Lecture Notes in Networks and Systems

Volume 337

The series "Lecture Notes in Networks and Systems" publishes the latest developments in Networks and Systems—quickly, informally and with high quality. Original research reported in proceedings and post-proceedings represents the core of LNNS.

Volumes published in LNNS embrace all aspects and subfields of, as well as new challenges in, Networks and Systems.

The series contains proceedings and edited volumes in systems and networks, spanning the areas of Cyber-Physical Systems, Autonomous Systems, Sensor Networks, Control Systems, Energy Systems, Automotive Systems, Biological Systems, Vehicular Networking and Connected Vehicles, Aerospace Systems, Automation, Manufacturing, Smart Grids, Nonlinear Systems, Power Systems, Robotics, Social Systems, Economic Systems and other. Of particular value to both the contributors and the readership are the short publication timeframe and the world-wide distribution and exposure which enable both a wide and rapid dissemination of research output.

The series covers the theory, applications, and perspectives on the state of the art and future developments relevant to systems and networks, decision making, control, complex processes and related areas, as embedded in the fields of interdisciplinary and applied sciences, engineering, computer science, physics, economics, social, and life sciences, as well as the paradigms and methodologies behind them.

Indexed by SCOPUS, INSPEC, WTI Frankfurt eG, zbMATH, SCImago.

All books published in the series are submitted for consideration in Web of Science.

For proposals from Asia please contact Aninda Bose (aninda.bose@springer.com).

More information about this series at https://link.springer.com/bookseries/15179

Ernesto León-Castro · Fabio Blanco-Mesa ·
Victor Alfaro-García · Anna Maria Gil-Lafuente ·
José M. Merigó · Janusz Kacprzyk
Editors

Soft Computing and Fuzzy Methodologies in Innovation Management and Sustainability

Springer

Editors
Ernesto León-Castro
Faculty of Economics and Adminsitrative Sciences
Universidad Catolica de la Santisima Concepcion
Concepcion, Bio Bio, Chile

Victor Alfaro-García
Facultad de Contaduría y Ciencias Administrativas
Universidad Michoacana de San Nicolas de Morelia, Michoacán, Mexico

José M. Merigó
School of Information, Systems and Modelling
University of Technology
Sydney, NSW, Australia

School of Economics and Business
University of Chile
Santiago, Chile

Fabio Blanco-Mesa
Facultad de Ciencias Económicas y Administrativas Escuela de Administración de Empresas
Universidad Pedagógica y Tecnológica d
Tunja, Colombia

Anna Maria Gil-Lafuente
Department of Business Administration
University of Barcelona
Barcelona, Spain

Janusz Kacprzyk
Systems Research Institute
Polish Academy of Sciences
Warsaw, Poland

ISSN 2367-3370 ISSN 2367-3389 (electronic)
Lecture Notes in Networks and Systems
ISBN 978-3-030-96149-7 ISBN 978-3-030-96150-3 (eBook)
https://doi.org/10.1007/978-3-030-96150-3

This Springer imprint is published by the registered company Springer Nature Switzerland AG
The registered company address is: Gewerbestrasse 11, 6330 Cham, Switzerland

Preface

This volume constitutes a new and original contribution to the development of new modeling, analytic and implementation approaches, and tools and techniques which can not only be useful for but also give a new boost to dealing with various problems faced by the world in which sustainability is a key concern. Various aspects of sustainability are discussed in the contributions in the volume but, basically, sustainability is here meant as an ability to satisfy present needs of the world without jeopardizing, or endangering, the satisfaction of needs of the future generation. Of course, these needs can be very diverse, and sustainability should be considered from various perspectives, notably: economic, environmental and social.

It is obvious that sustainability needs innovation. There is a simple reason for this because the satisfaction of growing needs of the humans, in view of both a steep population growth and growing needs and aspirations of both individuals and human groups who get richer and richer in most countries, implies a need for some new solutions. This calls both for new, more effective and efficient production and manufacturing technologies, and new solutions related to, for instance, management or organization, at all levels.

In this volume there is a collection of papers which show deep analyses of various aspects of some interesting projects and case studies related to sustainability, notably related to some economic pillar of sustainability. In particular, the context of tourism or agriculture which are relevant for countries in Latin America is what the volume focuses on. Here, again, relevant innovative approaches are needed. For the analysis of these problems, novel formal and algorithmic tools and techniques are proposed, for instance, some advanced aggregation operators exemplified by the OWAs (ordered weighted averaging operators), fuzzy logic based classifiers, etc.

An interesting and original part of the volume is Part II in which a conceptually new approach is used to recognize and study various aspects of sustainability and innovation, and also of related topics, by analyzing the scientific literature on this topic. Scientific papers in journals, books, research reports or articles in conference proceedings can easily be downloaded from various bibliometric databases, notably the Web of Science and Scopus, and then can be analyzed by using data analytic and data mining tools and techniques complemented by visualization. In such a way one can obtain information on which topics are popular, who are the most productive or cited researchers, which research groups or institutions are the most active, etc. This information can be useful for both some present day analyses and strategic, or even policymaking related analyses and decisions.

Part I "Analyses of Aspects of Economic, Social and Technological Development" is concerned with the use of various tools and techniques of computational intelligence, artificial intelligence, data sciences, data mining and related fields for the formulation, analysis and solution of various problems related to the economic, social and technological development, with an emphasis on both theoretical analyses and practical problems, notably related to case studies of Latin America.

Martín Isimayrt Huesca-Gastélum, Alicia Delgadillo-Aguirre, Martín León-Santiesteban and Ernesto León-Castro ("Ranking of Innovation and Sustainability of Tourist Destinations in Sinaloa: An Analysis with the Ordered Weighted Average Operator") are concerned with the classification of innovation and sustainability of tourist destinations in the State of Sinaloa in Mexico. They use the ordered weighted average operator (OWA), a powerful and widely employed aggregation operator. The use of the OWA operator makes it possible to order the recreation sites according to a relative importance of each criterion assumed. The results indicate the city of Culiacán as the best recreation place from the point of view of these criteria, while the municipality of Cosalá is found to be the worst one. This result can be useful for local authorities and policymakers by suggesting where to allocate resources.

A. J. Villa Silva, L. A. Pérez Domínguez, E. Martínez Gómez and R. Romero López ("Dimensional Analysis under Pythagorean Fuzzy Set with Hesitant Linguists Term Entropy Information") deal with the method of dimensional analysis (DA) which considers an association of all the criteria involved in a problem and makes it possible to capture interrelations that usually occurs in multi-criteria decision-making problems. The authors use the so-called Pythagorean fuzzy sets (PFSs), a recently introduced tool for handling fuzziness and vagueness that is viewed to yield a greater flexibility for decision-makers who are to provide evaluations and assessments and also may have difficulty for the specification of weights when information available is unknown and incomplete which can notably happen while applying tools and techniques of multi-criteria decision-making (MCDM). The use of a combination of three important tools is proposed: the dimensional analysis, Pythagorean fuzzy sets and entropy measure for hesitant fuzzy linguistic term sets for problems with a lack of, or scarcity of information on qualitative criteria, interrelationship among the multiple criteria, and weights. The analyses are illustrated in examples.

Ingrid Nineth Pinto López, Cynthia M. Montaudon Tomas and Alicia L. Yáñez Moneda ("Conditions of Technology Access for Remote Work in the Quaternary Sector in Mexico in Times of COVID-19") perform an analysis of conditions of an access to technology for teleworking that professionals in the quaternary sector in Mexico face. The quaternary sector is meant, as usually, by businesses and industries providing information related services, such as computing, ICT (information and communication technologies), consultancy, R&D (research and development, notably in scientific fields), and related types of activities. Data collection for the analyses has been performed between five and seven months after the Covid-19 related lockdown measures have been ordered in Mexico, so that professional activities have been moved to telework (distant) mode. The data have been collected for 966 participants in 27 of the 32 states of Mexico. Valuable results are obtained on the conditions of access to technology, training and tools required to work remotely.

Pavel Anselmo Alvarez ("Study of the Geographical Marginality in a Mexican Region Using the MR-sort Method") studies the geographical marginality in the State of Sinaloa in Mexico. The marginality evaluation of municipalities is developed by employing a multi-criteria decision-making based sorting method. More specifically, the MR-sort method is used to classify the municipalities into four categories of marginality. The results obtained indicate three municipalities with the highest marginality level due to low performance in education and income factors. The analysis suggests that low performance has a negative impact on the marginalization and leave the population in the state of a higher value of marginality.

Rubén Chávez, Federico González, Victor Alcaraz and Jesús Ricard Ramos ("Strategic Diagnostics of Stress and Impulse Control for Second Order Change: Inclusion of Forgotten Effects in Diffuse Cognitive Maps") are concerned with the determination of strategic cognitive factors that impact the so-called Second Order Change (SOC) which, roughly speaking, boils down to doing something fundamentally different than done before, and which can be decisive for a transformation of organizational systems. A methodology based on the Fuzzy Cognitive Maps (FCM) and the Forgotten Effects (FE) model is proposed. The inclusion of the FEs makes it possible to reduce inference errors in the inference matrix with an objective to provide certainty to the inference matrix when applying the FE model. First, the elements are separated into two matrices containing positive and negative elements. The negative element matrix is then temporarily transformed into a state of positive elements so that both matrices can undergo a fuzzification-inferential-defuzzification process in the FE model. Then, the matrix of temporarily positive elements returns to its original (negative) state. Finally, both matrices are added to obtain an adjusted inference matrix to be applied in the fuzzy cognitive map model. In the case study presented the cognitive factors that impact the behavior of the personnel of four companies are found, starting with two highly correlated factors: the stress tolerance and impulse control, using the absolute Hamming distances.

Walayat Hussain, José M. Merigó, Fethi Rabhi and Honghao Gao ("Aggregating Fuzzy Sentiments with Customized QoS Parameters for Cloud Provider Selection Using Fuzzy Best Worst and Fuzzy TOPSIS") are concerned with a very important problem of dealing with the hesitancy of consumers who are faced with an abundance of services (and the same for products) amplified even more when multiple service providers offer the same type and quality of services. To make an informed choice the decision-makers have to take into account and combine multiple factors and aspects. Sentiment mining is here increasingly popular as one of the key techniques to determine the service quality and get an insight into business. It can help service providers precisely deduce a consumer's emotions to then find an optimal service provider matching the request. Though there is much literature on the Quality of Service (QoS) of the offered services in cloud service selection, there are a very few of them which consider a user's experience of a consumer in the decision-making process. Moreover, a very few works are on the combination of the Quality of Experience (QoE) and the customized Quality of Service (QoS). This problem is dealt with here by aggregating the consumer's sentiments with customized QoS parameters to choose an optimal service provider. The fuzzy Best Worst Method (BWM) is employed to determine the weights of selection criteria and then the fuzzy TOPSIS is used to handle uncertain linguistic preferences. The results obtained confirm the applicability and effectiveness of the approach.

Jhesus Wilson Panca Galindo, Blanca Roldán-Clarà and Alfredo Pelayo Calatayud Mendoza ("Economic Benefits to Conserve the Tourism Potential of Chifron Beach in the District of Capachica, Puno, Peru") are concerned with the Chifron beach in the Capachica district, Puno, Peru, which is a popular tourist attraction. It should be subject to a critical conservation due to environmental pollution by solid wastes, visual contamination from graffiti painted on the rocks, and alteration of beaches by the extraction of sand for construction that threaten the quality of the beach and water and have a negative impact on neighboring households. This work deals with an attempt to stimulate economic benefits for conserving the Chifron beach by trying to identify a socioeconomic profile of visitors and to determine a proper entrance fee to the beach by using a survey and contingent valuation method. The survey concerns 252 visitors among whom 54% are male, 53% are married, 76% have higher education, 66% work in private or public institutions, 34% have their own businesses, 27% are residents of rural areas and 73% live in urban areas of different cities. To summarize the results obtained, 94% of the tourists agreed that the beach should be cleaned and sanitary services be available. However, 66% of the surveyed tourists rated the cleanliness and hygiene of the sand as poor, and 64% considered the cleanliness and hygiene of the shoreline as poor while 59% of the tourists rated the landscape at the beach as very good. As found, 63% of the visitors would be willing to pay an entrance fee for the conservation of the beach. This means that is though the conservation condition of the Chifron beach is critical, there is a willingness to pay for visiting the beach as long as it is in optimal conditions, that is, with a clean sand and shoreline, and with appropriate sanitary facilities. Moreover, the analysis showed that the monthly income and educational level of visitors is directly proportional to the willingness to pay for a conservation fee, and the hypothetical cost of visiting the beach is inversely proportional to that willingness to pay.

Feng Chung Wu, Huei Diana Lee, Newton Spolaôr, Moacir Fonteque Junior, Weber Shoity Resende Takaki, Claudio Saddy Rodrigues Coy, João José Fagundes, Raquel Franco Leal, Renato Bobsin Machado and Maria de Lourdes Setsuko Ayrizono ("Monitoring Videocolonoscopy Examinations in Real-time via Internet") are concerned with the use of computational technologies in medicine, more specifically with the support of the establishment of collaboration networks among professionals in medical centers, clinics and hospitals, notably after the advent and expansion of the Internet. These networks can help the experts located in different locations to monitor and discuss medical procedures in real time, as well as to collect, store and share multimedia data associated with these procedures. An innovative method in telemedicine and a computational system is proposed that allows experts to interact in real time by text, audio and video, as well to perform a real-time gathering and sharing of videocolonoscopic images and videos with authorized users via the Internet. These multimedia data are stored and published into a centralized server implemented taking into account security concerns. The results obtained confirm that the proposed system can be an alternative to support collaboration among experts and institutions that offer videocolonoscopy or other types of video-based examinations and medical procedures to patients.

Antonio Kido Cruz and M. C. Felipe Andoni Luna Campos ("Economic Growth and Universities: Empirical Evidence for Mexico") discuss and estimate the relationship between the economic growth in the states of Mexico (EGM) and the number of colleges and universities (NU). In order to prove the existence of a long-term association among the variables considered, yearly data from 1995 to 2015 are used and a fixed effects data panel model is employed. The results obtained show that a 100% increase in the number of public universities is positively and significantly related to the growth of 38.2% in the GDP per capita. The proposed analysis can potentially be useful for the design of public policies of higher education in various countries.

Juan C. Ruiz-Torres and Gina P. Fonseca-Cifuentes ("ABC Costing System Applied to a SME specializing in Dairy Production in Colombia") discuss how to develop a methodology for the implementation of management accounting associated with the Activity Based Costing (ABC) which is a method fo reassigning overhead and indirect costs to products and services. The work is focused on an example of a Colombian Small and Medium Enterprize (SME) specializing in dairy products.

Nine stages for the construction of the cost system are assumed. As a result of the analyses, the determination of the cost of the agricultural products as well as the measurement of biological assets are performer according to the international IAS 41 standard which basically states that a biological asset is any living plant or animal owned by business, and their measurement is at a fair value minus selling costs. An important conclusion obtained is that the system costs make it possible to make appropriate decisions and optimize processes and to increase profitability.

María Eugenia Estrada-Chavira, Sylja Viridiana Guerrero-García, Maribel Rocío Hernández-Velázquez and Guillermo Arredondo-García ("Did Mexico Lose the Fight of Tomato Exports in the Times of USA Tariff, Facing the Umsca?") are concerned with the problem implied by introducing in 2019 a 17.5% tariff on the tomato imports from Mexico which has strongly affected the tomato market. The objective of this work is to measure the demand price elasticity of tomatoes for two periods: with the tariff and without the tariff. The price elasticity is calculated using the logarithmic regression. It is shown that the tomato price elasticity is elastic in both periods but is higher in the period with tariff. That means that if the tomato price changes, then the volume of the demand diminishes more than proportionally. In the beginning the producers, exporters and both the American and Mexican economies lose because the consumer price of the product increases. As a conclusion, it is stated that since Mexico does not have a strong international presence on the tomato market around the world, except for the USA, then it is highly advisable to extend the exportation to other markets, notably the Asian markets.

Miriam Edith Pérez-Romero, Martha Beatriz Flores-Romero, and Victor G. Alfaro-García ("Tourism Competitiveness Theory Perspectives Through the Application of Counter-expertise Methods") deal with a comparison of opinions of two groups of experts regarding the relationship between causes and effects of the competitiveness of tourist destinations. Formally, the formation of two groups of experts, those who investigate the touristic activity and those who work in the tourism sector are first undertaken. Both groups evaluate cause-effect relationships using the forgotten effects theory. The results are then compared using the counter-expertise technique. Finally, the results are grouped via a frequency distribution to facilitate its interpretation and analysis. The application of the counter-expertise technique shows that the distance between the opinions of the two groups of experts is from 0.00% to 18.75% in direct incidences and from 0.00% to 16.25% in indirect incidences. The frequency distribution shows that the data is highly skewed to the right so that most of the results are in the lowest values, that is, the distance between the opinions of the two groups of experts is small in most cases, and therefore, information provided by the experts appears to be valid. In both cases of using the direct incidents and forgotten effects, the environmental commitment cause appears to be important.

Norma Laura Godínez-Reyes, Rodrigo Gómez-Monge, Gerardo Gabriel Alfaro-Calderón, and Argelia Calderón-Gutiérrez ("Sustainable Value: An Empirical Research in Large Firms") deal with the determination of the extent to which the environmental, social and governance indicators can explain corporate efficiency which is gauged by the return on assets (ROA), return on equity (ROE) and return on sales (ROS). To find the validity of the corporate efficiency measurement first a model of three linear regressions with panel data is developed. This model tests whether the environmental (E), social performance (S) and corporate governance (G) variables for the companies considered can explain the profitability of these companies for the period 2014–2017. The results obtained show that the sustainable value (S) explains the profitability—or return on sales (ROS)—with a high degree of significance, whereas it explains the corporate efficiency to a moderate extent. The corporate governance (G) is the most significant variable for the generation of a sustainable value. As a result it is shown that the generation of sustainable values makes it possible for the companies to meet their profitability goals and to mitigate environmental and social impacts. That is, the model can be useful for measuring the corporate efficiency in sustainable companies.

Part II: Bibliometric analyses of main research directions

This part of the volume is quite novel and is rarely included in research reports, books and volumes on the topics considered. Basically, the authors use bibliometric data from well-known systems which include scholarly data on publications, authors, etc. exemplified by the Web of Science, Scopus or Google Scholar. These data are analyzed by advanced data mining methods which makes it possible to extract much information about various important aspects like the main areas and topics of research and their impact on the community as shown by, for instance, the number of citations, the main researchers and research groups working on specific topics, an application potential of particular papers and research directions, etc. Needless to say that these analyses can yield much information that can be very useful for both usual productivity analyses of research efforts and also for scientific policymaking to support promising fields.

Marlenne Velázquez Cázares, Sergio Alvarado-Altamirano, Ernesto León-Castro, and Fabio Blanco-Mesa ("A Bibliometric Analysis of Corporate Social Responsibility and Competitiveness") are concerned with the corporate social responsibility and its relationship to the competitiveness which is an important field of study, both for companies and institutions, and more generally the society. The paper focuses on a comprehensive presentation of the main contributions in these subjects via a bibliometric analysis. Specifically, a wide range of bibliometric indicators is employed using the Web of Science (WoS) Core Collection and complemented with the Scopus database. The leading journals, authors, research groups and institutions, countries, and even articles are considered. The results confirm a growing relevance of the research field and find both the most influential authors and the most productive authors. Moreover, leading countries, in the sense of the largest number of publications and citations, are shown. As a conclusion, it is stated that the relationship between the corporate social responsibility and competitiveness is a more and more relevant field of study that contributes to the development of new theories, approaches and applications.

Pavel López-Parra, Anselmo Alvarez-Carrillo, Ernesto León-Castro, Marlenne Velázquez Cázares and Manuel Muñoz Palma ("A Bibliometric Analysis of Robustness and MCDA") are concerned with some important aspects, both for theory and practice, of Multicriteria Decision Aiding (MCDA) which can be described as an important area of operations research that offers the methodology, algorithms and implementations for the formulation and solution of complex decision-making problems with multiple conflicting criteria. An important aspect is here robustness, and its related analyses are used to find a more precise solution within the methodologies of MCDA, as well as those less vulnerable to uncertainties, imprecision or imperfect and missing data. The purpose of this work is to present an overview of this topic from the perspective of bibliometric analysis using the Scopus and Web of Science databases. Using a large number of indicators, relevant information about journals, authors, countries and institutions is derived. The results obtained list the most productive and influential authors, research groups and countries, as well as indicate the usefulness of these results for various aspects of research analyses and policymaking.

Denisse Ballardo-Cárdenas, Ernesto León-Castro, Fabio Blanco-Mesa, and Ramón Martínez-Huerta ("A Bibliometric Analysis on Innovation Management Research") deal with an analysis of information on scientific research publications related to innovation management available using the Web of Science Core Collection database, more specifically published from 1982 to 2018. The documents analyzed included articles, reviews and notes. The approach proposed graphically maps the bibliographic material by using the VOS viewer software, a popular software used for such bibliometric analysis, to obtain a more illustrative and in depth analysis of data and relations between them. The paper identifies leading and most inspiring publications, the most productive authors, the most productive countries, the most productive institutions and analyzes the evolution of their output and activities over time.

Rodrigo Gómez Monge, Víctor G. Alfaro-García, Irma C. Espitia-Moreno, Dalia García-Orozco, and Manuel Ricardo Romo de Vivar Mercadillo ("Environmental Sustainability: A 10-year Bibliometric Analysis of the Developments in Management, Business, Finance and Economics") are concerned with a much talked about and extremely relevant general problem of sustainability which is understood basically as the satisfaction of needs of people living in the present times without jeopardizing the resources for future generations. This general goal calls for various systems, economic, social, technological, etc., to be designed for a proper transformation, consumption, strategic coordination and allocation of resources. A bibliometric analysis proposed is designed to find core academic developments indexed in the Web of Science scientific database from 2010 to 2019. The results obtained show the leading authors, papers, journals, organizations, countries, etc. which are relevant for the field of sustainable management, business, finance and economics. Trends, connections and networks using analytical and statistical tools are obtained. The use of the VOS viewer software provides a fast and comprehensible visual representation of the retrieved core documents.

Abraham Nuñez Maldonado and Martha Beatriz Flores Romero ("A Bibliometric Analysis of Sustainable Tourism Research for the Period 1991–2019") present a bibliometric analysis of sustainable tourism by analyzing 28 years of research published between 1991 and 2019, using the Web of Science database. The analysis focuses on the results obtained by using the bibliometric review software, Rstudio. This makes it possible to obtain in a comprehensible form various useful results exemplified by the most cited papers and authors, the most productive and influential researchers, institutions and countries, to just mention a few. The results obtained indicate that sustainable tourism attracts much attention of the research community reflected by an exponential growth of scientific papers on the topic.

Adriana Paulina Aranzolo-Sánchez, Donaji Jiménez-Islas, and Miriam Edith Pérez-Romero ("Research Growth on Bioethanol: A Bibliometric Analysis") provide a bibliometric analysis of research on bioethanol using the Web of Science and Scopus databases for the period of 1990 to 2019. The logistic equation is used to quantitatively describe the growth of publication volume reported in both databases. The results show that the Scopus database exhibits the rate of publications of 0.31 per years and for the WoS database this rate is 0.29 per year, and—in general—the logistic equation is found appropriate for the problem considered. The method also makes it possible to find the leading countries who produce the highest number of publications.

We strongly believe that the high quality, interesting and inspiring contributions included in this volume will be of much interest and use for a wide research community. The contributions concern many questions that are important not only for researchers and scholars, but also practitioners, notably economic analysts and policymakers. All these people can find in the volume very much of useful material that can help in their operational, strategic and policy related activities.

We wish to thank all the contributors for their great works. Special thanks are due to anonymous referees whose deep and constructive remarks and suggestions have helped to greatly improve the quality and clarity of the contributions.

And last but not least, we wish to thank Dr. Tom Ditzinger, Dr. Leontina di Cecco and Mr. Holger Schaepe for their dedication and help to implement and finish this important publication project on time, while maintaining the highest publication standards.

Baja California, Mexico	Ernesto Leon-Castro
Tunja, Colombia	Fabio Blanco-Mesa
Morelia, Mexico	Victor Alfaro-Garcia
Barcelona, Spain	Anna Maria Gil-Lafuente
Sydney, Australia	José M. Merigó
Warsaw, Poland	Janusz Kacprzyk
Summer 2021	

Contents

Analyses of Aspects of Economic, Social and Technological Development

Ranking of Innovation and Sustainability of Tourist Destinations in Sinaloa: An Analysis with the Ordered Weighted Average Operator

Huesca-Gastélum Martín Isimayrt, Delgadillo-Aguirre Alicia, León-Santiesteban Martín, and León-Castro Ernesto

Abstract The objective of this work is to classify the innovation and sustainability of tourist destinations in Sinaloa through the ordered weighted average operator (OWA). The application of this method allows to order the recreation sites according to the relative importance of each criterion. The results indicate the city of Culiacán as the recreation place with the best levels in its evaluation, while the municipality of Cosalá turned out to be the lowest valued establishment. This information is useful for policy makers because they will be able to allocate their resources based on their areas of opportunity. Finally, the document demonstrates the application of the OWA operator to measure innovation and sustainability of tourist attractions in Sinaloa, Mexico.

Keywords Innovation · Sustainability · Tourist destinations · OWA operator

1 Introduction

According to [33], in recent years tourism has experienced rapid growth around the world, where the extensive diversification of recreational spaces has managed to turn it into a sector it is an important economic factor for the host communities, but it has also caused those responsible for planning it to include certain innovative and sustainable strategies within the administration of these leisure spaces.

In the same way, due to globalization, innovation and sustainability have been positioned as relevant aspects for the competitiveness of tourist attractions [25]. Therefore, in such a globalized and competitive context, it is up to the management of entertainment venues to implement increasingly innovative and sustainable actions to guarantee the best possible results over time [33].

H.-G. M. Isimayrt · D.-A. Alicia (✉) · L.-S. Martín
Universidad Autónoma de Occidente. Blvd, Lola Beltrán Km 1.5, Culiacán, Sinaloa, México

L.-C. Ernesto
Faculty of Economics and Administrative Sciences, Universidad Católica de La Santísima Concepción, Concepción, Chile
e-mail: eleon@ucsc.cl

E. León-Castro et al. (eds.), *Soft Computing and Fuzzy Methodologies in Innovation Management and Sustainability*, Lecture Notes in Networks and Systems 337,
https://doi.org/10.1007/978-3-030-96150-3_1

Hence, innovation and sustainability are considered by tourist attractions to improve their competitiveness and remain active in an increasingly diversified world [2, 5, 26, 27, 35]. For these reasons, in recent years tourist destinations have been studied from different disciplines [36].

However, there is still little empirical evidence that considers these factors in the questions of recreation places [23]. In addition, a large number of academics argue that research on these dimensions is limited, so there is a need for further elaboration of documents focused on the evaluation of these phenomena, as well as on the theoretical contribution regarding innovation and sustainability studies tourism [5, 6, 30].

Hence, the objective of this study is to classify the innovation and sustainability of tourist destinations in Sinaloa through the ordered weighted average (OWA) operator. The use of this technique allowed not only the classification of recreation sites according to their innovative capacity and sustainable competence, but also managed to highlight the relative importance of the criteria based on the expectations, knowledge and aptitude of the decision maker.

To achieve the purpose of this research, it was decided to use the approach proposed by [6], as well as the framing presented by [8] because they are the best options to measure and characterize the innovation and sustainability of recreational spaces, but also because they have been used in various publications in recent years.

Finally, this article begins with the presentation of the theoretical framework on the aforementioned dimensions. Subsequently, the OWA operator is defined and the use of this method is presented to classify recreational sites in Sinaloa according to their innovation and sustainability. Finally, the main conclusions of the document are summarized and the references used are indicated.

2 Theoretical Framework

2.1 A Literature Review of the Innovation and Sustainability of Tourist Destinations

As pointed out by [32], due to globalization, tourism managed to position itself as a fast-growing industry worldwide and according to [7], different countries have considered it as one of their main economic activities due to its great capacity to generate benefits in the host communities.

For [41], it is due to this relevance that competition between tourist destinations around the world has increased, but more regions have also decided to resort to travel proclivity to incorporate them as an important element in their economic work, as they recognize the benefits at stake [10, 13]. Therefore, the proper management of this industry could make it a key element to achieve broader social objectives [9].

On the other hand, recreational sites are in constant competition with each other due to the growth in tourist mobility. Consequently, recreation venues should strive to be more competitive [14, 41] because as travel increases, competition among tourist attractions also increases.

Hence, the potential of the recreational vocation of any host community strongly depends on its ability to maintain an advantage in the delivery of goods and services to its visitors [11]. According to [28], for this reason entertainment venues need more flexible approaches to be effective in this competitive process [28].

On the one hand, they intend to implement new technologies to be more innovative [16], but on the other, they strive to promote more sustainable entertainment venues [17]. Otherwise, if recreational sites do not pay attention to these elements, all efforts to be more competitive would be in vain [1].

However, it is difficult to evaluate and compare their degree of innovation and sustainability because there is no consensus on the ideal indicators to make this comparison [16, 17]. In addition, most studies have decided to address these factors to estimate good practices, but they have also shown the limitations faced in these aspects and which, in turn, limit the competitiveness of tourist destinations [3].

In this sense, the reflection begins with the exploration of [34], these authors propose a model to evaluate these criteria in some destinations in Portugal through the entropy of information and through different weights that were calculated with alternative approaches such as the Fuzzy Rasch and the hierarchical analytical process (AHP). The results of this application not only showed that there is a heterogeneity between Portuguese attractions in terms of innovation and sustainability, but also evidenced the relationship between these tourist practices.

On the other hand, [31] carried out an investigation to determine the places of recreation that implement sustainable and innovative strategies in the province of Fars, Iran. To achieve this purpose, the authors used ordered weighted average (OWA) algorithms and fuzzy quantifiers. The results showed that almost the entire study area applied few innovative and sustainable instruments.

In the same way, [16] applied a methodology that allowed them not only to compare the innovation and sustainability of tourist destinations in Spain, but also facilitated the generation of a ranking of destinations according to these criteria. This analysis was done through the hierarchical analytical process (AHP) and the ordered weighted average (OWA). Among the most relevant findings is the fact that leisure sites can develop innovative and sustainable ideas with little resources, but with more efficient management processes.

Similarly, [40] applied the induced ordered weighted average (IOWA) operator to determine the level of innovation and sustainability presented by 13 cities in China. The results of this evaluation indicated that the innovative and sustainable development of most of the localities was poor. Where the performance values were below 0.5 and only Beijing and Tianjin managed to be above this parameter.

Finally, [21] used the analytical hierarchy process (AHP) and the ordered weighted average operator (OWA) to evaluate and prioritize tourist sites based on their innovative and sustainable capabilities. The study was carried out in the area of La Vera, Spain and its results showed the application of these fuzzy logic techniques to compare these geographical areas. Furthermore, this method of analysis proposes an approach to improve the attitudes of local residents regarding the management of tourist destinations.

However, the utility approaches are the one proposed by [8], as well as that presented by [6], since these authors have developed some indicators that allow assessing the innovation and sustainability of the recreation sites at the local and regional levels. For this reason, this document opts to use these theoretical perspectives. In addition, with the proposal of this study the theoretical and methodological lag is reduced, therefore, it contributes to the generation of knowledge from this perspective in question.

2.2 Innovation and Sustainability in the Tourist Destinations of Mexico

Currently, tourism is experiencing significant changes that are the product of globalization and a variety of new trends in tourism demand [12]. Hence, tourist destinations around the world must compete with each other to maintain their participation in the tourism market [15].

This implies that innovation and sustainability strategies are key factors to retain this competitive validity, but they have also become essential criteria for evaluating the competitiveness of recreational sites [22]. On the other hand, around the world there is a growing demand related to the analysis of this question, since different methodological proposals are visualized that try to evaluate the competition of certain tourist destinations through indexes and ranking made up of these two dimensions [16].

For example, the [37] measures the competitiveness of the travel and tourism sector every two years through 14 main dimensions. Among these pillars is that of innovation, as well as the criterion of sustainability. In the same way, said evaluation is carried out by means of an index that is calculated as an average of all the pillars considered.

The results show that Mexico ranks 19 in this competition index (see Table 1). Where its level is characterized by having exceptional natural and cultural resources that, effectively combined, manage to offer relatively strong price competitiveness, but also favors the high percentage achieved [37].

However, if the innovation criterion is analyzed, the Mexican case is located in position 81 with a total of 4.4 points. Now, looking at the pillar related to sustainability, this nation is affected in the ranking, since it reaches 108th place with a 3.9 score [37]. Consequently, the Aztec government will have to pay more attention to these issues, otherwise, its performance in the competitiveness index could be affected.

Similarly, the [20] carried out a measurement on these aspects in the Mexican states through an index made up of 6 components and 160 indicators. In this way,

Table 1 Travel and tourism competitiveness index

Rank	Country	Score
1	Spain	5.4
2	France	5.4
3	Germany	5.4
19	Mexico	4.7
20	Norway	4.6
70	Ecuador	3.9
71	Azerbaijan	3.8
138	Liberia	2.6
139	Chad	2.5
140	Yemen	2.4

these parameters offer the states the possibility of analyzing and evaluating their competitiveness through a ranking.

The results indicate that the best evaluated states according to this index are Mexico City, Nuevo León and Querétaro. In contrast, the territories of Chiapas, Oaxaca and Guerrero constitute the worst evaluated entities in their performance. In this sense, with regard to the question of innovation and sustainability, it is possible to affirm that Mexico City remains with the first place in the ranking, this tourist destination being the best compared to the rest [20].

Along the same lines, the [18] developed a state competitiveness index to analyze the 32 Mexican states according to various criteria, among which innovation and sustainability stand out. This classification was made up of 100 indicators that were categorized into 10 sub-indices and made it possible to identify the strengths as well as the weaknesses of each state. The analysis shows Mexico City, Querétaro and Chihuahua as the most competitive entities.

Likewise, the results based on the sustainable criterion also show that Mexico City obtained the first place in this criterion, while Oaxaca was the entity that was located in the last position. On the other hand, in relation to the innovation aspect, the state of Querétaro remains number one in the ranking. For its part, Tabasco is positioned as the worst federal entity [19].

Finally, the [18] developed an index to classify the 73 most important cities in Mexico according to 70 indicators that were grouped into 10 dimensions. This parameter was created with the intention of identifying the opportunities and challenges of each urban area to create and attract talent and investment. In general, the results show the Ciudad del Carmen and the Piedras Negras municipality as the best-evaluated cities.

On the contrary, the towns of Tecomán and La Piedad-Pénjamo were evaluated as the worst tourist destinations based on this global index. However, analyzing the criteria related to sustainability, the best city evaluated was Aguascalientes, while Toluca was ranked as the worst performing one. However, the ranking changes if the innovation factor is observed, since in the first position is the Valle de Mexico and as last place is Tuxtla Gutiérrez [19].

3 The Ordered Weighted Average Operator

An operator that can be used to aggregate information is the OWA operator introduced by [38]. This operator let you aggregate information between the maximum and the minimum and since then many applications have been made [4, 39]. The definition is as follows.

Definition 1
An OWA operator of dimension n is a mapping OWA : $R^n \rightarrow R$ with an associated weight vector W of dimension n such that $\sum_{j=1}^{n} w_j = 1$ and $w_j \in [0, 1]$, according to the following formula:

$$\mathrm{OWA}(\mathrm{a}_1, \mathrm{a}_2, \ldots, \mathrm{a}_\mathrm{n}) = \sum_{\mathrm{j}=1}^{\mathrm{n}} \mathrm{w}_\mathrm{j} \mathrm{b}_\mathrm{j}, \tag{1}$$

where b_j is the jth largest element of the collection a_i.

Note that we can distinguish between the descending OWA (DOWA) and the ascending OWA (AOWA) operator. This difference is related by $\mathrm{w}_\mathrm{j} = \mathrm{w}^*_{\mathrm{n-j+1}}$, where w_j is the jth weight of the DOWA operator and $\mathrm{w}^*_{\mathrm{n-j+1}}$ the jth weight of the AOWA operator.

Decisions within OWA operators can be generated under different criteria, the most important of which are the following.

1. Optimistic criterion. It assumes that the most favorable state will be presented, in such a way that the most favorable result of each alternative must be selected and from the results obtained select the most favorable of all. In such a way that this criterion is based on a maxim that is formulated

$$Decision = Max\{E_i\} = Max[Max\{Max\{a_j\}\}]$$

2. Pessimistic or Wald criterion. It maintains that the decision maker must select the alternative that provides the highest level of security, in such a way that our decision must be the most favorable result among the most unfavorable for each alternative. This method is known as max min and its formula is

$$Decision = Max\{E_i\} = Max[Min\{a_j\}]$$

3. Hurwics criterion. It consists of weighing the best and worst case respectively with an optimistic coefficient and another pessimistic one, subsequently the

values are added and the alternative that proposes a better result is chosen. The formula for this criterion is

$$Decision = Max\{E_j\} = Max[\alpha Max\{a_j\} + (1-\alpha)Min\{a_j\}]$$

where $\alpha + (1-\alpha) = 1$.

4. Laplace's criterion. It is based on the principle of insufficient reason, in such a way that the same degree of probability is associated with the different scenarios, provided that there are no indications to the contrary. The formula is

$$Decision = Max\{E_j\} = Max\left[(1/n)\sum_{j=1}^{n} a_j\right]$$

4 Measurement of the Innovation and Sustainability in the Tourist Destinations in Sinaloa with the OWA Operator

To measure a ranking of tourist destinations based in Sinaloa based on the OWA operator, there are some steps that must be follow.

Step 1. Because the purpose is to rank the tourist destinations based on innovation and sustainability, these two items will be the main criterions and for each of them different indicators have been defined based on the theoretical framework presented in Section II. This information is presented in Table 2.

Step 2. For each of the indicators proposed in Table 1, the information for each of the 18 municipalities for Sinaloa were obtained (See Table 2). The data is presented in Table 3.

Step 3. To give an actual value to each indicator, it was decided that the highest value will have the value of 1 and the others one will be evaluated in function of that. In this sense, the value for each indicator will be $Result = \frac{Value}{MaxValue}$. The information is presented in Table 4.

Step 4. The next step will be giving importance to the criterions by a weighting vector. This vector will be the one used to calculate the results using the OWA operator. For the criterions and because innovation presents many municipalities does not present results in the indicators the weights will be 30% and sustainability will be 70%. The sum of all indicators will be the evaluation for each criterion (See Table 5).

Step 5. The OWA operator is calculated based on the information provided. For this article, the weighted average, the OWA maximum and OWA minimum will be calculated. The results are presented in Tables 6, 7 and 8.

Table 2 Indicators to measure the innovation and sustainability in tourism destinations

Identifier	Description	Category
INN	Innovation	Criteria
NEI	Number of educational institutions dedicated to the training of human resources for innovation purposes	Indicator
PII	Public investment for innovation	Indicator
NCR	Number of companies that receive public resources for innovation issues	Indicator
ARD	Amount of R&D expenditure	Indicator
NRP	Number of registered patents	Indicator
SUS	Sustainability	Criteria
NTR	Number of tourism related jobs	Indicator
PIC	Public investment aimed at various cultural aspects	Indicator
NIW	Number of inhabitants with access to water	Indicator
CRE	Complaints received on environmental matters	Indicator
PDW	Percentage of dwellings with waste disposal	Indicator

Table 3 Municipalities of Sinaloa, Mexico

Identifier	Municipality
A1	Ahome
A2	Angostura
A3	Badiraguato
A4	Choix
A5	Concordia
A6	Cosalá
A7	Culiacán
A8	El Fuerte
A9	Elota
A10	Escuinapa
A11	Guasave
A12	Mazatlán
A13	Mocorito
A14	Navolato
A15	Rosario
A16	Salvador Alvarado
A17	San Ignacio
A18	Sinaloa

Table 4 Results of each identifier for each municipality

Identifier	A1	A2	A3	A4	A5	A6	A7	A8	A9
INN									
NEI	6	0	0	0	0	0	19	0	0
PII	0	0	0	0	0	0	0	0	0
NCR	3	0	0	0	0	0	0	2	0
ARD	0	0	0	0	0	0	0	0	0
NRP	0	0	0	0	0	0	3	0	0
SUS									
NTR	170,982	14,964	6104	7238	8220	3556	366,171	33,752	23,070
PIC	30,700,072	0	2,200,000	0	1,200,000	0	10,025,000	1,660,000	9,549,000
NIW	449,033	47,203	31,819	32,987	27,127	16,292	905,152	100,445	53,856
CRE	20	4	0	0	0	0	13	0	10
PDW	124,209	13,246	8167	8698	7459	3840	247,718	26,435	13,721
Identifier	A10	A11	A12	A13	A14	A15	A16	A17	A18
INN									
NEI	0	0	10	0	0	0	0	0	0
PII	0	0	6,500,000	0	0	0	0	0	0
NCR	0	0	1	6	0	0	0	0	0
ARD	0	0	0	0	0	0	0	0	0
NRP	0	0	2	0	0	0	0	0	0
SUS									
NTR	22,002	101,770	209,914	13,421	63,709	18,640	29,804	6,609	22,023

(continued)

Table 4 (continued)

Identifier	A1	A2	A3	A4	A5	A6	A7	A8	A9
PIC	1,536,000	9,000,000	94,500,000	–	–	–	2,200,000	6,494,000	1,600,000
NIW	59,436	295,353	502,282	45,351	153,937	53,763	81,101	21,438	88,655
CRE	1	3	18	–	2	–	–	7	3
PDW	15,186	77,005	146,485	12,223	38,965	14,540	22,569	6,046	22,342

Table 5 Values for each indicator for each municipality

Identifier	A1	A2	A3	A4	A5	A6	A7	A8	A9
INN									
NEI	0.32	0.00	0.00	0.00	0.00	0.00	1.00	0.00	0.00
PII	0.00	0.00	0.00	0.00	0.00	0.00	0.00	0.00	0.00
NCR	0.50	0.00	0.00	0.00	0.00	0.00	0.00	0.33	0.00
ARD	0.00	0.00	0.00	0.00	0.00	0.00	0.00	0.00	0.00
NRP	0.00	0.00	0.00	0.00	0.00	0.00	1.00	0.00	0.00
SUS									
NTR	0.47	0.04	0.02	0.02	0.02	0.01	1.00	0.09	0.06
PIC	0.32	0.00	0.02	0.00	0.01	0.00	0.11	0.02	0.10
NIW	0.50	0.05	0.04	0.04	0.03	0.02	1.00	0.11	0.06
CRE	1.00	0.20	0.00	0.00	0.00	0.00	0.65	0.00	0.50
PDW	0.50	0.05	0.03	0.04	0.03	0.02	1.00	0.11	0.06
Identifier	A10	A11	A12	A13	A14	A15	A16	A17	A18
INN									
NEI	0.00	0.00	0.53	0.00	0.00	0.00	0.00	0.00	0.00
PII	0.00	0.00	1.00	0.00	0.00	0.00	0.00	0.00	0.00
NCR	0.00	0.00	0.17	1.00	0.00	0.00	0.00	0.00	0.00
ARD	0.00	0.00	0.00	0.00	0.00	0.00	0.00	0.00	0.00
NRP	0.00	0.00	0.67	0.00	0.00	0.00	0.00	0.00	0.00
SUS									
NTR	0.06	0.28	0.57	0.04	0.17	0.05	0.08	0.02	0.06
PIC	0.02	0.10	1.00	0.00	0.00	0.00	0.02	0.07	0.02
NIW	0.07	0.33	0.55	0.05	0.17	0.06	0.09	0.02	0.10
CRE	0.05	0.15	0.90	0.00	0.10	0.00	0.00	0.35	0.15
PDW	0.06	0.31	0.59	0.05	0.16	0.06	0.09	0.02	0.09

Table 6 Value for each criterion

Identifier	A1	A2	A3	A4	A5	A6	A7	A8	A9
INN	0.82	0.00	0.00	0.00	0.00	0.00	2.00	0.33	0.00
SUS	2.79	0.35	0.11	0.09	0.10	0.04	3.76	0.33	0.78
Identifier	A10	A11	A12	A13	A14	A15	A16	A17	A18
INN	0.00	0.00	2.36	1.00	0.00	0.00	0.00	0.00	0.00
SUS	0.25	1.16	3.62	0.14	0.60	0.17	0.29	0.48	0.42

Table 7 Weighted average results

A1	A2	A3	A4	A5	A6	A7	A8	A9
2.20	0.24	0.08	0.06	0.07	0.03	3.23	0.33	0.55
A10	A11	A12	A13	A14	A15	A16	A17	A18
0.18	0.81	3.24	0.40	0.42	0.12	0.20	0.34	0.29

Table 8 OWA maximum results

A1	A2	A3	A4	A5	A6	A7	A8	A9
2.20	0.24	0.08	0.06	0.07	0.03	3.23	0.33	0.55
A10	A11	A12	A13	A14	A15	A16	A17	A18
0.18	0.81	3.24	0.74	0.42	0.12	0.20	0.34	0.29

Table 9 OWA minimum results

A1	A2	A3	A4	A5	A6	A7	A8	A9
1.41	0.10	0.03	0.03	0.03	0.01	2.53	0.33	0.23
A10	A11	A12	A13	A14	A15	A16	A17	A18
0.08	0.35	2.74	0.40	0.18	0.05	0.09	0.15	0.12

Step 6. Finally, the ranking and the analysis of the information it is done. The ranking is presented in Table 9.

With the results obtained in Table 10, it is interesting that the Top 3 of the municipalities are always presented and in the same order that is Mazatlan, Culiacan and Ahome. This is important because it is possible to visualize that these three municipalities have an specific rank even when the weights are changed. After that, some interesting changes are presented such is the case of Mocorito, that can go from 7 to 4th depending on the operator that it is used, another similar municipality that have some important changes are Navolato, El Fuerte, Elota and San Ignacio. Finally, from ranking 10–18th there is no important change in the ranking.

With this analysis can be seen that the way that information is analyzed and the importance that is given to each indicator can change the ranking that each municipality has. Also, because of the information from some municipalities are 0, the results from tank 10–18th does not change at all, presenting an important challenge for that municipalities to obtain some results in the medium term to increase their results and increase their position in the ranking.

Table 10 Ranking based on different aggregation operators

Ranking	WA	OWA_{max}	OWA_{min}
1	Mazatlán	Mazatlán	Mazatlán
2	Culiacán	Culiacán	Culiacán
3	Ahome	Ahome	Ahome
4	Guasave	Guasave	Mocorito
5	Elota	Mocorito	Guasave
6	Navolato	Elota	El Fuerte
7	Mocorito	Navolato	Elota
8	San Ignacio	San Ignacio	Navolato
9	El Fuerte	El Fuerte	San Ignacio
10	Sinaloa	Sinaloa	Sinaloa
11	Angostura	Angostura	Angostura
12	Salvador Alvarado	Salvador Alvarado	Salvador Alvarado
13	Escuinapa	Escuinapa	Escuinapa
14	Rosario	Rosario	Rosario
15	Badiraguato	Badiraguato	Badiraguato
16	Concordia	Concordia	Concordia
17	Choix	Choix	Choix
18	Cosalá	Cosalá	Cosalá

5 Conclusions

This article presents a ranking of innovation and sustainability of tourist destinations in Sinaloa, Mexico through the operator OWA. A fuzzy systems tool that made it possible to compare the different levels of innovation and sustainability based on different weighting vectors. This analysis showed that with the OWA operator it is possible to obtain a ranking based on the importance of each criterion.

The results indicate the city of Culiacán as the recreation place with the best levels in its evaluation, while the municipality of Cosalá turned out to be the lowest valued establishment. In this sense, the use of this instrument represents an opportunity for decision makers because it enables the recognition of those factors that require greater attention, but also makes it possible to identify the areas with the best performance.

For future research, the application of more complex aggregation operators based on the OWA operator should be carried out, such as the heavy OWA [24] or the prioritized OWA [29]. In addition, it is necessary to make a broader analysis using information from different tourist destinations in Mexico and around the world.

References

1. Andrades L, Dimanche F (2017) Destination competitiveness and tourism development in Russia: issues and challenges. Tour Manage 62:360–376
2. Bagiran Ozseker D (2019) Towards a model of destination innovation process: an integrative review. Serv Ind J 39(3–4):206–228
3. Batle J, Orfila-Sintes F, Moon C (2018) Environmental management best practices: towards social innovation. Int J Hosp Manag 69:14–20
4. Beliakov G, Pradera A, Calvo T (2007) Aggregation functions: a guide for practitioners, vol 221. Springer, Berlin
5. Booyens I, Rogerson C (2016) Tourism innovation in the Global South: evidence from the Western Cape, South Africa. Int J Tour Res 18(5):515–524
6. Camisón C, Monfort-Mir V (2012) Measuring innovation in tourism from the Schumpeterian and the dynamic-capabilities perspectives. Tour Manage 33(4):776–789
7. Carayannis E, Ferreira F, Bento P, Ferreira J, Jalali M, Fernandes B (2018) Developing a socio-technical evaluation index for tourist destination competitiveness using cognitive mapping and MCDA. Technol Forecast Soc Change 131:147–158
8. Choi H, Sirakaya E (2006) Sustainability indicators for managing community tourism. Tour Manage 27(6):1274–1289
9. Crouch G, Ritchie J (1999) Tourism, competitiveness, and societal prosperity. J Bus Res 44(3):137–152
10. Drakulić Kovačević N, Kovačević L, Stankov U, Dragićević V, Miletić A (2018) Applying destination competitiveness model to strategic tourism development of small destinations: the case of South Banat district. J Destin Mark Manag 8:114–124
11. Dwyer L, Forsyth P, Rao P (2000) The price competitiveness of travel and tourism: a comparison of 19 destinations. Tour Manage 21(1):9–22
12. Gallardo-Vázquez D, Hernández-Ponce O, Valdez-Juárez L (2019) Impact factors for the development of a competitive and sustainable tourist destination. Case: Southern Sonora Region. Euro J Tour Hosp Recreat 9(2):3–14
13. Goffi G, Cucculelli M, Masiero L (2019) Fostering tourism destination competitiveness in developing countries: the role of sustainability. J Clean Prod 209:101–115
14. Hanna P, Font X, Scarles C, Weeden C, Harrison C (2018) Tourist destination marketing: from sustainability myopia to memorable experiences. J Destin Mark Manag 9:36–43
15. Huber G, Mungaray A (2017) Los índices de competitividad en México. Gestión y Política Pública 26(1):167–218
16. Huertas A, Moreno A, Ha My T (2019) Which destination is smarter? Application of the (SA)6 framework to establish a ranking of smart tourist destinations. Int J Inf Syst Tour 4(1):19–28
17. Iniesta-Bonillo M, Sánchez-Fernández R, Jiménez-Castillo D (2016) Sustainability, value, and satisfaction: model testing and cross-validation in tourist destinations. J Bus Res 69(11):5002–5007
18. Instituto Mexicano para la Competitividad (2020a). Índice de competitividad estatal 2020, que no vuelva a pasar: estados prevenidos valen por dos. Instituto Mexicano para la Competitividad. Ciudad de México
19. Instituto Mexicano para la Competitividad (2020b). Índice de competitividad urbana 2020, ciudades resilientes. Instituto Mexicano para la Competitividad. Ciudad de México
20. Instituto Tecnológico y de Estudios Superiores de Monterrey (2017) Índice de competitividad sostenible de los estados mexicanos (ICSEM 2017). Instituto Tecnológico y de Estudios Superiores de Monterrey. Monterrey
21. Jeong J (2016) A GIS-Supported approach with AHP & OWA for site suitability evaluation of sustainable rural housings towards ecotourism. Int J Fuzzy Syst Adv Appl 3:54–61
22. Knežević Cvelbar L, Dwyer L, Koman M, Mihalič T (2016) Drivers of destination competitiveness in tourism: a global investigation. J Travel Res 55(8):1041–1050

23. Kuščer K, Mihalič T, Pechlaner H (2017) Innovation, sustainable tourism and environments in mountain destination development: a comparative analysis of Austria, Slovenia and Switzerland. J Sustain Tour 25(4):489–504
24. León-Castro E, Avilés-Ochoa E, Merigó J, Gil-Lafuente A (2018) Heavy moving averages and their application in econometric forecasting. Cybern Systemas 49(1):26–43
25. Martínez-Pérez Á, Elche D, García-Villaverde P (2019) From diversity of interorganizational relationships to radical innovation in tourism destination: the role of knowledge exploration. J Destin Mark Manag 11:80–88
26. Martínez-Román J, Tamayo J, Gamero J, Romero J (2015) Innovativeness and business performances in tourism SMEs. Ann Tour Res 54:118–135
27. Pappas N (2015) Achieving competitiveness in Greek accommodation establishment during recession. Int J Tour Res 17(4):375–387
28. Pearce D (1997) Competitive destination analysis in Southeast Asia. J Travel Res 35(4):16–24
29. Pérez-Arellano L, León-Castro E, Avilés-Ochoa E, Merigó J (2019) Prioritized induced probabilistic operator and its application in group decision making. Int J Mach Learn Cybern 10(3):451–462
30. Rodríguez I, Williams A, Hall C (2014) Tourism innovation policy: implementation and outcomes. Ann Tour Res 49:76–93
31. Ronizi S, Mokarram M, Negahban S (2020) Utilizing multi-criteria decision to determine the best location for the ecotourism in the east and central of Fars province, Iran. Land Use Policy 99:105095
32. Shahzad S, Shahbaz M, Ferrer R, Kumar R (2017) Tourism-led growth hypothesis in the top ten tourist destinations: new evidence using the quantile-on-quantile approach. Tour Manage 60:223–232
33. Sigalat-Signes E, Calvo-Palomares R, Roig-Merino B, García-Adán I (2020) Transition towards a tourist innovation model: the smart tourism destination: reality or territorial marketing? J Innov Knowl 5(2):96–104
34. Teixeira S, Ferreira J, Wanke P, Moreira Antunes J (2019) Evaluation model of competitive and innovative tourism practices based on information entropy and alternative criteria weight. Tourism Econ XX(X):1–22
35. Thomas R, Wood E (2015) The absorptive capacity of tourism organisations. Ann Tour Res 54:84–99
36. Van der Zee E, Gerrets A, Vanneste D (2017) Complexity in the governance of tourism networks: balancing between external pressure and internal expectations. J Destin Mark Manag 6(4):296–308
37. World Economic Forum (2019) The travel & tourism competitiveness report 2019: travel and tourism at tipping point. World Economic Forum. Switzerland
38. Yager RR (1988) On ordered weighted averaging aggregation operators in multicriteria decisionmaking. IEEE Trans Syst Man Cybern 18(1):183–190
39. Yager, R. R., & Kacprzyk, J. (2012). *The ordered weighted averaging operators: theory and applications*. Springer Science & Business Media
40. Yi P, Li W, Zhang D (2019) Assessment of city sustainability using MCDM with interdependent criteria weight. Sustainability 11(6):1632
41. Zainuddin Z, Radzi S, Zahari M (2016) Perceived destination competitiveness of Langkawi Island, Malaysia. Procedia Soc Behav Sci 222:390–397

Dimensional Analysis Under Pythagorean Fuzzy Set with Hesitant Linguists Term Entropy Information

A. J. Villa Silva, L. A. Pérez Domínguez, E. Martínez Gómez, R. Romero López, and D. J. Valles Rosales

Abstract Dimensional Analysis (DA) is a method that consider an association of all the criteria involved in a problem, able to capture the interrelationship usually presents in multi-criteria problems. At the same time Pythagorean Fuzzy Set (PFS), is a recent tool used for handling fuzziness and vagueness, due is able to provide greater flexibility for decision makers to give their assessments. In addition, Multi-criteria decision making (MCDM) problems involves criteria predetermined weights and difficulty when information given is unknown or incomplete. This paper proposes the application and combination of three important tools: Dimensional Analysis, Pythagorean fuzzy sets and entropy measure for hesitant fuzzy linguistic term sets (HFLTSs) in order to solve the qualitative criteria, the interrelationship among the multiple criteria, and weights calculation when are unknown. Finally, an example case is given in order to show the functioning of the proposed hybrid method, and comparison with other weight methods.

Keywords Multi-Criteria Decision Making (MDCM) · Pythagorean Fuzzy Set (PFS) · Analysis Dimensional (DA) · Entropy

1 Introduction

Since theory of fuzzy sets was introduced by Zadeh in 1965 has reached an important success in several fields and became an important approach to handle uncertainty and inaccurate information that appears in several real life problems [1–3]. Since then, different versions of fuzzy set have been studied and proposed by some researchers [1]. Under this context, researchers are working with Pythagorean Fuzzy Set (PFS) in order to improve and developing studies concerning decision-makers are truly familiar with the criteria and alternatives evaluated [4–6].

A. J. Villa Silva (✉) · L. A. Pérez Domínguez · E. Martínez Gómez · R. Romero López
Universidad Autónoma de Cd. Juárez, Ciudad Juárez, Mexico

D. J. Valles Rosales
New Mexico University, Albuquerque, NM, USA

E. León-Castro et al. (eds.), *Soft Computing and Fuzzy Methodologies in Innovation Management and Sustainability*, Lecture Notes in Networks and Systems 337,
https://doi.org/10.1007/978-3-030-96150-3_2

After introducing PFS, Yager and Abbasov studied the relationship between the Pythagorean fuzzy numbers (PFNs) and the complex number [7]. In addition other proposed works has been introduced: Zhang and Xu [8] extended the TOPSIS approach concerning to hand the Multi-criteria decision-making (MCDM) problems in terms of Pythagorean fuzzy environment. Peng and Yang [9] proposed the division and subtraction equations for PFNs. Zhang [10] developed a closeness index for Pythagorean fuzzy number (PFN) and for interval-valued Pythagorean fuzzy number (IVPFN) based on distance measures of PFNs and IVPFNs. Garg [11] developed a new approach of Pythagorean fuzzy, using information aggregation by Einstein equations and applied it to decision making.

PFS is symbolized by three values: membership, non-membership and indeterminacy [3], but the main difference consist: the addition of the degree of membership and non-membership given by experts it can be more than unit, but its square sum is the same to or less than unit [12]. Particularly, if decision makers gives their valuations or perceptions information where membership grade is 0.9 and degree of non-membership is 0.5, you can know that the Intuitionistic Fussy Sets (IFS) does not address adequately this problem because $0.9 + 0.5 > 1$ their sum exceeds 1, IFS fail to handle such situations [13]. However, $0.9^2 + 0.5^2 < 1$ therefore the PFS has capacity to represent evaluation and characterize better the uncertainty by lack of clarity than IFS [14], this advantage provides a more powerful representation of uncertainty established by the Fuzzy intuitionist and therefore Fuzzy sets are best and proved tools for modeling uncertainty [3, 15].

In other hand Dimensional Analysis (DA) is a method with capacity to consider the mutual influence between several criteria [16], then, makes it properly in multi-criteria decision making (MCDM) problems in different scales of measuring, within of a single dimensionless index [16–18]. The most remarkable advantage of DA is concerning to join the valuations or perceptions of a group of decision makers (DM) based on different information, such as alternatives, criteria and their importance [16]. It should be noted that the DA is widely mentioned and applied in different industries, but there is a vast literature on its application in the agricultural and automotive sector [19]. Nevertheless, DA presents the weakness to operate with qualitative information usually involved in MCDM problems [19].

In other hand, entropy measure for hesitant fuzzy linguistic term sets (HFLTSs), is applied when information is missing, incomplete, or lots of information are lost [20, 21], due expressing decision maker´s opinion in uncertainty (caused by subjective weights) is hard to provide accurate values, therefore real numbers would change to linguistic terms which are closer to the human cognitive processes thru a proper predefined linguistic evaluation scale, is more adequate reasonable and applicable in real life problems [20, 22, 23].

However, in the literature [24–27] has been found a great amount of methods including fuzzy versions and others, but almost the majority of them has limitations, basically our purpose try to overcome the next limitations on MCDM problems:

- Consider that all the input criteria are independent and cannot consider the interrelationship among input criteria [28–30]. Few researches are concerned about consider the interrelationship among criteria in MDCM [28, 30, 31].
- There are limitations about the handling of subjective/ uncertainly information [31–33] regarding to MCDM problems.
- Weight/preference of the decision makers'. In several situations, performance ratings and weights cannot be given precisely [34], in some methods it is difficult to determine if the weights are used as trade-offs or importance coefficients [20, 35].

Based on the considerations mentioned above, this paper proposes the application and combination of three important tools: Dimensional Analysis, Pythagorean fuzzy sets and entropy measure for hesitant fuzzy linguistic term sets, in order to solve the qualitative criteria, the interrelationship among the multiple criteria, and weights calculation when are unknown. The structure of the paper is summarized as follows: in Sect. 2, basic concepts of Pythagorean Fuzzy sets (PFS), Dimensional Analysis (DA) and entropy measure for hesitant fuzzy linguistic term sets (HFLTSs) are described. In Sect. 3, description of integration of (PFS) and (DA) proposes a hybrid method, and an algorithm is given. In Sect. 4 an application is presented numerical example to illustrate our technique Dimensional Analysis-Pythagorean fuzzy (DA-PF), we present a comparison between DA-PFS with calculated weights and Entropy weights. Finally, the conclusion is provided in Sect. 5.

2 Preliminaries

In the following sections some fundamental concepts of PFS [9–15], DA [16–19] and Entropy [20–22] are described.

2.1 Pythagorean Fuzzy Sets

Definition 1 [35–46], if S, R ∈ PFSs equations are described as follows:

$$R \oplus S = \{< T, \sqrt{\mu_R^2(T) + \mu_S^2(T) - \mu_R^2(T)\mu_S^2(T)}, \nu_R(\mathrm{T})\nu_S(\mathrm{T})\}; \tag{1}$$

$$R \ominus S = \{< T, \sqrt{\frac{\mu_R^2(T) - \mu_S^2(T)}{1 - \mu_S^2(T)}}, \frac{\nu_R(T)}{\nu_S(T)} > |T \in X\} \tag{2}$$

$$R \oslash S = \{< T, \frac{\mu_R(T)}{\mu_S(T)}, \sqrt{\frac{\upsilon_R^2(T) - \upsilon_S^2(T)}{1 - \upsilon_S^2(T)}} > |T \in X\} \tag{3}$$

$$R \otimes S = \{< T, \mu_R(T)\mu_S(T), \sqrt{\upsilon_R^2(T) + \upsilon_S^2(T) - \upsilon_R^2(T)\upsilon_S^2(T)} > |T \in X\} \tag{4}$$

$$p^{\lambda} = (\mu^{\lambda}, \sqrt{1-(1-\nu^2)^{\lambda}}) \tag{5}$$

$$\lambda p = (\sqrt{1-(1-\mu^2)^{\lambda}}, \nu^{\lambda}) \tag{6}$$

$$s(p) = (\mu)^2 - (\nu)^2 \tag{7}$$

2.2 Dimensional Analysis

DA is a MCDM method applied in the decision-making process, that operates with an optimal solution better or chosen in a set of alternatives. DA operates with a comparison of each alternative in evaluation against optimal alternative and calculate an index of similarity, where the highest index of similarity is consider as the best alternative [16, 18].

Definition 1
Be $a_l^N (N = 1, \ldots, n)(M = 1, \ldots m)$ and $S_l^* = a_j^* (M = 1, \ldots, m)$ represent a data base of crisp values. DA equation is described as follows:

$$IS_i\left(a_1^i, a_2^i, \ldots, a_m^i\right) = \prod_{j=1}^{m} \left(\frac{a_j^i}{S_l^*}\right)^{w_j} \tag{8}$$

where IS_i represents the index of similarity for alternative i.

Where a_l^k represents crisp values of criterion l for alternative i.

Where S_l^* represents crisp values of the optimal alternative for criterion l.

Where $w_j(z = 1, \ldots, m)$ represents crisp weight value for criterion 1.

2.3 Entropy with Unknown Weights in Hesitant Fuzzy Linguistic Term Setting

According with Gou et al. [21], usually MCDM concerning two important steps: first: calculate criteria weights, and second: obtain an adequate ranking of alternatives. In accordance with Farhadinia [20]; described entropy measures, are applied to treat with the MCDM problems where information concerning criteria weights is missing or lack of data. The following equations stand for entropy measure based on generalized distance:

$$E_{dg}(H_\xi) = 1 - \frac{2}{N}\sum_{i=1}^{N}\left[\left(\frac{1}{L}\sum_{l=1}^{L}\left(\frac{|\delta_l|}{2\tau}\right)^\lambda\right)^\lambda\right], \lambda > 0 \tag{9}$$

Then, to calculate the entropy weights as follows:

$$W_j = \frac{1 - E_j}{m - \sum_{j=1}^{m} E_j}, j = 1, \ldots, m \tag{10}$$

Then we use linguistic labels that represents the preferences given by decision makers therefore a predefined linguistic evaluation scale is needed. For this, in accordance with Xu [47] a discrete linguistic term set is described as: $\vartheta = \{S_\alpha | \alpha = -T, \ldots, -1, 0, 1, \ldots, T\}$ where S_α represents a linguistic variable.

3 DA-PFS with Hesitant Entropy

In this section, we propose the hybrid method of DA and PFSs is given for dealing with both types of information, and an algorithm is proposed.

3.1 Dimensional Analysis Under Pythagorean Fuzzy Set (DA-PFS)

Based on Eq. (8) of DA described in Sect. 2, the definition of DA-PFS is described as follows: Let $\omega_j^i = \left(\mu_{w_j^i}, \nu_{w_j^i}\right)(i = 1, 2, \ldots n)(j = 1, 2 \ldots m)$ and $S_j^i = \left(\mu_{w_j^i}, \nu_{w_j^i}\right)(j = 1, 2 \ldots m)$ be a collection of P F_s, if:

$$PFIS_i\left(\omega_1^i, \omega_2^i, \ldots, \omega_m^i\right) = (\otimes_{j=1}^m (\frac{\omega_j^i}{S_j^*})^{T_j}) = (\otimes_{j=1}^m (\psi_j^i)^{T_j} \tag{11}$$

According with Eqs. (3–6) of the PFS given in Sect. 2 and Eq. (11), we deduct the next results.

Theorem 1 Let $\psi_j^i = (\mu_{\psi_j^i}, \nu_{\psi_j^i})(i = 1, 2, \ldots, n)(j = 1, 2, \ldots, m)$ be a set of PFS. Therefore, the aggregated value, by using $PFIS$, is also an IFN, and

$$PFIS_i\left(\omega_1^i, \omega_2^i, \ldots., \omega_m^i\right) = \left(\underset{j=1}{\overset{m}{\otimes}} \left(\psi_j^i\right)^{T_j}\right)$$

$$= \left(\prod_{j=1}^{m} \left(\mu_{\psi_j^i}\right)^{T_j}, \sqrt{1 - \prod_{j=1}^{m} \left(1 - \left(\nu_{\psi_j^i}\right)^2\right)^{T_j}} \right) \tag{12}$$

3.2 Algorithm for DA-PFS with Hesitant Entropy

According with above analysis, DA-PFS is described as follows:

- Step 1: Build Pythagorean decision matrix, preferences given by decision makers.
- Step 2: Choose optimal solution according to (BN) or (C) criteria values.
- Step 3: Calculate hesitant entropy criteria weights, by Eq. (9 and 10)
- Step 4: Calculate standardized matrix: according to (BN) criteria, or (C) criteria
- Step 5: Standardized matrix elevated according to entropy criteria weights, use Eq. (5)
- Step 6: Calculate Pythagorean fuzzy index, by Eq. (12)
- Step 7: Calculate the highest similarity index or defuzzy, by score Eq. (7)
- Step 8: The index similarity must be organized in descending order and choose the alternative with the highest value.

4 Application

4.1 Numerical Example

A Company, needs to evaluate and select the most properly Forklift machine, to choose the best forklift that offers maximum performance at lowest cost. There are five alternatives or brands to select (A1, A2, A3, A4, and A5), and eight criteria to consider: C1: Load capacity (pounds), C2: Maximum travel speed full load (mph), C3: Maximum lift speed full load (fpm), C4: Maximum grade ability full load (%), C5: Basic right angle stack (in), C6: More Intelligent, C7: Safer and C8: Robuster. Criteria C1-C5 are quantitative data that can be treated with simple AD, in other hand criteria C6-C8 are qualitative data, in this case AD-PFS is applied (Table 1).

According to algorithm proposed in Sect. 3, it is important to note that they should be treated separately criteria C1 to C5 using simple DA, and criteria C6-C8 using DA-PFS, then steps are the following:

In this part DA-PFS is applied: Step 1: In accordance with DM evaluations, the Pythagorean fuzzy decision matrix is defined as follows:

Table 1 Alternative and criteria Forklift selection

Options	C1	C2	C3	C4	C5	C6	C7	C8
A1	3000.491	10.9	110	43	110	{0.7,0.6}	{0.8,0.44}	{0.5,0.8}
A2	3999.185	10.9	110	35	87.6	{1.0,0}	{0.8,0.44}	{0.7,0.60}
A3	3999.185	10.6	120	36	92.3	{0.5,0.8}	{0.7,0.6}	{0.6,0.71}
A4	5000.084	11	125	31	95	{0.8,0.44}	{1.0,0}	{0.8,0.44}
A5	5511.557	14	130	55	115	{0.8,0.44}	{0.8,0.44}	{1.0,0}

$$\begin{bmatrix} \{0.70, 0.60\} & \{0.80, 0.44\} & \{0.50, 0.44\} \\ \{1.00, 0.00\} & \{0.80, 0.44\} & \{0.70, 0.60\} \\ \{0.50, 0.80\} & \{0.70, 0.60\} & \{0.60, 0.71\} \\ \{0.80, 0.44\} & \{1.00, 0.00\} & \{0.80, 0.44\} \\ \{0.80, 0.44\} & \{0.80, 0.44\} & \{1.00, 0.00\} \end{bmatrix}$$

Step 2: Establish ideal solution in accordance to criteria values:

$$S^{+} : \{1.00, 0.00\}\{1.00, 0.00\}\{1.00, 0.00\}$$

Step 3: Establish the entropy criteria weights use Eq. (9) and (10):

$$W_{\{C_6, C_7, C_8\}} = \begin{bmatrix} \{0.1345\} \\ \{0.1578\} \\ \{0.0935\} \end{bmatrix}$$

Step 4: In order to standardized matrix, use Eqs. (3) in accordance to BN or C:

$$\begin{bmatrix} \{0.70, 0.60\} & \{0.80, 0.44\} & \{0.50, 0.80\} \\ \{1.00, 0.00\} & \{0.80, 0.44\} & \{0.70, 0.60\} \\ \{0.50, 0.80\} & \{0.70, 0.60\} & \{0.60, 0.71\} \\ \{0.80, 0.44\} & \{1.00, 0.00\} & \{0.80, 0.44\} \\ \{0.80, 0.44\} & \{0.80, 0.44\} & \{1.00, 0.00\} \end{bmatrix}$$

Step 5: Then, each criteria column in standardized matrix is elevated with entropy criteria weights, use Eq. (5):

$$\begin{bmatrix} \{0.9532, 0.2414\} & \{0.9650, 0.1828\} & \{0.9370, 0.3019\} \\ \{1.0000, 0.0000\} & \{0.9650, 0.1828\} & \{0.9670, 0.2022\} \\ \{0.9110, 0.3583\} & \{0.9450, 0.2608\} & \{0.9530, 0.2520\} \\ \{0.9704, 0.1689\} & \{1.0000, 0.0000\} & \{0.9790, 0.1412\} \\ \{0.9704, 0.1689\} & \{0.9650, 0.1828\} & \{1.0000, 0.0000\} \end{bmatrix}$$

Step 6: Then, to generate an index of similarity $PFIS_i$ use Eq. (12):

$$\begin{bmatrix} 0.8624 & 0.4980 \\ 0.9337 & 0.3330 \\ 0.8209 & 0.5360 \\ 0.9504 & 0.2590 \\ 0.9368 & 0.2470 \end{bmatrix}$$

Step 7: To get the highest index of similarity use Eq. (7), however for this case we need to get the IS from the simple DA using Eq. (8) in order to solve criteria C1-C5 due they are quantitative data. Then we have the following matrix in accordance with forklift specifications:

$$\begin{bmatrix} 3000.491 & 10.9 & 110 & 43 & 110 \\ 3999.185 & 10.9 & 110 & 35 & 87.6 \\ 3999.185 & 10.6 & 120 & 36 & 92.3 \\ 5000.084 & 11 & 125 & 31 & 95 \\ 5511.557 & 14 & 130 & 55 & 115 \end{bmatrix}$$

In this part we use simple DA: Step 7.1: Establish ideal solution in accordance to criteria values:

$$S^+ : [5511.557, 14, 130, 55, 115]$$

Step 7.2: Establish the criteria weights, use Eqs. (9) and (10) from Hesitant Entropy:

$$W_{\{C_1,C_2,C_3,C_4C_5\}} = \begin{bmatrix} 0.0935 \\ 0.1481 \\ 0.1403 \\ 0.1091 \\ 0.1228 \end{bmatrix}$$

Step 7.3: Normalized matrix use Eq. (8):

$$\begin{bmatrix} 0.544 & 0.799 & 0.846 & 0.782 & 0.957 \\ 0.726 & 0.799 & 0.846 & 0.636 & 0.762 \\ 0.726 & 0.757 & 0.923 & 0.655 & 0.803 \\ 0.907 & 0.786 & 0.962 & 0.564 & 0.826 \\ 1.000 & 1.000 & 1.000 & 1.000 & 1.000 \end{bmatrix}$$

Table 2 Rankings with entropy-based weights of criteria

	AD	AD	PF IS	ΛP		SCORE	RANK
IS 1	0.861	0.862	0.498	0.831	0.549	0.389	4
IS 2	0.841	0.934	0.333	0.907	0.397	0.665	3
IS 3	0.856	0.821	0.536	0.785	0.587	0.272	5
IS 4	0.872	0.950	0.259	0.933	0.307	0.775	2
IS 5	1.000	0.937	0.247	0.937	0.247	0.817	1

Step 7.4: Normalized matrix weight elevated:

$$\begin{bmatrix} 0.945\ 0.964\ 0.977\ 0.973\ 0.995 \\ 0.970\ 0.964\ 0.977\ 0.952\ 0.967 \\ 0.970\ 0.960\ 0.989\ 0.955\ 0.973 \\ 0.991\ 0.965\ 0.995\ 0.939\ 0.977 \\ 1.000\ 1.000\ 1.000\ 1.000\ 1.000 \end{bmatrix}$$

Step 7.5: Establish AD index of similarity IS_i:

$$\begin{bmatrix} 0.8609 \\ 0.8409 \\ 0.8558 \\ 0.8725 \\ 1.0000 \end{bmatrix}$$

Step 8: Establish the ranking, use Eq. (6) and (7), and then we got the following calculations (Table 2):

Where: $A_4 > A_3 > A_5 > A_2 > A_1$, therefore A_4 is selected as the best Forklift machine, according to the highest index of similarity.

4.2 *Hesitant Entropy Weight*

In accordance with Farhadinia [20]; calculations for entropy measure based on generalized distance, are based in the hesitant fuzzy linguistic judgment matrix is given by decision makers in the Tables 3 and 4:

Using Eq. (9 and 10), we have entropy-based weights of criteria as follows:

$w1 = 0.093$, $w2 = 0.148$, $w3 = 0.140$, $w4 = 0.109$, $w5 = 0.122$, $w6 = 0.134$, $w7 = 0.157$, $w8 = 0.093$

Table 3 Hesitant fuzzy linguistic matrix given by the decision makers. Part 1

ALT	C1	C2	C3	C4
A1	{s−1, s−1, s−1}	{s−1, s0, s−1}	{s0, s0, s−1}	{s1, s1, s1}
A2	{s0, s0, s0}	{s1, s0, s0}	{s1, s1, s1}	{s0, s0, s0}
A3	{s0, s0}	{s0, s1, s1}	{s1, s1, s1}	{s1, s0, s0}
A4	{s1, s1, s1}	{s1}	{s1, s1, s2}	{s−1, s−1, s−1}
A5	{s2, s2, s2}	{s1, s2, s2}	{s3, s2, s2}	{s2, s2, s2}

Table 4 Hesitant fuzzy linguistic matrix given by the decision makers, part 2

ALT	C5	C6	C7	C8
A1	{s1, s2}	{s1, s0}	{s1, s1, s1}	{s−1, s−1, s−1}
A2	{s0, s0, s0}	{s2, s2}	{s1, s1, s1}	{s0, s0, s0}
A3	{s0, s0}	{s0}	{s1, s0}	{s0, s0, s0}
A4	{s1, s1, s1}	{s1, s1, s1}	{s2, s2}	{s1, s1, s1}
A5	{s2, s2, s2}	{s1, s1, s1}	{s1, s1, s1}	{s2, s2, s2}

4.3 *Sensitivity Analysis*

Sensitivity analysis is commonly used to ensure robustness of solutions [48]. In other words the sensitivity analysis can be described as stability or behavior when a solution is subjected to small changes by decision makers, or change the parameters values, and these small changes do not affect the result is consider an efficient multi-criteria decision method referring Pamučar and Ćirović [49].

4.3.1 Entropy

In accordance with Farhadinia [20]; calculations for entropy measure based on generalized distance, are based in the hesitant fuzzy linguistic judgment matrix provided by decision makers (Table 5).

Step 1: Using Eq. (9), we have the following (Table 6):

Then, using Eq. (9) we get that:

$$E_{1dg} 1 - \frac{2}{5} \sum_{i=1}^{5} E_{dg}\left(h_{\xi}^{il}\right) = 0.733$$

Table 5 The hesitant fuzzy linguistic judgment matrix provided by the decision organization

Alt	C1	C2	C3	C4	C5	C6	C7	C8
A1	{s−1, s−1, s−1}	{s−1, s0, s−1}	{s0, s0, s−1}	{s1, s1, s1}	{s1, s2}	{s1, s0}	{s1, s1, s1}	{s−1, s−1, s−1}
A2	{s0, s0, s0}	{s1, s0, s0}	{s1, s1, s1}	{s0, s0, s0}	{s0, s0, s0}	{s2, s2}	{s1, s1, s1}	{s0, s0, s0}
A3	{s0, s0}	{s0, s1, s1}	{s1, s1, s1}	{s1, s0, so}	{s0, s0}	{s0}	{s1, s0}	{s0, s0, s0}
A4	{s1, s1, s1}	{s1}	{s1, s1, s2}	{s−1, s−1, s−1}	{s1, s1, s1}	{s1, s1, s1}	{s2, s2}	{s1, s1, s1}
A5	{s2, s2, s2}	{s1, s2, s2}	{s3, s2, s2}	{s2, s2, s2}	{s2, s2, s2}	{s1, s1, s1}	{s1, s1, s1}	{s2, s2, s2}

Table 6 Determining the entropy-based weights of criteria by generalized distance

Alt	C1	C2	C3	C4	C5	C6	C7	C8
A1	0.167	0.111	0.056	0.167	0.375	0.125	0.167	0.167
A2	0.000	0.056	0.167	0.000	0.000	0.500	0.167	0.000
A3	0.000	0.111	0.167	0.056	0.000	0.000	0.125	0.000
A4	0.167	0.500	0.222	0.167	0.167	0.167	0.500	0.167
A5	0.333	0.278	0.389	0.389	0.333	0.167	0.167	0.333

In addition, we have:

$E_{2dg} = 0.578$, $E_{3dg} = 0.600$, $E_{4dg} = 0.689$, $E_{5dg} = 0.650$, $E_{6dg} = 0.617$, $E_{7dg} = 0.550$ & $E_{8dg} = 0.733$

Step 2: Consequently, the entropy-based weights of criteria using Eq. (10): c_j(j = 1, 2, 3, 4, 5, 6, 7 and 8) are achieved as:

$$W_1 = \frac{1 - 0.733}{8 - (0.733 - 0.578 - 0.600 - 0.689 - 0.650 - 0.617 - 0.550 - 0.733)}$$

Therefore, we have entropy-based weights of criteria as follows:

$W_1 = 0.093$, $W_2 = 0.148$, $W_3 = 0.140$, $W_4 = 0.109$, $W_5 = 0.122$, $W_6 = 0.134$, $W_7 = 0.157$, $W_8 = 0.093$

4.3.2 Fuzzy Weighted

Step 1: Establish a team of DM and capture preferences. If the $DM_k = \{\mu_{k,} v_{k,} \pi_k\}$ is a Pythagorean fuzzy number for DM, then we have the following:

$$\delta_k = \frac{\left(\mu_k + \pi_k\left(\frac{\pi_k}{\mu_k+v_k}\right)\right)}{\sum_{k=1}^{l}\left(\mu_k + \pi_k\left(\frac{\pi_k}{\mu_k+v_k}\right)\right)} \tag{13}$$

Apply Table 5 for DM preferences (Table 7):

Then we have three DM (Table 8):

Using Eq. (46), we get the following:

$DM_1 = 0.35$, $DM_2 = 0.35$, $DM_3 = 0.30$

Step 2: Establish preferences of criteria. Apply Table 5 now for criteria preferences (Table 9):

Step 3: Using Eq. (14), preferences must be gathered and mixed in just one, we have the following (Table 10):

$$PFWA_W = \left[\left(1 - \prod_{j=i}^{n}(1 - \mu_{\alpha_j}^2)^{w_j}\right)^{\frac{1}{2}}, \prod_{j=1}^{n} v_{\alpha_j}^{w_j}\right], \tag{14}$$

Step 4: in addition, we use again Eq. (13):

Table 7 Linguistic scale for DM preferences

Meaning	PFNs (μ, ν)
Apprentice (Ap)/Very Insignificant (VI)	(0.10, 0.90)
Leaner (Lr)/Insignificant (I)	(0.35, 0.60)
Capable (Ct)/Average (A)	(0.50, 0.45)
Skillful (S)/Imperative (Im)	(0.75, 0.40)
Dominant (D)/Very Significative (VS)	(0.90, 0.10)

Table 8 DM preferences

	DM1			DM2			DM3	
μ	ν	π	μ	ν	π	μ	ν	π
0.9	0.1	0.42	0.9	0.1	0.42	0.75	0.4	0.53

Table 9 Criteria preferences

C1	C2	C3	C4	C5	C6	C7	C8
0.50, 0.45	0.90, 0.10	0.75, 0.40	0.50, 0.45	0.90, 0.10	0.75, 0.40	0.90, 0.10	0.75, 0.40
0.90, 0.10	0.35, 0.60	0.90, 0.10	0.90, 0.10	0.35, 0.60	0.90, 0.10	0.35, 0.60	0.90, 0.10
0.75, 0.40	0.90, 0.10	0.90, 0.10	0.75, 0.40	0.90, 0.10	0.90, 0.10	0.90, 0.10	0.90, 0.10

Table 10 Criteria preferences gathered in one

C1	C2	C3	C4	C5	C6	C7	C8
0.78, 0.26	0.82, 0.19	0.86, 0.16	0.78, 0.26	0.82, 0.19	0.86, 0.16	0.82, 0.19	0.86, 0.16

Table 11 Hesitant entropy weight versus fuzzy weighted

C1	C2	C3	C4	C5	C6	C7	C8
0.093	0.148	0.14	0.109	0.122	0.134	0.157	0.093
0.12	0.13	0.13	0.12	0.13	0.13	0.13	0.13

$W_1 = 0.12, W_2 = 0.13, W_3 = 0.13, W_4 = 0.12, W_5 = 0.13, W_6 = 0.13, W_7 = 0.13,$ $W_8 = 0.13$

Then have calculated the weights of criteria in both methods we got the same result:

$$A_4 > A_3 > A_5 > A_2 > A_1$$

In other hand we compare Hesitant Entropy weight and Fuzzy weighted calculated in the previous sections (Table 11).

Results reveal that: DA-PFS with entropy or fuzzy weighted, the alternative A_4 is selected as the best Forklift machine, according to the highest index of similarity and the rankings are consistent.

5 Conclusion

In this paper we have introduced a method DA-PFS with entropy measure for hesitant fuzzy linguistic term sets, in order to solve the qualitative criteria where exist uncertainty and lack of clarity [14], the interrelationship among the multiple criteria [16], and weights calculation when are unknown and prevent the loss of lots or sets information when the process is being carried out [50]. This method combines the best features of DA which consists the capacity to consider the mutual influence between several criteria [16], PFS, has capacity to represent evaluation and characterize better the uncertainty by lack of clarity [15], and HFLTSs by using linguistic labels instead numbers due are closer to the human cognitive processes [20, 22, 23].

As a result of this combination of tools, we obtain a more robust tool capable of considering information not involved in classical methods or their fuzzy extensions, mainly given in TOPSIS and AHP.

In addition we have compared different results concerning to weight concepts: a comparison, between Hesitant Entropy weight against Fuzzy weighted, then we got the same result, however, Hesitant Entropy weight requires less steps, therefore it's more efficient.

For the near future, we advise apply the conjugation of these tools in different fields where exist uncertainty, unknown criteria weights and interrelationship between criteria be an important factor.

References

1. Li D, Zeng W (2018) Distance measure of pythagorean fuzzy sets. Int J Intell Syst 33(2):348–361
2. Zhou J, Su W, Baležentis T, Streimikiene D (2018) Multiple criteria group decision-making considering symmetry with regards to the positive and negative ideal solutions via the Pythagorean normal cloud model for application to economic decisions. Symmetry (Basel) 10(5)
3. Zeng S, Mu Z, Baležentis T (2018) A novel aggregation method for Pythagorean fuzzy multiple attribute group decision making. Int J Intell Syst 33(3):573–585
4. Garg H (2017) A novel improved accuracy function for interval valued pythagorean fuzzy sets and its applications in the decision-making process. Int J Intell Syst 32(12):1247–1260
5. H. Garg H (2017) Confidence levels based Pythagorean fuzzy aggregation operators and its application to decision-making process. Comput Math Organ Theory 23(4):546–571
6. Bolturk E (2018) Pythagorean fuzzy CODAS and its application to supplier selection in a manufacturing firm. J Enterp Inf Manag 31(4):550–564
7. Yager RR, Abbasov AM (2013) Pythagorean membership grades, complex numbers, and decision making. Int J Intell Syst 28(5):436–452
8. Zhang X, Xu Z (2014) Extension of TOPSIS to multiple criteria decision making with pythagorean fuzzy sets. J Int Syst Intell
9. Peng X, Yuan H, Yang Y (2017) Pythagorean fuzzy information measures and their applications. Int J Intell Syst 32(10):991–1029
10. Zhang X (2016) A novel approach based on similarity measure for Pythagorean fuzzy multiple criteria group decision making. Int J Intell Syst 31:593–611
11. Peng X, Yuan H, Yang Y (2017) Pythagorean fuzzy information measures and their applications. I J I Syst 32:991–1029
12. Mandal P, Ranadive AS (2018) Decision-theoretic rough sets under Pythagorean fuzzy information. Int J Intell Syst 33(4):818–835
13. Hussain A, Mahmood T, Irfan M (2019) Rough Pythagorean fuzzy ideals in semigroups. Comput Appl Math 38(2):1–15
14. Wan SP, Jin Z, Dong YJ (2018) Pythagorean fuzzy mathematical programming method for multi-attribute group decision making with Pythagorean fuzzy truth degrees. Knowl Inf Syst 55(2):437–466
15. Çoban V, Onar SÇ (2018) Pythagorean fuzzy engineering economic analysis of solar power plants. Soft Comput 22(15):5007–5020
16. Pérez-Domínguez L, Alvarado-Iniesta A, García-Alcaraz JL, Valles-Rosales DJ (2018) Intuitionistic fuzzy dimensional analysis for multi-criteria decision making. Iran J Fuzzy Syst 15:17–40

17. García JL, Ju C (2010) Evaluación y Selección de Tractores Agrícolas con Análisis Dimensional. Academia J.com Rev Ing Ind 4:1–8
18. Pérez-Domínguez L, Alvarado-Iniesta A, García-Alcaraz JL, Valles-Rosales DJ (2015) Análisis dimensional difuso Intuicionista para la selección de personal. Proc Congreso Internacional de Investigación Academia J 2:20–24
19. Cristina I, Rendón A (2015) Selección y certificación de proveedores : un camino hacia el mejoramiento de la gestión de. 4(4):205–230
20. Farhadinia B (2016) Multiple criteria decision-making methods with completely unknown weights in hesitant fuzzy linguistic term setting. Knowl Based Syst 93:135–144
21. Gou X, Xu Z, Liao H (2017) Hesitant fuzzy linguistic entropy and cross-entropy measures and alternative queuing method for multiple criteria decision making. Inf Sci (NY) 388–389:225–246
22. Peng XD, Yang Y, Song JP, Jiang Y (2015) Hesitant fuzzy linguistic decision method based on combination weight. Comput Eng 41(9):190–198
23. Chen N, Xu ZS, Xia MM (2013) Interval-valued hesitant preference relations and their applications to group decision making. Knowl Based Syst 37:528–540
24. Makui A, Gholamian MR (2016) A hybrid intuitionistic fuzzy multi-criteria group decision making approach for supplier selection. J Optim Ind Eng 20:61–73
25. Jafarnejad A, Aghasi E (2015) An integration between fuzzy Promethee and fuzzy linear program for supplier selection problem: case study. 4(1):79–88
26. Zhu B, Xu Z (2016) Extended hesitant fuzzy sets. Technol Econ Dev Econ 22(1):100–121
27. Teng F, Liu Z, Liu P (2018) Some power Maclaurin symmetric mean aggregation operators based on Pythagorean fuzzy linguistic numbers and their application to group decision making. Int J Intell Syst 33(9):1949–1985
28. Hung Y, Huang T, Hsieh J, Tsuei H, Cheng C (2012) Online reputation management for improving marketing by using a hybrid MCDM model. Knowl-Based Syst 35:87–93
29. Garg H (2018) SS symmetry multi-criteria decision-making method based on prioritized muirhead mean aggregation operator under neutrosophic set environment. Symmetry 10:280
30. Sun L, Dong H, Liu AX (2018) Aggregation functions considering criteria interrelationships in fuzzy multi-criteria decision making : state-of-the-art. IEEE Access 6:68104–68136
31. Shen K (2018) Advances in multiple criteria decision making for sustainability : modeling and applications. Ml:1–7
32. Chatterjee K, Zavadskas EK, Adhikary K, Kar S (2018) A hybrid MCDM technique for risk management in construction projects. Symmetry 10:46
33. Ahmed R, Kim S (2018) Uncertainty-aware Integrated Fuzzy AHP-WASPAS model to evaluate public cloud computing services. Procedia Comput Sci 130:504–509
34. Kannan D, Beatriz A, Sousa LD, José C, Jabbour C (2014) Selecting green suppliers based on GSCM practices: using fuzzy TOPSIS applied to a Brazilian electronics company. Eur J Oper Res 233(2):432–447
35. Zhang X, Xu Z (2014) Extension of TOPSIS to multiple criteria decision making with pythagorean fuzzy sets. J. Int Syst Intell 29:1061–1078
36. Naz S, Ashraf S, Akram M (2018) A novel approach to decision-making with pythagorean fuzzy information. Mathematics 6(6):95
37. Peng X, Yuan H, Yang Y (2017) Pythagorean fuzzy information measures and their applications. Int J Intell Syst 32:991–1029
38. Wei G, Lu MAO (2017) Dual hesitant pythagorean fuzzy Hamacher aggregation operators in multiple attribute decision making. 27(3):365–395
39. Liang W, Zhang X, Liu M (2015) The maximizing deviation method based on interval-valued pythagorean fuzzy weighted aggregating operator for multiple criteria group decision analysis. Discret Dyn Nat Soc
40. Wei G (2018) Models for green supplier selection in green supply chain management with pythagorean 2-Tuple linguistic information. IEEE 6:18042–18060
41. Yager R (2013) Pythagorean fuzzy subsets. Proc 2013 Jt. IFSA World Congr NAFIPS Annu Meet IFSA/NAFIPS 2013 2(x):57–61

42. Xu Q, Yu K, Zeng S, Liu J (2017) Pythagorean fuzzy induced generalized owa operator and its application to multi-attribute group decision making. Int J Innov Comput Inf Control 2017(13):1527–1536
43. Rahman K, Ali A (2018) Some interval-valued pythagorean fuzzy weighted averaging aggregation operators and their application to multiple attribute decision making. J Math 50:113–129
44. Liang D, Xu Z (2017) The new extension of TOPSIS method for multiple criteria decision making with hesitant Pythagorean fuzzy sets. Appl Soft Comput J 60:167–179
45. Sajjad M, Khan A (2018) Pythagorean fuzzy prioritized aggregation operators and their application to multi-attribute group decision making. Granul Comput
46. Biswas A, Sarkar B (2018) Pythagorean fuzzy multicriteria group decision making through similarity measure based on point operators. Int J Intell Syst 33:1731–1744
47. Xu ZS (2012) Linguistic decision making: theory and methods. Science Press, Beijing
48. Yazdani M, Zavadskas EK, Ignatius J, Abad MD (2016) Sensitivity analysis in MADM methods: application of material selection. Eng Econ 27(4):382–391
49. Pamučar D, Ćirović G (2015) The selection of transport and handling resources in logistics centres using Multi-attributive Border Approximation area Comparison (MABAC). Expert Syst Appl 42:3016–3028
50. Biswas A (2019) Pythagorean fuzzy TOPSIS for multicriteria group decision—making with unknown weight information through entropy measure. Int J Intell Syst 1–21

Wages Returns in Mexico: A Comparison Between Parametric and Nonparametric Approaches

Berenice Mendoza, Salvador Cruz Aké, and Fernando Ávila Carreón

Abstract This work studies the existence of salary differences with parametric and non-parametric tools, from a gender perspective using the National Household Income and Expenditure Survey (ENIGH). We use Mincer's equation (parametric) to control for traditional factors in wage analysis; we also use decision trees to verify the gender gap while controlling for educational level and work experience in Mexico (non-parametric) and discriminant analysis to observe other elements that influence salary. We have found that a man can have better returns on his wages than a woman, other factors such as remittances were also analyzed: scholarships, current monetary expenditure, overtime, independent income, property income, among others, all this, based on in the INEGI Survey. With this research it is concluded that women will have to have more work experience and better studies to achieve better salaries, that is, they will have to try harder than they do.

Keywords Gender gap · Wage return

1 Introduction

Human capital is the knowledge that each individual possesses. As he increases his knowledge of him his potential will grow. Attitudes are found on the basis of all human capital; in them you can see the behaviors of people and how all behavior can be modified [51].

B. Mendoza (✉)
Faculty of Accounting and Administrative Sciences, Michoacan University of San Nicolás de Hidalgo, Morelia 58004, México

S. C. Aké
National Polytechnic Institute, Mexico City 11350, México
e-mail: scruza@ipn.mx

F. Á. Carreón
Department of Basic Sciences, Technological Institute of Morelia, Morelia 58120, México
e-mail: fernando.ac@morelia.tecnm.mx

E. León-Castro et al. (eds.), *Soft Computing and Fuzzy Methodologies in Innovation Management and Sustainability*, Lecture Notes in Networks and Systems 337,
https://doi.org/10.1007/978-3-030-96150-3_3

Since the issue of human capital began, it has been having more and more boom in various investigations, and performance models, it begins as the income received by workers, now it has its own identity within the branch of the economy. This topic has been perceived by economists as a phenomenon that varies rapidly within the Theory of Supply and Demand (we can understand it as the interaction of the consumer and the producer of a good in relation to its price and sale), in our case and as an example education can be seen as a tangible and intangible good for the consumer. Research on education and wages, the formation of human capital, is a classical subject in Economics and Social Sciences. Authors as [32, 40, 47] analyzed the relation between the human capital and wages controlling by variables such as race, ethnicity, gender, social environment and others.

The concept of human capital is a relevant topic in Economics since the middle of the past century. Authors as [3, 4, 41, 48] emphasize its importance and formation process in developed economies. The human capital formation process is so crucial that International organizations decided to study its formation closely, stock, classes and its role as a trigger for economic development, *V.g.* [35], [20, 36, 33] or [50].

Despite its lengthy study, the topic is far from exhausted. Authors as [8, 29, 43] analyzed the subject with a fresh perspective incorporating issues as the new technologies accessibility and the public policy role on the education and capability acquisition process. Authors as [25, 7, 30, 45, 31, 37, 42, 21] are updating and expanding the field using novelle econometrics techniques, economic theories or new databases.

The subject is also relevant for developing economies such the Mexican. Authors as [16, 46, 28, 34] applied and broadened the classic papers in Economics of Education to the Latinamerican and Mexican context, where the social class segmentation, differences between urban and rural environments and culture create a different context for the human capital formation process.

The human capital formation in México, as a particular case, is a previously studied topic. Authors as [23, 1, 27] studied the education's rentability and its unequal returns due to social factors.

The objective of this research is to study human capital, but seen from the existence of salary differences by gender between Mexico City and Michoacán using the National Survey of Household Income and Expenditure (ENIGH) of 2016,taking into consideration factors such as education and work experience, the Mincer equation is developed. In addition to the above, through data mining, Mincer is analyzed with decision trees. Subsequently, observing factors such as scholarships, remittances, current monetary expenditure, overtime, independent income, property income, among others, it is analyzed which of them influences the salary the most, through a matrix of structures.

Data mining uses large sets of information and observes certain similar behavior patterns, it has been used as a tool for different investigations, but in particular for the analysis of the gender gap in different countries [24, 13, 14, 38, 17].

The methodology used for this article is first a selection of data from the survey through data mining, then regressions that are used with the support of the Eviews-10 program, the regression of the Mincer equation is calculated. To make the decision trees, data mining is taken into consideration first and then the trees are made with the SPSS Statistical program. Likewise, for the case of the matrix of structures, we use Statistical SPSS.

Therefore, with this research, we first begin with data mining, to continue with the Mincer regressions followed by the decision trees. Based on the information from the decision trees, an analysis of outliers is carried out to determine the educational level that most influences wages in two states of the Mexican Republic, Mexico City and Michoacán. We close with the calculation of various factors for the matrix of structures.

Among the important conclusions that were obtained is that education influences men more to improve their salary than women (using Mincer's equation) and that they influence them for a complete professional good, instead they have to have a higher level of studies, that is to say, postgraduate. Specifically, human capital influences men more than women. Using the outliers, it is observed that in Mexico City the fact of having a higher degree of studies is better evaluated than in Michoacán.

Next we will talk about the database on which we based this study, we will also talk about the Mincer equation, as a parametric analysis. Later we will start with the non-parametric studies that are decision trees with Mincer data, followed by outliers and discriminant analysis.

2 Preliminaries

The ENIGH[1]is a biannual strata survey conducted by the INEGI[2] to analyze the expenses, income, occupation and social features (*V.g* house characteristics and home equipment) from selected households. The households are not the same in each survey, but the methodology tries to represent the same population mix. For more details, see [19].

The ENIGH consists of depth analysis of the amounts, origin and distribution of income among each economic unit (home). The survey has its origins in the 1956 survey performed by the General Directorate of Statistics (DGE). Due simplicity and to avoid comparability issues, we will focus our efforts on the 2016 ENIGH survey [19].

[1] National Survey of Household Income and Expenditure.

[2] National Institute of Geography and Statistics, the Mexican equivalent to de Census Bureau in the USA.

2.1 Mincer Equation Analysis for Mexico

With its long academic tradition, several authors, with a great variety of data and techniques, have studied the Mincer equation. As examples of this pleiad of works are the papers of [26, 10, 11, 49, 6, 34, 40, 41, 12, 15, 2, 18, 22]. The most common representation of the Mincer equation is:

$$\ln Y = \beta_0 + \beta_1 s + \beta_2 t + \beta_3 t^2 + \varepsilon$$

where:

Y Wage.

s Schooling.

t Years of labour experience.

Departing from the [32] paper, we analyze the household head's income, controlling by education, sex and labour experience using the ENIGH 2016 database. We performed the econometric analysis using the EViews-10.

$$\ln sal = c(1) + c(2) * s + c(3)w + c(4) * w^2$$

where:

$\ln sal$ Natural logarithm of salaries of the heads of family.

s schooling of the head of the family.

w work experience of the head of the family.

w^2 work experience of the head of the family squared.

The base model shows that schooling is the most significant regressor in this model, although the R^2 is small, the coefficients' signs are correct. The low R^2 indicates that there are other explanatory variables not included in the regression, examples of those variables are sex, their city (a combination of cluster effect and physical capital availability) or social context (V.g. presence of children or elders that avoids formal labour).

2.2 Mincer with Decision Trees for Mexico

We use decision trees to explore hidden hierarchy in the data. This hidden information may appear using a different classification or segmenting the data. Using the decision tree's terminology, we can analyze the nodes (represented by a table) as the critical factor that classifies observations. In the same manner, we cut off some terminal nodes (in which all cases have the same value in the dependent variable), and branches that

Table 1 Regression from Mincer equation for Mexico

ln sal = c(1) + c(2)* s + c(3)*w + c(4)*w²				
Both genders				
	Coefficient	Std. Error	t-Statistic	Prob
C(1)	7.943488	0.02971	267.3699	0
Schooling	0.098671	0.001274	77.43293	0
Work_Experience	0.025331	0.000971	26.07429	0
Work_experience²	−0.000322	1.31E−05	−24.47134	0
R-squared	0.114403	Mean dependent var		9.87127
Adjusted R-squared	0.114356	S.D. dependent var		1.047612
S.E. of regression	0.985893	Akaike info criterion		2.809534
Sum squared resid	54,979.4	Schwarz criterion		2.810166
Log likelihood	−79,460.85	Hannan-Quinn criter		2.809731
F-statistic	2435.682	Durbin-Watson stat		0.043545
Prob(F-statistic)	0			
Men				
C(1)	7.93656	0.03256	243.7513	0
Schooling	0.100353	0.001408	71.25186	0
Work_Experience	0.027873	0.00111	25.1194	0
Work_experience²	−0.000374	1.55E−05	−24.10726	0
R-squared	0.127833	Mean dependent var		9.914724
Adjusted R-squared	0.127772	S.D. dependent var		1.025396
S.E. of regression	0.95765	Akaike info criterion		2.751425
Sum squared resid	38,946.18	Schwarz criterion		2.75224
Log likelihood	−58,423.88	Hannan-Quinn criter		2.751682
F-statistic	2074.795	Durbin-Watson stat		0.051419
Prob(F-statistic)	0			
Woman				
C(1)	7.799551	0.069703	111.8976	0
Schooling	0.096349	0.002877	33.49473	0
Work_Experience	0.024085	0.002068	11.6488	0
Work_experience²	−0.000245	2.58E−05	−9.493796	0
R-squared	0.082868	Mean dependent var		9.740353
Adjusted R-squared	0.082672	S.D. dependent var		1.101587

(continued)

Table 1 (continued)

ln sal = c(1) + c(2)* s + c(3)*w + c(4)*w^2				
Both genders				
	Coefficient	Std. Error	t-Statistic	Prob
S.E. of regression	1.055069	Akaike info criterion		2.945374
Sum squared resid	15,687.93	Schwarz criterion		2.947517
Log likelihood	−20,756.47	Hannan-Quinn criter		2.946087
F-statistic	424.4587	Durbin-Watson stat		0.084586
Prob(F-statistic)	0			

Note: Own elaboration Source: ENIGH 2016

do not contribute decisively to the wage's explanation. For more details on the tree's pruning see [5, 39] or [44].

In this paper, we use classification trees, using the SPSS Statistics 23, to detect the importance of each factor in the wage of each household head. Under the assumption of normality on regressors, we calculated the CHAID (Chi-square automatic interaction detector) to detect the most robust interactions between the dependent and independent variables. We state the model as follows:

- Dependent variable: Wage's natural logarithm.
- Independent variables: work experience, formal education and gender of the head of the household.

We base our analysis on the Mincer equation. We took as dependent variable the natural logarithm of wages. Also, the independent variables are schooling, work experience and squared work experience. The most important result is that schooling affects wages in a differenced way. The first years are not so influential, while the superior and postgraduate education heavily affects the wage. When we observe the post-graduated branch, we noted that the experience becomes relevant only for the highest salaries, reshaping the Mincer's equation.

We explain this because of the relative scarcity of highly educated personnel in México when compared to its demand. In this part of the analysis, we are still using the complete sample to get a baseline in which we can compare the differences when the sex is taken into account. We show the Mincer tree (Fig. 1).

On the analysis's next step, we take into consideration the sex of the household's head while using the same regressors. We show the results in the next illustration (Fig. 2).

2.3 *Outliers*

As noted previously on other papers, the wages follow a Pareto distribution, so it is common to find outliers on its distribution. With the idea of studying those pieces of

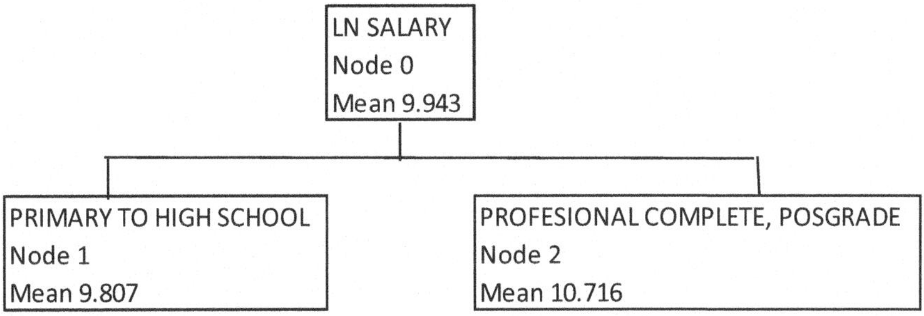

Fig. 1 Mincer decision tree in Mexico Own elaboration

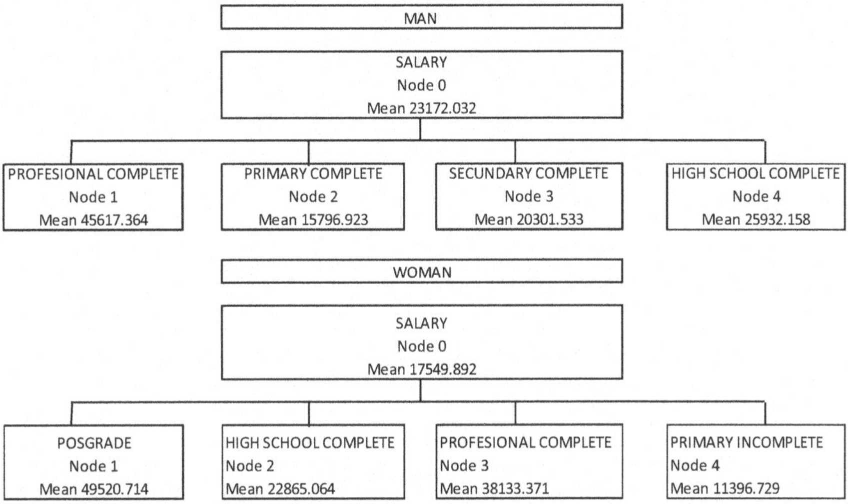

Fig. 2 Decision tree for Mexico Own elaboration

information, we observe the outliers in salaries emphasizing its sex while controlling for formal education. The most important conclusion in this paper remains that higher education affects salaries more than basic education.

Next figure shows the wage's median, which remains stable except for postgraduate levels, which are higher. It is also remarkable that most of the outliers are males and that wages tend to be higher in Mexico City than in Michoacán, possibly because of cluster effects on capital accumulation (Figs. 3 and 4).

2.4 *Discriminant Analysis*

The discriminant analysis is a statistical method that characterizes two or more objects, classifying them using dimensional reduction [9]. We use the discriminant analysis to identify any possible difference between the wages of both sexes.

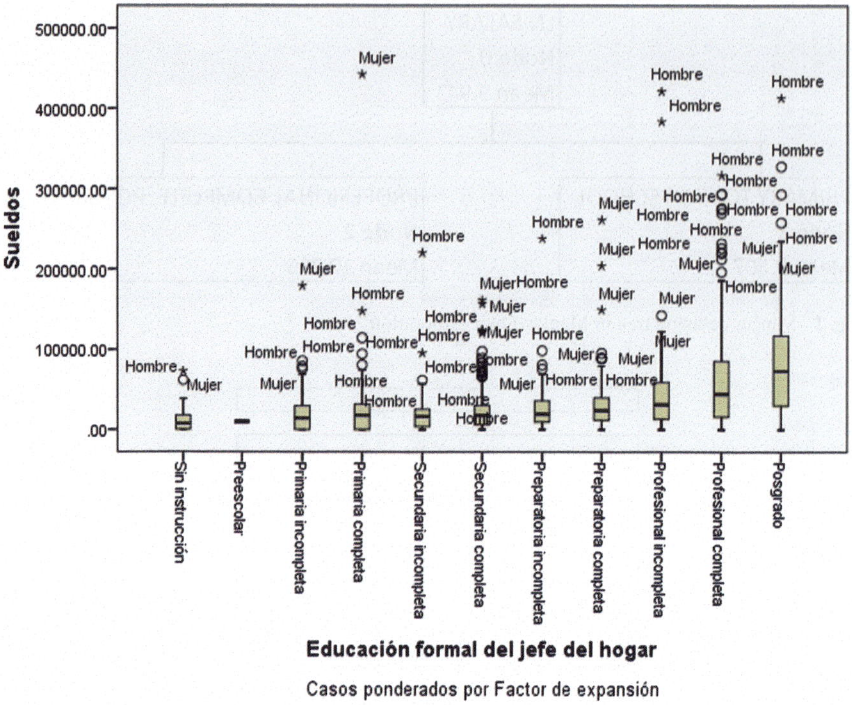

Fig. 3 Outliers Mexico City, 2016. *Note* Own elaboration with data from ENIGH 2016

To control for other variables that influence the wages, we added variables such as extra hours worked, independent income, property income, scholarships, remittances, living place, and home characteristics. We made the analysis using SPSS for the entire database, emphasizing the sex of the household's head, we show our results in Table 2.

Table 2 shows the structure matrix. In this matrix, we can observe the correlations within the combined groups; these variables are ordered by the absolute size of the correlation within the function (Table 3).

3 Conclusions

Calculating human capital through the Mincer equation, first for both genders and later for men and women, gives us a difference between them and women in schooling, which tells us that if men have a higher academic degree they will be able to get better salaries, but it is also better for them if they have more work experience than women.

In the analysis of salaries, they are below the average both in schooling and educational level, so they will probably have to make more effort in studies and work experience, since these factors favor men in explanation for their salary.

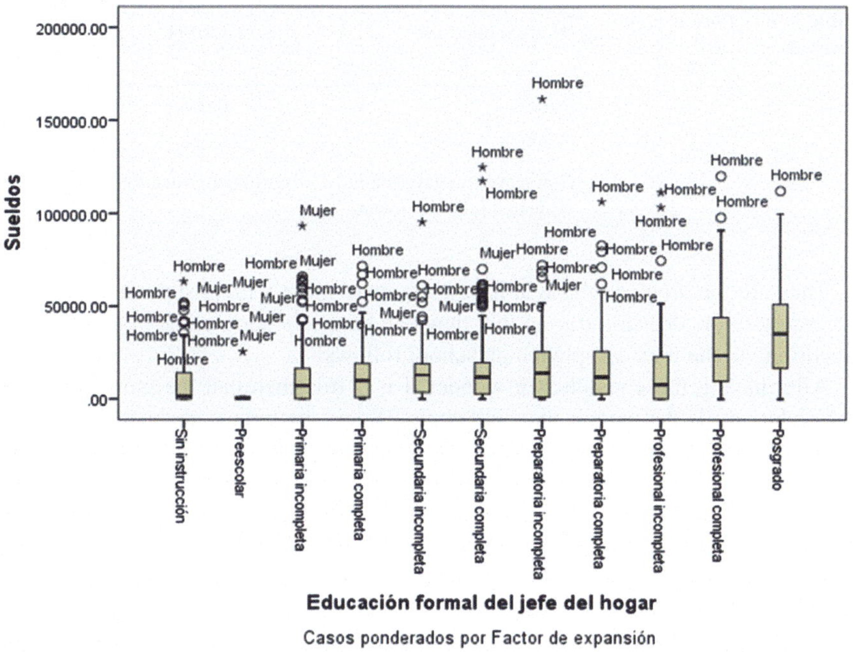

Fig. 4 Outliers Michoacan, 2016. *Note* Own elaboration with data from ENIGH 2016

Table 2 Standardized canonical discriminant function coefficients

	Función
	1
Salaries	0.485
Income recipients	0.401
Monetary current expenditure	0.343
Extra hours	0.045
Independent income	0.203
Property income	−0.052
Schoolarships	−0.017
Remitrances	−0.542
Living place	−0.332
Propoerty and installments	−0.06
Water	−0.082

When asked about the educational degree then that will help women or men to have a better income, decision trees were chosen, using the Mincer variables, obtaining that it is enough for them to have a full professional career, to despite the fact that the average of both sexes shows that it is from primary to secondary.

Table 3 Functions in group centroids

Householder sex	Function
	1
Men	0.088
Women	−0.228

Non-standardized canonical discriminant functions have been evaluated in group means

They are the women who will have to prepare more with postgraduate degrees according to the decision tree so that they can have a better income in their salary. According to the tree, complete high school follows.

After having these results, we wonder if it is true then that the more studies a person has, the better income they will obtain? So we continue with the outliers, here we had to perform data mining to select only two entities, it would be Mexico City and Michoacán those selected to analyze, finding that indeed the mean of the candle can be seen how it grows as the studies increase. Of course, it is in the capital of Mexico where a clear trend of salary improvement with better educational levels is observed. In parametric analysis it allowed us to know precisely how much education and work experience influence both sexes and later separately, but the non-parametric analysis defines us on which educational level influenced more on each gender (regressions and decision trees), therefore that one is a complement to the other.

To finalize the study, we decided to analyze other factors that may influence salaries, such is the case of scholarships, remittances, current monetary expenditure, overtime, independent income, property income, among others, noting that there are some that have a negative influence, such as in the case of rents, housing taxes, payment of water, among others, which will affect income.

At the end of this research we realized that it is interesting to know that women will have to make more effort in their studies and work experience in order to obtain better benefits in their salary.

References

1. Angulo H, Rojas M, Velázquez I (2000) Rentabilidad de La Inversión En Capital Humano En México Rentabilidad de La Inversión En Capital Humano En México. Economía Mexicana. Nueva Época IX (2000):113–42
2. Austria MA, Venegas-Martínez F (2011) Rendimientos Privados de La Educación Superior En México En 2006. El Trimestre Económico LXXVIII(2):441–68. http://www.eltrimestreeconomico.com.mx/index.php/te/article/view/39/41
3. Becker GS (1962) Investment in human capital: a theoretical analysis. J Political Econ. https://doi.org/10.1086/258724
4. Benhabib J, Spiegel M (1994) The Role of Human Capital in Economic Development. Evidence from Aggregate Cross-Country Data. J Monet Econ 144:117–143. https://doi.org/10.1016/0304-3932(94)90047-7

5. Berlanga V, Rubio MJ, Vila R (2013) Cómo Aplicar Árboles de Decisión En SPSS. Reire Revista D´Innovació i Recerca En Edicació 6:1–120. https://doi.org/10.1344/reire2013.6.1615
6. Cabañete A (1997) La Demanda de Educación Superior En El Sistema Público Catalán. Un Modelo Estructural de Previsión de La Demanda Universitaria Pública Presencial En Cataluña. Universitat Politécnica de Catalunya. Departament D´Organització D´Empreses.
7. Sergio C (2015) Tendencias Globales En Capital Humano 2015. Oderando En El Nuevo Mundo Del Trabajo. Deloitte University Press, pp 28–72.
8. Ehrenberg RG, Smith RS (2012) Modern labor economics theory and public policy - eleventh edition. journal of chemical information and modeling, vol 11. Prentice Hall. https://doi.org/10.1017/CBO9781107415324.004
9. Fernandez FS (2011) Análisis Discriminante. Universidad Autónoma de Madrid, Fac. Ciencias Económicas y Empresariales, p 45
10. Falgueras I (2008) La Teoría Del Capital Humano: Orígenes y Evolución. Temas Actuales de Economía 2:17–48
11. Francine D, Lawrence M (2016) The gender wage gap: extent, trends, and explanations. IZA Discussion Papers, Institute for the Study of Labor (IZA) 9656. https://www.econstor.eu/bitstream/10419/130341/1/dp9656.pdf
12. Becker GS, Murphy KM, Tamura R (1994) Economic Growth 1. Human capital: a theoretical and empirical analysis with special reference to education (3rd edn), pp 323–50. http://www.nber.org/chapters/c11239
13. Ghaljahi M, Rahdar S, Almasi SZ, Ahmadi S, Igwegbe CA (2018) Survey dataset on the externalizing self-esteem and gender effects on self-esteem subscales of students in Zabol University of Medical Sciences, Iran. Data Brief 21:407–413. https://doi.org/10.1016/j.dib.2018.10.019
14. Hajian S, Domingo-Ferrer J, Farràs O (2014) Generalization-based privacy preservation and discrimination prevention in data publishing and mining. Data Mining Knowl Discovery 28. https://doi.org/10.1007/s10618-014-0346-1
15. Heckman JJ, Lochner LJ, Todd PE (2013) Fifty years of mincer earnings regressions. J Chem Inf Model 53(9):1689–1699. https://doi.org/10.1017/CBO9781107415324.004
16. Herrero PP (2000) Economía de La Educación: Una Disciplina Pedagógica En Pleno Desarrollo. Ediciones Universidad de Salamanca 12:143–158
17. Hirsch B, Schnabel C (2012) Women move differently: job separations and gender. J Lab Res 33(4):417–442. https://doi.org/10.1007/s12122-012-9141-1
18. Humphreys J (2012) An alternative to the mincer model of education the basic mincer model. Univ Queensland 1954:1–14
19. Instituto Nacional de Estadística y Geografía (2016) Encuesta Nacional de Ingresos y Gastos de Los Hogares 2016 : ENIGH. Módulo de Las Condiciones Socioeconómicas 2010. http://www.inegi.org.mx/est/contenidos/Proyectos/Encuestas/Hogares/modulos/mcs/mcs2010/default.aspx%5Cn. http://www.inegi.org.mx/est/contenidos/espanol/metodologias/encuestas/hogares/enigh10_mcs_operativo.pdf
20. Cruz K, Antonio, and María Teresa Kido Cruz. (2015) Modelos Teóricos Del Capital Humano y Señalización: Un Estudio Para México. Contaduría y Administración 60(4):723–734. https://doi.org/10.1016/j.cya.2014.06.001
21. Koch A, Nafziger J, Nielsen HS (2015) Behavioral economics of education. J Econ Behav Organ 115:3–17. https://doi.org/10.1016/j.jebo.2014.09.005
22. Kredler M (2014) Vintage human capital and learning curves. J Econ Dyn Control 40:154–178. https://doi.org/10.1016/j.jedc.2014.01.003
23. Lächler U (1999) Education and Earnings Inequality in Mexico. The World Bank, Policy Research Working Paper Series: 1949, 1999, 1998. http://search.proquest.com/docview/56203232?accountid=17248
24. Lauwo S (2018) Challenging masculinity in CSR disclosures: silencing of women's voices in Tanzania's mining industry. J Bus Ethics 149(3):689–706. https://doi.org/10.1007/s10551-016-3047-4

25. Lee J-W, Lee H (2016) Human capital in the long run. J Dev Econ 122:147–169. https://doi.org/10.1016/j.jdeveco.2016.05.006
26. Lopez-Acevedo G (2001) Evolution of earnings and rates of returns to education in Mexico. The World Bank, Policy Research Working Paper, p 2691
27. López-Acevedo G, Tinajero M, Rubio M (2005) Mexico : human capital effects on wages and productivity. World Bank Policy Research Working Paper, p 3791. http://siteresources.worldbank.org/EXTLACREGTOPPOVANA/Resources/SkoufiasProgresaanditsimpactonhumancapitalandthewelfareofhh.pdf
28. López SL, Almagro AC (2005) Economía de La Educación: capital Humano y Rendimiento Educativo. Análisis Económico 79–106. http://www.analisiseconomico.com.mx/pdf/3604.pdf
29. Machin S (2014) Developments in economics of education research. Labour Econ 30:13–19. https://doi.org/10.1016/j.labeco.2014.06.003
30. Mason A, Lee R, Jiang JX (2016) Demographic dividends, human capital, and saving. J Econ Ageing 7:106–122. https://doi.org/10.1016/j.jeoa.2016.02.004
31. Miciuła I (2016) The measurement of human capital methods. Folia Oeconomica Stetinensia 16(1). https://doi.org/10.1515/foli-2016-0003
32. Mincer J (1974) The human capital earnings function. Human Capital Earn Funct I:83–96
33. Ministerio de Educación Cultura y Deporte (2015) Panorama de La Educación. Indicadores de La OCDE 2015. http://www.mecd.gob.es/dctm/inee/internacional/panorama-de-la-educacion-2015.-informe-espanol.pdf?documentId=0901e72b81ee9fa3
34. Morduchowicz A (2004) Discusiones de Economia de La Educación. http://unesdoc.unesco.org/images/0015/001507/150777so.pdf
35. OCDE (2007) Capital Humano: Cómo Moldea Tu Vida Lo Que Sabes. Resumen En Español, pp 1–7. http://www.oecd.org/insights/38435951.pdf
36. Organización para la Cooperación y el Desarrollo Económicos (OCDE) (2017) Estudios Económicos de La OCDE México. Enero, no. Visión general, p 53. https://doi.org/10.1787/9789264227682-es
37. Perri T (2016) Online education, signaling, and human capital. Inf Econ Policy 36:69–74. https://doi.org/10.1016/j.infoecopol.2016.06.001
38. Petreski M, Blazevski NM, Petreski B (2014) Gender wage gap when women are highly inactive: evidence from repeated imputations with macedonian data. J Lab Res 35(4):393–411. https://doi.org/10.1007/s12122-014-9189-1
39. Podgorelec V, Zorman M (2015) Decision tree learning. Encycl Complex Syst Sci. https://doi.org/10.1007/978-3-642-27737-5_117-2
40. Psacharopoulos G (1994) Returns to investment in education: a global update. World Dev 22(9):1325–1343. https://doi.org/10.1016/0305-750X(94)90007-8
41. Psacharopoulos G (1995) The profitability of investment in education: concepts and methods. Human Capital Develpment and Operations Policy. HCO. Working Papers, pp 1–22. https://doi.org/10.1017/CBO9781107415324.004
42. Rahayu D, Ismail M, Santoso DB, Pratomo DS (2015) Do natural resources and human capital matter to regional income convergence? (A Case Study at Regencies/Municipalities of Kalimantan Area—Indonesia). Procedia Soc Behav Sci 211:1112–1116. https://doi.org/10.1016/j.sbspro.2015.11.148
43. Refrigeri L, Aleandri G (2014) The economics of education as educational science. Procedia Soc Behav Sci 116:2059–2063. https://doi.org/10.1016/j.sbspro.2014.01.519
44. Rokach L, Maimon O (2013) Decision Trees. Data Mining Knowl Discovery Handbook 3:84–88. https://doi.org/10.4018/978-1-60960-557-5.ch006
45. Rustiadi S (2015) Creating better education system, building stronger human capital: a creative industries perspective. Procedia Soc Behav Sci 169:378–86. https://doi.org/10.1016/j.sbspro.2015.01.323
46. Salas M (2002) Cuatro Décadas En Economía de La Educación. Revista de Educación 328:427–49. http://www.educacionyfp.gob.es/dam/jcr:399c556a-58ea-467e-861d-f27e6fa96f1d/re3282310861-pdf.pdf

47. Schultz TW (1960) Capital formation by education. J Polit Econ 68(6):571–583. https://doi.org/10.1086/258393
48. Schultz TW (1961) Investment in human capital. Am Econ Rev
49. Seoane MJF, Álvarez MT (2010) Las Ecuaciones de Mincer y Las Tasas de Rendimiento de La Educación En Galicia. Asociación de Economía de La Educación, Chapter 14 in Investigaciones de Economía de La Educación vol 5, pp 285–304
50. Unesco (2009) Indicadores de La Educación Especificaciones Técnicas. UNESCO Institute for Statistics, p 10. http://www.uis.unesco.org/Library/Documents/eiguide09-es.pdf
51. Velázquez Valadez G (2004) Las Organizaciones y El Capital Humano. Liderazgo Del Tercer Milenio Opciones Para Aumentar La Productividad, pp 82–88. http://www.mundosigloxxi.ciecas.ipn.mx/pdf/v03/09/05.pdf

Study of the Geographical Marginality in a Mexican Region Using the MR-Sort Method

Pavel Anselmo Alvarez

Abstract The present work studies the geographical marginality in the Sinaloa region in Mexico. The marginality evaluation of municipalities was developed with a multicriteria decision-making sorting method. The MR-sort method classified the municipalities into four categories of marginality. The study showed three municipalities with the highest marginality level due to the low level of performance in education and income factors. The analysis suggests those performances are impacting negatively and leaves the population in more grades of marginality.

Keywords Geographical marginality · Ordered classification · Sorting · MR-sort

1 Introduction

The issue of inequality from the perspective of territorial units is a relevant field to study. The measurement of this inequality shows a panorama of lack of services and poverty in the marginalization of regions. Marginality is basically a social phenomenon, but the term marginal also has an important use in economy. Economic factors are also very important in the process of the marginalization of certain individuals and social groups besides the marginalization of areas or regions as spatial units [14].

Bock [1] describes some marginalizing factors related to the population in rural areas. Some of those factors are the following: losing specific population groups, consequential loss of social and cultural capital, reduction of community's capacity to act and regenerate, loss of socioeconomic and political power, deterioration of public services, decrease in profitability of private business and others.

The current research is focused on the spatial dimensions of a marginality phenomenon. Pelc and Nel [14] describe the geometrical or spatial dimension of marginality as the locational characteristics of marginality. Geographical location

P. A. Alvarez (✉)
Department of Management and Economic Sciences, Universidad Autónoma de Occidente, 80020 Culiacan, Mexico
e-mail: pavel.alvarez@uadeo.mx

E. León-Castro et al. (eds.), *Soft Computing and Fuzzy Methodologies in Innovation Management and Sustainability*, Lecture Notes in Networks and Systems 337,
https://doi.org/10.1007/978-3-030-96150-3_4

and the characteristics of the region may still be very demanding from the developmental perspective, but it is the problem of connectivity rather than location itself that causes marginalization [14].

The marginalization in Mexico is a current problem that remains. A broad spectrum of the impact of marginalization in Mexico can be visualized by the descriptive definition by Gatzweiler et al. [8]. Authors define marginalization as an involuntary position and condition of an individual or group on the margins of the social, political, economic, ecological, and physical system that prevents them from accessing resources, goods, and services. It limits their freedom of selection, reduces capacities, and eventually causing extreme poverty. It is not difficult to understand the negative impact of marginalization on some areas of the population. Analyzing data about the quality of homes, elementary services at homes, income, and educational level; helps us visualize the seriousness of the social phenomenon in some regions of any country.

The present work analyzes the spatial dimensions of a marginality in the Sinaloa region of Mexico. The marginality in Mexico is studied by National Council of Population in Mexico (from the Spanish, Consejo Nacional de Población (CONAPO)). The CONAPO identified four dimensions of expression of the phenomenon and, therefore, of action: education, housing, monetary income, and an affectation due to the spatial location. It see the marginality as a social problem associated with the lack of opportunities and access by the population to services such as education, health, and income as well as urban amenities such as drinking water, sewer, and electricity [15].

In this paper, the marginality of 18 municipalities are evaluated and assigned to four ordered categories of marginality. To represent the grade of marginality, we are using the term category and class indistinctly. The problem is addressed as a MultiCriteria Decision-Making (MCDM) problem with a sorting method (ordered classification) called MR-sort. The main contribution of the research is the identification of municipalities in certain category of marginalization. Due its spatial dimension is able to show in a graphical manner the location of that areas with significant level of marginalization and high level of marginalization.

The paper is structured as follow. In Sect. 2 is described the work developed about marginalization in Mexico. The MR-sort method is described in Sect. 3. Section 4 presents the data, preference information stated for MR-sort method and the classification of municipalities related with their marginalization level. Finally, concluding remarks are provided in Sect. 5.

2 Previous Work of Marginalization in Mexico

The National Population Council of Mexico defines marginalization as a structural process in relation to the socio-economic development achieved by the Mexican country [6]. Low socio-economic development will be reflected in lack and poverty in certain social groups, even if some progress is developed in other areas. This situation will generate some repercussions on the productive structure and be expressed in territorial inequalities [5].

In the study developed by Peña [15], it was found that despite good performances in economic indicators in border cities in Mexico, urban marginality continues. It is similar to the findings by Guillen López [9] in border cities, analyzing unemployment, salaries, and productivity indicators. We can say that good results in economic variables do not avoid enough marginality in some population sectors.

In Mexico, marginality condition is expanded in the whole territory, some studies developed in some areas of Mexico are presented in Collins and Ley García [4], Peña [15], Roldán et al. [16]. Moreover, it expresses how people are in societal positions and where they are and what services they have. In Mexico, marginalized people are disabled to traditional access and rights to use essential resources.

Collins and Ley García [4] considers the level of happiness and marginalization for born and migrated persons in two cities of Baja California, Mexico. Roldán et al. [16] analyzed the Mexico's public health problem across indigenous, rural and urban areas in relation to degree of marginalization and health service coverage. The study found the indigenous are the most affected population sector, followed by the rural sector. The sector that was least affected was urban.

In Soto-Perez-De-Celis et al. [18] studied the cancer-related mortality in The Mexico City metropolitan area. The authors wanted to know if this cause is influenced by the urban marginality status of the population. Statistical analysis for the years 2000 and 2010 showed that cancer-specific deaths and Global Marginality Index (GMI) present relation with high-GMI municipalities having a higher number of deaths.

Maldonado et al. [13] analyzed the creeping eruption of the skin caused by the cutaneous larva migrans (CLM). This case was found in a Mexican population of high marginalization in the state of Veracruz. It is found that the living environment, in addition to the patient's work in masonry, generates greater contact with contaminated soil with the larva.

The above studies developed about marginalization in Mexico show some of the inequalities evidencing by the different levels or marginalization on particular spatial units. The following will show the marginalization of 18 municipalities in a region in Mexico with the MCDM sorting method called MR-sort.

3 The MR-Sort for Ordered Classification

The ordered classification corresponds to the decision-making problematic called sorting. The multicriteria decision-making sorting consist in assign a set of alternatives $A = \{a_1, a_2, \ldots, a_m\}$ in a predefined ordered classes $C_1, C_2, \ldots, C_k$, where the class C_h is preferred to class C_l ($C_h \succ C_l$). Each alternative a_i is evaluated by each criterion in $G = \{g_1, g_2, \ldots, g_n\}$ and compared against a limit profile or central profile. When classes are defined by limiting profiles like MR-sort, the profile lp is the vector of attributes values $(b_{h,1}, \ldots, b_{h,j}, \ldots, b_{h,n})$.

The Sorting Method Based on a Majority Rule (MR-sort) is a noncompensatory sorting method studied by Bouyssou and Marchant [2, 3]. It is an outranking approach similar to the ELECTRE-TRI method. We use the notation of both [12, 17] to describe the method in more simple way. In the MR-Sort model the class C_h is delimited by its lower limit profile b_{h-1} and its upper limit profile b_h. The profile b_h is the vector of attribute values $(b_{h,1}, \ldots, b_{h,j}, \ldots, b_{h,n})$.

An alternative a_i is assigned to a category C_h if its attribute values are at least as good as the category lower profile values on a weighted majority of criteria and this condition is not fulfilled when the alternative's attribute values are compared to the category upper profile values [17].

MR-sort measures the strength of the coalition of criteria with the assertion a_i is at least as good as the profile b_h. It is called the coalition of criteria $\sigma(a_i \succ b_h)$, $\sum_{g_j(a_i)\geq b_{h,j}} w_j$. The assignment rule of MR-sort is the following.

$$\sum_{j\in N:g_j(a_i)\geq b_{h-1,j}} w_j \geq \lambda \text{ and } \sum_{j\in N:g_j(a_i)\geq b_{h,j}} w_j < \lambda \quad (1)$$

where $N = \{1, \ldots, n\}$, and it is assumed that $b_{h-1,j} \leq b_{h,j}$, $h = 1, \ldots, k$; $h < k$. The majority threshold λ tells which coalitions are strong enough to conclude that indeed a_i is at least as good as b_h. The MR-Sort assignment rule described above involves $kn + 1$ parameters. It is n weights, $(k - 1)$ profiles evaluations, and one majority threshold (λ).

4 Results

4.1 Data Marginalization in Mexico

The marginalization data is able in the survey developed by INEGI [10, 11]. Most of the indicators are obtained from the basic questionnaire tabulations, and income information is obtained from the expanded questionnaire (census sample). The set of criteria to evaluate the marginalization of each municipality are the following. Criteria g1 and g2 correspond to Education factors. The decision criteria g3, g4, g5, g6 and g7 are evaluating house condition and services. Criterion g8 measures the level of distribution of the population, and g9 measures the incomes for working. Table 1 list the municipalities and the performance on each criterion.

- Criteria (g1): Percentage of the illiterate population aged 15 years or over
- Criteria (g2): Percentage of the population without complete primary education aged 15 years or over
- Criteria (g3): Percentage of occupants in private homes without drainage or sanitary service
- Criteria (g4): Percentage of occupants in private homes without electricity

Table 1 Evaluation of municipalities in each criterion

Label	Municipality	g1	g2	g3	g4	g5	g6	g7	g8	g9
A1	Ahome	2.75	12.99	1.36	0.44	1.8	29.88	2.02	30.36	36.55
A2	Angostura	4.39	23.67	1.98	0.32	1.93	28.94	1.63	73.71	40.72
A3	Badiraguato	10.87	32.47	23.84	6.32	12.11	37.29	16.63	100	45.18
A4	Concordia	6.85	22.03	6.34	1.57	2.67	34.47	6.96	70.77	35.33
A5	Cosalá	10.23	28.88	8.32	0.44	8.7	35.95	6.34	60.61	37.09
A6	Culiacán	3.13	12.89	1.16	0.21	1.4	26.22	1.78	14.73	21.59
A7	Choix	11.12	36.09	8.86	3.82	18.18	36.21	10.08	71.8	52.1
A8	Elota	8.49	27.7	2.21	0.16	2.86	47.13	3.47	63.51	34.82
A9	Escuinapa	6.81	19.64	2.46	1.38	8.8	39.19	3.88	32.37	32.62
A10	El Fuerte	6.03	23.5	3.34	1.05	5.45	35.75	6.91	63.31	59.04
A11	Guasave	5.69	22.25	2.6	0.23	5.38	32.42	3.47	51.25	46.92
A12	Mazatlán	1.95	10.61	0.45	0.1	0.83	27.24	1.18	8.48	29.48
A13	Mocorito	8.69	31.81	6.87	0.91	13.54	34.91	6.3	74.33	48.86
A14	Rosario	5.53	19.32	4.08	1.31	4.71	34.81	5.28	67.6	37.58
A15	Salvador Alvarado	3.33	14.59	0.88	0.09	0.86	29.91	1.46	12.47	30.71
A16	San Ignacio	7.09	27.62	7.34	2.16	4.67	34.44	6.08	100	43.77
A17	Sinaloa de Leyva	10.07	33.5	5.33	2.8	7.55	36.16	7.75	86.92	58.58
A18	Navolato	6.73	26.32	10.93	0.21	3.76	42.48	3.07	53.43	39.65

*The data were generated in the survey developed by INEGI [11]

- Criteria (g5): Percentage of occupants in private homes without piped water
- Criteria (g6): Percentage of private homes with some level of overcrowding
- Criteria (g7): Percentage of occupants in private dwellings with dirt floors
- Criteria (g8): Percentage of the population residing in towns with less than 5,000 inhabitants
- Criteria (g9): Percentage of the employed population with incomes of up to two times the minimum wage

4.2 Preference Information

The MR-sort method for ordered classification requires some parameters from the analyst. It is required the limits profiles between classes (or categories). Due the marginality in Sinaloa region stated four classes, three profiles are defined with nine values each. Also, it is needed nine importance of weights (one weight for each criterion). An additional parameter is the majority threshold $\lambda = 0.5$. Table 2 show the complete set of parameters defined for the municipalities classification.

Table 2 Parameters of the MR-sort

Criteria	Min/Max	Weight	Profile classes		
			b1	b2	b3
g1	Min	0.14	3.5	7	9
g2	Min	0.14	13	24	29
g3	Min	0.07	1.5	5	7.5
g4	Min	0.09	0.5	1.7	2.5
g5	Min	0.09	2	6	10
g6	Min	0.11	30	34	35
g7	Min	0.1	2.5	5	7
g8	Min	0.13	30	60	75
g9	Min	0.13	30	40	49

The number of parameters is 37. It corresponds to $kn + 1$ parameters, where k is the number of classes, n is the number of criteria, plus one threshold. For the current problem test, no veto threshold was needed.

A schematic representation of the profiles and classes defined for the level of marginalization in the municipalities is shown in Fig. 1. The ordered classes correspond to C_1, C_2, C_3, C_4, where C_4 is considered the best class with the lowest level of marginality and C_1 is the worst class with the highest level of marginality ($C_{h-1} < C_h$).

Each municipality a_i is compared against each class C_h to define that a_i is at least as good as b_h, $\sigma(a_i \succ b_h)$. If this assertion is validated a_i is assigned to class C_h. Else, a_i is assigned to a lower class.

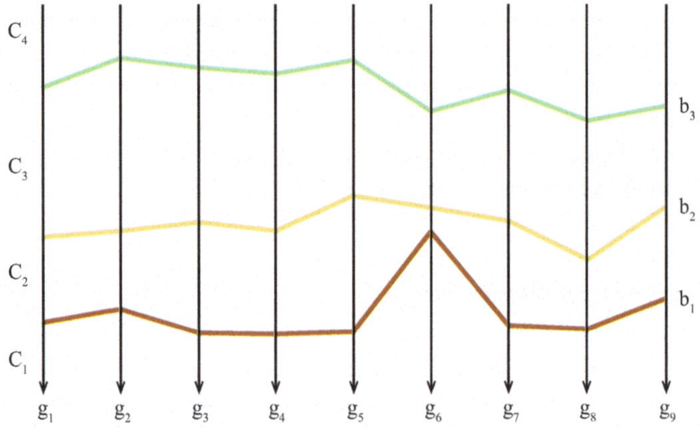

Fig. 1 Profiles and classes of the geographical marginality

4.3 Result Analysis

For the problem of assignment of the municipalities of the Region of Sinaloa based on their marginalization, four classes (or categories) of marginalization were defined and ordered. The class C_4 is the best class because groups municipalities with the lowest level of marginality. C_3 is constituted with municipalities with moderate level of marginality. C_2 is grouping municipalities with significant level of marginality. And, C_1 is the worst class because it contains the highest level of marginality in the region.

Table 3 shows the result of the classification of municipalities. As we stated above, the elements belonging to C_4 correspond to the four municipalities with the lowest degree of marginalization, Ahome (A1), Culiacán, (A6), Mazatlán (A12), Salvador Alvarado (A15). Class C_3 consist in moderate level of marginality. Here, seven municipalities are assigned Angostura (A2), Concordia (A4), Escuinapa (A9), El Fuerte (A10), Guasave (A11), Rosario (A14), Navolato (A18). Class C_2 correspond to a significant level of marginality. For municipalities are assigned to this class, Cosalá (A5), Elota (A8), Mocorito (A13), San Ignacio (A16). Three municipalities correspond to the highest level of marginalization, Badiraguato (A3), Choix (A7) and Sinaloa (A17).

The municipalities categorized in the highest marginality level (C_1) present the worst education and income from work indices. They are categorized by the main

Table 3 Ordered classification of municipalities

Label	Municipality	Class
A1	Ahome	C4
A6	Culiacán	C4
A12	Mazatlán	C4
A15	Salvador Alvarado	C4
A2	Angostura	C3
A4	Concordia	C3
A9	Escuinapa	C3
A10	El Fuerte	C3
A11	Guasave	C3
A14	Rosario	C3
A18	Navolato	C3
A5	Cosalá	C2
A8	Elota	C2
A13	Mocorito	C2
A16	San Ignacio	C2
A3	Badiraguato	C1
A7	Choix	C1
A17	Sinaloa	C1

four-factors (criteria) influencing marginalization. In Table 1, Badiraguato (A3), Choix (A7), and Sinaloa de Leyva (A17) present the highest percentage of the population lacking education. Each municipality shows more than 10% of the illiterate population aged 15 years or over (see values from criteria g1 in Table 1). Each present more than 32% population without complete primary education aged 15 years or over (see criteria g2). The current situation on those municipalities show a high number of the population that do not count with an expected level of education.

On the other hand, analyzing the monetary incomes for working. The same municipalities count with the lowest income level, between 45 and 58% of the employed population present incomes of up to two times the minimum wage (see evaluation on criteria g9 from Table 1). The majority of the population of these municipalities reside in towns with less than 5,000 inhabitants (criteria g8 from Table 1). The four-factors' performance describing the population of class C_1 are the main factors impacting the marginality condition of the municipalities. They are municipalities with low opportunity for development because of the performances of those factors.

The west of the region counts with the coast from the Pacific Ocean. The Mountain range goes through from south to north, and it is more pronounced inside the east side. This Mountain range is called *the Sierra Madre Occidental* and reach heights of more than 2000 km (Cuentame INEGI, n.d.). The analysis of the geographical marginality shows the municipalities in a more critical marginality situation. Figure 2 shows the colored municipalities based on their marginality level. The illustration helps to identify the geographical locations and some limitations based on its location.

There are some municipalities in the lower level of marginality (C_4). It is observable they are on the coast or next to the coast. This factor can help to more work opportunities related to fishing or related actives. The municipalities categorized in the highest level of marginality (C_1) are far away from the cost. Some rural locations in those municipalities are even in the Mountain range with limited access to roads to communicate the population. The three municipalities with the most marginality are located in this Mountain range.

The applied sorting method shows the municipalities assigned in the predefined ordered classes of marginality. The analysis clearly evidences the level of marginality of the municipalities. The Sinaloa region's geographical marginality in Mexico is very high in those municipalities with a geographical location in the Mountain range (A7, A17, A3). The municipalities with a significant marginality level must improve their performance to avoid a high level of marginality. The municipalities with significant marginality level share both the plain and mountain reliefs (A13, A5, A8, A16). The lowest marginality is in municipalities far away from this Mountain range (A1, A15, A6, A12).

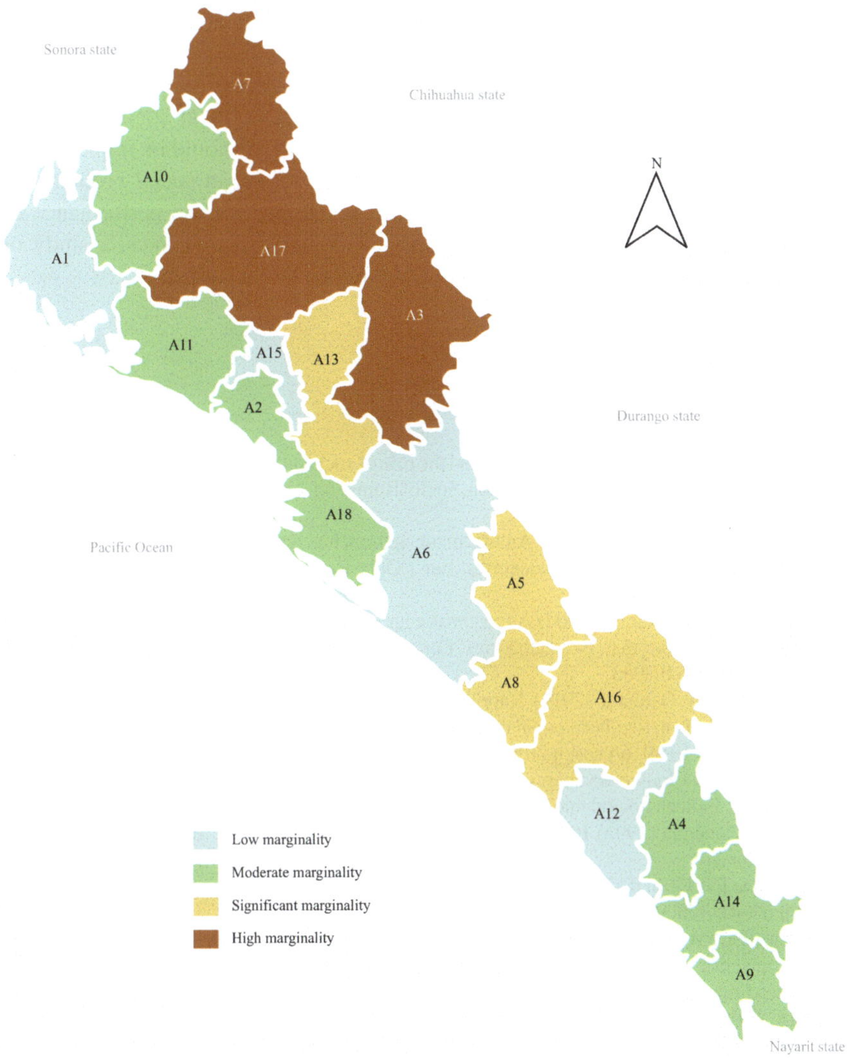

Fig. 2 The graphical marginality of the Sinaloa region in Mexico

5 Conclusions

The multicriteria sorting process applied, make it possible to observe in the classification of the municipalities those that presented a lowest degree of marginalization and those that presented a highest degree of marginalization.

The analysis of the multicriteria decision-making approach with the sorting method for ordered classification helps to understand the interaction between complex characteristics of population in spatial units, some of them rural locations.

This understanding helps to explain the marginality problem and identify locations with the greatest urgency for government attention and decision-making.

The analysis of geographical marginality should be further studied and analyzed by the developers of social politics, due the region needs critical attention to help the population in their own development. The fuzzyfication should be implemented in some decision criteria. It can be supported by an expert in social problems to improve understanding of marginalization and the impact of the performance of the municipalities in the population. As part of future research lines, a study of geographical marginality of each region in Mexico could be studied.

References

1. Bock BB (2016) Rural Marginalisation and the role of social innovation; a turn towards nexogenous development and rural reconnection. Sociol Rural 56(4):552–573. https://doi.org/10.1111/soru.12119
2. Bouyssou D, Marchant T (2007) An axiomatic approach to noncompensatory sorting methods in MCDM, I: the case of two categories. Eur J Oper Res 178(1):217–245. https://doi.org/10.1016/j.ejor.2006.01.027
3. Bouyssou D, Marchant T (2007) An axiomatic approach to noncompensatory sorting methods in MCDM, II: more than two categories. Eur J Oper Res 178(1):246–276. https://doi.org/10.1016/j.ejor.2006.01.033
4. Collins K, Ley García J (2019) Happiness and marginalization rates for internal Mexican migrants and the native-born population in Baja California, Mexico. Soc Sci J 51(4):598–606. https://doi.org/10.1016/j.soscij.2014.07.004
5. Conapo (2012) Índice de marginación por entidad federativa y municipio 2010. México: Colección: Índices Sociodemográficos
6. Conapo, Conagua (1993) Indicadores Socioeconómicos e Índice de Marginación Municipal 1990. México
7. Cuentame INEGI (n.d.) Relieve, from http://cuentame.inegi.org.mx/monografias/informacion/sin/territorio/relieve.aspx?tema=me&e=25
8. Gatzweiler FW, Baumüller H, Husmann C, von Braun J (2011) Marginality: addressing the root causes of extreme poverty. SSRN Electron J. https://doi.org/10.2139/ssrn.2235654
9. Guillen López TG (1990) Servicios públicos y marginalidad social en la frontera norte. Frontera Norte 2(4). https://doi.org/10.17428/rfn.v2i4.1630
10. INEGI (2010) Censo de Población y Vivienda 2010. Accessed 5 Nov 2020, from INEGI https://www.inegi.org.mx/
11. INEGI (2015) Encuesta Intercensal 2015. Accessed 5 Oct 2020, from INEGI https://www.inegi.org.mx/
12. Leroy A, Mousseau V, Pirlot M (2011) Learning the parameters of a multiple criteria sorting method. In: Brafman RI, Roberts FS, Tsoukias A (eds) Algorithmic decision theory, vol 6992, pp 219-+
13. Maldonado IA, Duarte SC, Velázquez FG, Aguilera AA (2014) A case report of cutaneous larva migrans in a Mexican population of high marginalization. Asian Pac J Trop Biomed 4(9):755–756. https://doi.org/10.12980/apjtb.4.2014apjtb-2014-0119
14. Pelc S, Nel E (2020) Social innovation and geographical marginality 5:11–21. https://doi.org/10.1007/978-3-030-51342-9_2
15. Peña S (2005) Recent developments in urban marginality along Mexico's northern border. Habitat Int 29(2):285–301. https://doi.org/10.1016/j.habitatint.2003.10.002

16. Roldán J, Álvarez M, Carrasco M, Guarneros N, Ledesma J, Cuchillo-Hilario M, Chávez A (2017) Marginalization and health service coverage among Indigenous, rural, and urban populations: a public health problem in Mexico. Rural Remote Health 17(4). https://doi.org/10.22605/rrh3948
17. Sobrie O, Mousseau V, Pirlot M (2018) Learning monotone preferences using a majority rule sorting model. Int Trans Oper Res. https://doi.org/10.1111/itor.12512
18. Soto-Perez-De-Celis E, Govezensky T, Chavarri-Guerra Y (2014) Does urban marginality influence cancer mortality rates? an analysis of the Mexico City Metropolitan Area. Ann Oncol 25:iv488. https://doi.org/10.1093/annonc/mdu353.9

Strategic Diagnostics of Stress and Impulse Control for Second Order Change: Inclusion of Forgotten Effects in Diffuse Cognitive Maps

Rubén Chávez, Federico González, Víctor Alcaraz, and Jesús Ricardo Ramos

Abstract The determination of the strategic cognitive factors that intervene in the Second Order Change (SOC) are fundamental to achieve the transformation of organizational systems [6]. This work proposes a methodology based on the Fuzzy Cognitive Maps (FCM) model and the Forgotten Effects (FE) model, for the formalization and identification of these factors that intervene in the inference matrix. The inclusion of FE allows reducing inference errors in the matrix. The objective being to give certainty to the inference matrix in its state of reliability when applying FE, with the trade-off that this model only handles positive elements in the analysis matrix. This condition requires, first, to separate the elements into two matrices, one containing positive elements and the other, negative ones. The negative-element matrix is then temporarily transformed into a state of positive elements so that both matrices can operate the fuzzification-inferential-defuzzification process in the FE model [34, 16]. Once the matrices have been defuzzified, the matrix of temporarily positive elements returns to its original (negative) state and, finally, both matrices are added to obtain an adjusted inference matrix so it can be applied to the FCM model. A case study is presented where the main objective is to find the cognitive factors that intervene in the behavior of four companies' staff, starting the procedure with two factors highly correlated: Stress Tolerance and Impulse Control. The main results represent the metrics of the absolute Hamming distances, laying the foundations for the formalization of the cognitive factors that hinder a SOC.

R. Chávez (✉)
QFB, UMSNH, Morelia, México
e-mail: ruben.chavez@umich.mx

F. González
FCCA, UMSNH, Morelia, México
e-mail: fegosa@inidem.edu.mx

V. Alcaraz
ININEE, UMSNH, Morelia, México
e-mail: jorge.alcaraz@umich.mx

J. R. Ramos
UTNL, Nuevo Laredo, México
e-mail: jramos@utnuevolaredo.edu.mx

E. León-Castro et al. (eds.), *Soft Computing and Fuzzy Methodologies in Innovation Management and Sustainability*, Lecture Notes in Networks and Systems 337,
https://doi.org/10.1007/978-3-030-96150-3_5

1 Introduction

Human resources in organizations are collectively responsible for decision-making, renegotiation and reaching consensus on decisions that will have repercussions on working habits, routines and existing communication channels, associated with commitment, loyalty and programmatic staff levels [5]. The change in strategy first affects the managers' personal interactions, which represents a paradigm shift in the organization [14, 16]. As a consequence, success depends on the effort to learn to produce the change and on the cognitive availability [12, 31]. Several authors point to the need to intensify the forces of change in designing intelligent and knowledge-based organizations, as well as to the importance of individual and group learning, [1, 9, 22, 23, 26–28].

Emotional intelligence includes several traits such as independence, represented by the contribution of the people, in performing their daily tasks; interdependence, where job performance depends, to some extent, on others; hierarchy, in which emotional capacities are mutually reinforcing [7, 8]. As for needs, as goals are met, new ones arise and allow the individual to grow. Generic needs are complemented by those in which different competences are required [2, 11, 12, 29, 30]

The need to assess the commitment, knowledge and implicit competences in the formulation of strategies, in addition to cognitive factors, all of them possessing subjective characteristics, lead to establish the main objective for using Forgotten Effects (FE) in the allocation matrix of the Fuzzy Cognitive Maps model (FCM) is to strengthen it and avoid omissions and errors, as well as to prevent the allocation matrix to be applied in the causalities of the cognitive factors included in the network system that hinder Second-Order Change (SOC). The objective of the diagnosis performed in this work is to identify critical factors that prevent change and, thus, establish new strategies on factors that actually promote the transformation of the organizational system.

2 Strategy Change: Second Order Change

The environment needs demand the design of new strategies, in order to generate a change in the system [20]. The process of modifying routines and work comfort, in exchange for structures capable to achieve a transformation in the system integrates two implicit aspects: inertia (I), represented by those who favor the *status quo*, and tension (S), represented by those who desire a transformation to be operated in the system [6, 14, 21, 32, 33]. Decision-making implies profound changes, with a perspective of stability and development in organizational behavior. Thus, it is not enough to merely incorporate new routines or staff into the structure, but rather to modify the system so as to generate a SOC, [6, 14, 32].

2.1 Fuzzy Relations on the Second-Order Change

The SOC occurs when the accumulated stress $S_{(i+1)}$ is greater than the accumulated inertia $I_{(i+1)}$. This step influences the homeostatic-cognitive process with little information [14]. In each of these dichotomous forces, relationships with a cognitive interface are required to allow the creation of two independent fuzzy sets with a common denominator: a cognitive set through the homeostatic-cognitive process (h_i). Then, the relationship between the matrix h_i and the diffuse tension matrix (S_i) is defined by the *maxmin* operator [16, 17].

2.2 Fuzzy Cognitive Maps

In the globalization era, information is the fundamental element that fuels the communication networks, with these ranging from the most basic to the extremely complex. Interconnected nodes using arcs that carry the necessary information load to positively or negatively activate other nodes from those networks. Nevertheless, the cognitive characteristic often complicates its formal presentation as the management of concepts and their relationships are usually linked through cognitive maps [3, 18, 19, 24].

Now, the network-composed systems receive two kinds of information: one that is internally and naturally generated within the system and the other that is externally provided. Then, through processing and transformation, the acquired information can be translated into knowledge. This condition promotes the existence of qualified advisers and experts in every organized system, so as to facilitate the knowledge management for the development of innovative processes (Castells 2002; Riesgo 2006). The intensities, represented by linguistic means, describe the existing relationships between concepts in the FCM and the corresponding arcs' orientation that connect the nodes, allowing to simulate the phenomenon using consecutive iterations, thus being fully predictive [18].

3 Methodology

The FCM causal graphs can easily integrate several different models: cognitive maps theory, fuzzy logic, neural networks, dynamic systems, non-linear dynamic systems, among others [15]. In this sense, the purpose for incorporating the FE model in the allocation matrix is to reinforce the confidence in all the weights portrayed and, at the same time, to take into account those factors that have been disregarded by mistake or as a result of underestimations [16]. The methodological proposal is described as follows:

3.1 Application of FCM Model Without FE to Know Its Base State

The FCM model is applied to the allocation matrix φ that was generated by an expert panel. Based on the responses of each one of the experts, it is possible to create the allocation matrix φ as:

$$\varphi = \begin{bmatrix} \cdots\cdots\cdots\cdots \\ \cdots\cdots\cdots\cdots \\ \cdots\cdots \varphi_{ij} \cdots \\ \cdots\cdots\cdots\cdots \end{bmatrix} \tag{1}$$

The gathering of the experts' opinions has led to the construction of the aggregated matrix φ_{ij} where each element has resulted from the expression (3), corresponding to the generalized mean [17].

$$\varphi_{ij}^{p} = \sum_{1}^{S} \left(\lambda_s \varphi_{ij}^{p}\right)^{1/p} \tag{2}$$

where S is the number of experts, p is a parameter that can take any value (1, would represent the arithmetic mean, 2, the quadratic mean, and so forth). The weighting vector fulfills the following condition: $\sum_{s=1}^{S} \lambda_s = 1$, $\lambda_s \in [0, 1]$.

The adaptation among variables in the FCM consists of an iterative process using linguistic concepts and their orientations of causal connections in the network of elements until stability is achieved [3, 13, 18, 24].

$$C_{t+1} = f(C_t, \varphi) = f(R) \tag{3}$$

where:

$$R = C_t * \varphi_k \tag{4}$$

The identity function allows visualizing the oscillations until an equilibrium is reached, making it possible to modify the elements of the matrix φ [4].

3.2 The FCM Model is Intervened with the FE Model

The matrix φ is divided into two parts, one of them gathers all the positive elements and the other, all the negative elements; in absolute value [25]. Both matrices are fuzzified in order to apply the FE model [10, 16, 17]. Subsequently, they are defuzzified so as to obtain a new allocation matrix.

$$\tilde{M}^{+}_{i,j} = \begin{cases} \varphi_{i,j}\, if\, \varphi_{i,j} \geq 0 \\ 0\, otherwise \end{cases} \tag{5}$$

Negative elements temporarily become positive items using their absolute value:

$$\tilde{M}^{-}_{i,j} = \begin{cases} |\varphi_{i,j}|\, if\, \varphi_{i,j} < 0 \\ 0 \text{ otherwise} \end{cases} \tag{6}$$

$$\Big\{$$

The fuzzification of those values is then performed and the FE model, applied. A convolution of the matrix $\tilde{M}^{+}$ results in causal $\tilde{A}^{+}$ and FE $\tilde{B}^{+}$ matrices:

$$\tilde{M}^{*+} = \tilde{A}^{+\circ}\tilde{M}^{+\circ}\tilde{B}^{+} \tag{7}$$

The FE model for positive elements is calculated using:

$$\tilde{O}^{+} = \tilde{M}^{*+} - \tilde{M}^{+} \tag{8}$$

Analogously, the FE model can be applied for the negative elements where: $\tilde{M}^{-}$, so the negative incidence matrix can be expressed as:

Analogously, for negative elements, we can apply FE, where to $\tilde{M}^{-}$. so we can express the negative incidence matrix as:

$$\tilde{M}^{*-} = \tilde{A}^{-\circ}\tilde{M}^{-\circ}\tilde{B}^{-} \tag{9}$$

While the FE model can be expressed as follows:

$$\tilde{O}^{-} = \tilde{M}^{*-} - \tilde{M}^{-} \tag{10}$$

Since matrix $\tilde{O}^{-}$ includes positive values, they ought to be reconverted into negative values, then added to the matrix $\tilde{O}^{+}$ and multiplied by the matrix φ, so as to obtain the new FE model:

$$\varphi^{new} = \varphi\left(\tilde{O}^{+} - \tilde{O}^{-}\right) \tag{11}$$

The matrix φ^{new} is then operated using the FCM model and Eq. (3). The result is: ${C_{t+1}}^{new} = f(C_t, \varphi^{new}) = f(R^{new})$ and, from Eq. (4): $R^{new} = C_t * {\varphi_k}^{new}$.

The new identity function allows the visualization of the oscillations, until an equilibrium is reached, which includes the FE model, concerning the identity function's original concept.

3.3 Identification and Distance of the Critical Factors

Identification of cognitive factors can be done through of the existing distance between vector R and vector R^{new}, once an equilibrium is reached. Both results are then contrasted using the Hamming distance equation [34]:

$$d(R, R^{new}) = \sum_{i=1}^{n} [|R_i(t)| - |R_i^{new}(t)|] \tag{12}$$

4 Results

Emotional intelligence is considered to be the basis for strategic change and organizational learning processes. A survey concerning the emotional intelligence, the implementation of new rules in the workplace and the formalization of ISO-9000 processes, was conducted in four fat-acids-producing organizations in Mexico, namely, QUIMIC, AAK, TRON HNOS and QUIMIKAO, including a total of 60 workers. The Bar-On model includes five components: intrapersonal, interpersonal, adaptability, stress management, and general mood [30]. The proposal in this work begins by doing a correlation diagnosis concerning stress tolerance and impulse control, as the most influential elements in work performance [12]. The results show that, from all five dimensions, stress management correlations of the indicators were the most significant. The analysed variables for this case are Stress Management and Impulse Control, as well as their impact on SOC. The Stress Management variable can be assigned the following values or indicators: "I know how to stay calm" (st1), "When I'm upset with someone, I feel upset for a long time" (st2), "It's hard for me to wait for my turn" (st3), "I get annoyed easily" (st4), and "When I get upset, I act without thinking" (st5). On the other hand, Impulse Control can adopt the following values: "I can stay calm when I'm upset" (pc1), "It is difficult for me to control my anger" (pc2), "I get too annoyed for anything" (pc3), "I fight with people" (pc4), "I have a bad temper" (pc5), "I annoy easily" (pc6), and "I'm mad at bothering" (pc7).

Taking into account these indicators, the experts have a formal knowledge base that enables them to make decisions in assigning values to the elements of matrix φ (Eq. 13) and the corresponding cognitive map (Fig. 1), on five dimensions (Tables 1 and 2).

A high correlation, at 5% significance level, can be observed. As there might exist a co-linearity, a multiple linear regression is required. The Model summaries are presented as follows: model 1:$R = 0.639^a, R^2 = 0.408$, $R^2 tight = 0.398$ and standard error = 0.642; model 2:$R = 0.639^a, R^2 = 0.471$, $R^2 tight = 0.452$, and standard error = 0.612 a. Predictors: (Constant), "I annoy easily" b. Predictors: (Constant), "I annoy easily", "I get too annoyed for anything". From these models, the following Stress Tolerance regression equation is obtained:$Y_{TE} =$

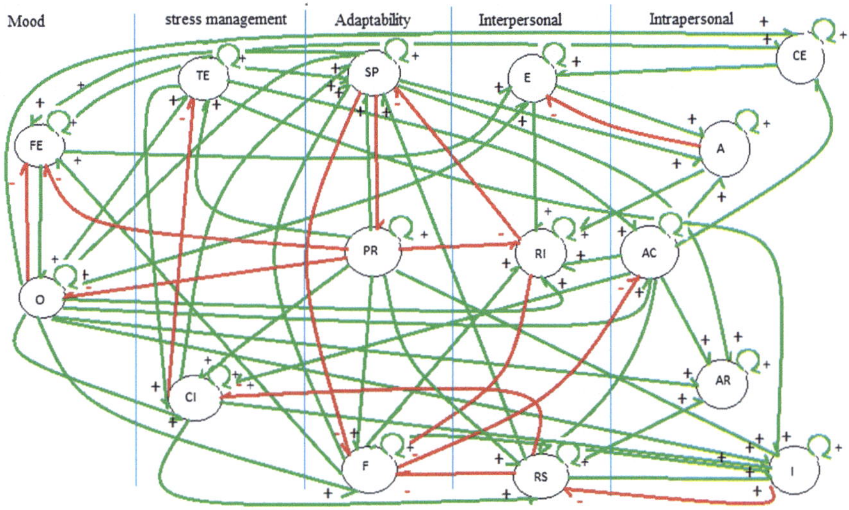

Fig. 1 Emotional Intelligence Cognitive Map. *Source* Own elaboration (2020)

Table 1 Stress Tolerance: indicator correlation

	Pearson correlation	st1	st2	st3	st4	st5
		1	−0.197	−0.228	−0.096	−0.235
st1	Significance		0.134	0.083	0.476	0.073
	N	59	59	59	58	59
	Pearson correlation	−0.197	1	0.173	**0.289***	0.196
st2	Significance	0.134		0.186	0.027	0.133
	N	59	60	60	59	60
	Pearson correlation	−0.228	0.173	1	0.064	**0.330***
st3	Significance	0.083	0.186		0.63	0.01
	N	59	60	60	59	60
	Pearson correlation	−0.96	**0.289***	0.064	1	0.176
st4	Significance	0.476	0.027	0.63		0.183
	N	58	59	59	59	59
	Pearson correlation	−0.235	0.196	**0.330***	0.176	1
st5	Significance	0.073	0.133	0.01	0.183	
	N	59	60	60	59	60
All tests are bilateral (2 tails), *The correl. sig at the 0.05 level in SPSS						

Source Own elaboration (2020)

Table 2 Impulse Control: indicator correlation

		pc1	pc2	pc3	pc4	pc5	pc6	pc7
pc1	Pearson correlation	1	−0.134	−0.605**	0.357**	−0.448**	−0.631**	0.118
	Significance		0.313	0	0.005	0	0	0.369
	N	60	59	60	60	60	60	60
pc2	Pearson correlation	−0.134	1	0.133	−0.024	0.236	0.292*	−0.270*
	Significance	0.313		0.314	0.86	0.072	0.025	0.039
	N	59	59	59	59	59	59	59
pc3	Pearson correlation	**−0.605****	0.133	1	**0.385****	**0.564****	**0.665****	−0.12
	Significance	0	0.314		0.002	0	0	0.363
	N	60	59	60	60	60	60	60
pc4	Pearson correlation	−0.357**	−0.024	**0.385****	1	0.295*	**0.407****	−0.087
	Significance	0.005	0.86	0.002		0.022	0.001	0.511
	N	60	59	60	60	60	60	60
pc5	Pearson correlation	**−0.448****	0.236	**0.564****	0.295*	1	**0.651****	−0.187
	Significance	0	0.072	0	0.022		0	0. 153
	N	60	59	60	60	60	60	60
pc6	Pearson correlation	**−0.631****	0.292*	**0.665****	**0.407****	**0.651****	1	−0.253
	Significance	0	0.025	0	0.001	0		0.051
	N	60	59	60	60	60	60	60
pc7	Pearson correlation	0.118	−0.270*	−0.12	−0.087	−0.187	−0.253	1
	Significance	0.369	0.039	0.363	0.511	0.153	0.051	
	N	60	59	60	60	60	60	60

All tests are bilat.(2 tails); * Correl. sig. at level 0,05;** Correl. sig. at level 0,01

Source Own elaboration (2020)

$1.587 + 0.299X_{1TE}$, Where: X_{1TE} = "When I'm upset with someone, I feel upset for a long time". On the other hand, the regression equation for Impulse Control is:$Y_{CI} = 4.315 - 0.401X_{1CI} - 0.301X_{2CI}$. Where: X_{1CI} = "I get annoyed easily" and X_{2CI} = "I get too annoyed for anything".

The regression equations shown above represent the first approximation for the elements of the allocation matrix, for the Stress Management dimension. The next step will consist on integrating all dimensions and following the Bar-On model: *Intrapersonal*: Depends on Emotional Empathy (EE), Assertiveness (A), Auto-concept (AC), Auto-realization (AR); *Independence* (I): Interpersonal, depends on Empathy (E), Interpersonal Relationships (IR), and Social Responsibility (SR);

Adaptability: Depends on Problem Solution (PS), Reality Test (RT), and Flexibility (F); *Stress Management*: Depends on Stress Tolerance (ST) and Impulse Control (IC). Finally, *Mood*: Depends on Happiness (H) and Optimism (O).

4.1 Application of the FCM Model, Without Integrating FE

At time $t = 0$, the Stress Tolerance (ST) and Impulse Control (IC) variables' activation represent initiating concepts, which integrate the system $C_0 = \{ST, IC\}$, from the following variable set: $C = \{ST, A, AC, AR, I, E, RI, SR, PS, RT, F, ST, IC, H, O\}$. The initial vector, when only these variables are active, is $C_0 = [000000000001100]$.

The relational variable matrix, constructed according to the experts' opinions is presented below:

φ=

	EE	A	AC	AR	I	E	IR	SR	PS	RT	F	ST	IC	H	O
EE	1	0	0	0	0	0.4	0	0	0	0	0	0	0	0	0
A	0	1	0	0	0	−1	0	0	1	0	0	0	0	0	0
AC	0.6	0.4	1	0.6	0	0	0.4	0.7	0	0	0	0	0.8	0	0
AR	0	0	0	1	0	0	0	0	0	0	0	0.8	0	0	0
I	0	0	0	0	1	0	0	−1	0	0	0	0	0	0	0
E	0	1	0	0	0	1	0.9	0	0	0	0	0	0	0	0
IR	0	0	0	0	0	0	1	0	−0.7	0	0	0	0	−0.6	0
SR	0	0	0	0.4	0.9	0	0	1	0	0	0	0	−0.6	−0.5	0
PS	0	0.7	0	0.7	0	0	0	0	1	−0.7	−1	0	0	0.7	1
RT	0	0	0	0	0.3	0	−0.2	1	0.7	1	0.1	0.2	0.5	−1	−1
F	0	1	0	0	0	0	0.9	0	0	0	1	0	0	0	0
ST	0	0	0.8	0	0.8	0	0	0	1	0	0	0	0.9	0	0
IC	0	0	0	0	0.7	0	0	0.7	1	0	0	−0.5	1	0	0
H	0	0	0	0	0	1	0	0	0	0	0	0	0	1	1
O	0.8	0	0.8	0.8	0.7	0.9	0.8	0	0	0	0.3	0.8	0.4	−0.5	1

(13)

Once the incidence matrix φ and the initial vector C_0 have been obtained, the product R can be calculated so as to find: $C(t+1) = C(0+1) = [111110001001100]$. This vector becomes the initial value for an iterative process, until an equilibrium is reached, where: $C_n = C_{n-1}$.

The oscillation of the defined variables is presented in Table 3 and Fig. 2.

As shown in Table 3, the variables Problem Solution (PS), Reality Test (RT) and flexibility (F) reached an early equilibrium, that is, in the 0 to 1 period. As for Assertiveness (A) and Problem Solution (PS), they reached it in the 1 to 2 period. From these results, it can be drawn the need to pay more attention to the critical

Table 3 Results of the fuzzy cognitive maps (initiators: TE and CI)

t	C(t)															
0	0	0	0	0	0	0	0	0	0	0	0	1	1	0	0	
1	1	1	1	1	1	0	0	0	1	0	0	1	1	0	0	
2	1	1	1	1	1	0	1	1	1	0	0	1	1	0	1	
3	1	1	1	1	1	1	1	1	1	0	0	1	1	0	1	
$R = C'\varphi$																
0	0	0.8	0	1.5	−1	−0.2	−0.3	2	−0.7	−1	0.5	1.3	−2.6	−1		
1.6	2.1	1.8	2.3	2.5	−0.6	1	0.4	3	−0.7	−1	0.5	2.1	−1.3	0		
2.4	2.1	2.6	3.5	4.1	0.3	2.8	1.4	3.7	−0.7	−0.7	1.3	2.5	−1. 3	1		
2.4	3.1	2.6	3.5	4.1	1.3	3.7	1.4	3.7	−0.7	−0.7	1.3	2.5	−1.3	1		
$C(t + 1) = f(R)$																
1	1	1	1	0	0	0	1	0	0	1	1	0	0			
1	1	1	1	0	1	1	1	0	0	1	1	0				
1	1	1	1	1	1	1	1	0	0	1	1	0				
1	1	1	1	1	1	1	1	0	0	1	1	0				

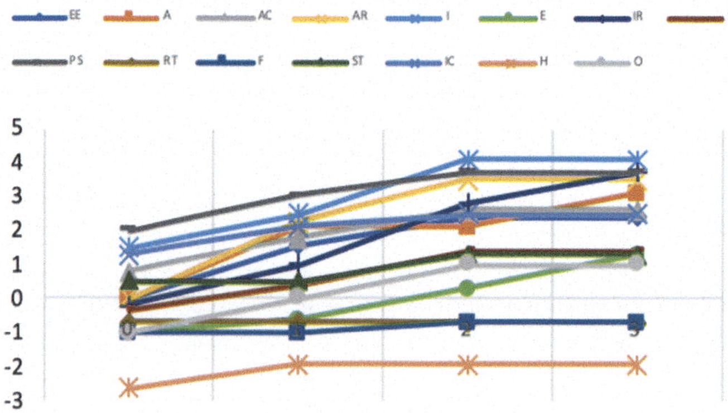

Fig. 2 Variable oscillation. *Source* Own elaboration (2020)

factors that reached equilibrium at the late stages of the process, in order to develop future strategies that could lead to faster changes.

4.2 Application of FCM Model Integrating FE

In order to include Forgotten Effects in the FCM model, some considerations must be made. In this case study, only two variables are being considered: Stress Tolerance

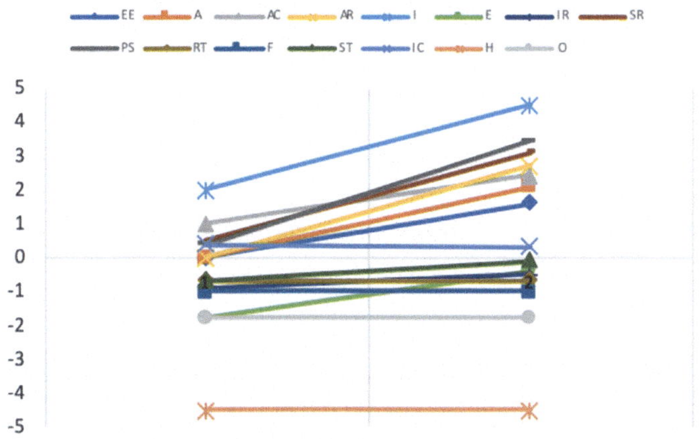

Fig. 3 The oscillation with FE. *Source* Own elaboration (2020)

(ST) and Impulse Control (IC), which were selected according to the discomfort level shown by the Bar-On survey.

The experts assigned a minimum level of incidence (pessimistic) after conducting the first attempt to induce a SOC within the staff: 0.2 for ST and 0.3 for IC, in matrix ${}^{+}_{1}\tilde{A}$ (Eq. 15). As the Forgotten Effects model considers only the positive values of matrix φ, the negative values have been set to zero. The matrix φ, is turned into the fuzzy allocation matrix ${}^{+}_{1}\tilde{M}$ (Estimated Positive Incidences):

$$\left[{}^{+}_{1}\tilde{M}\right] = \begin{bmatrix} 1 & 0 & 0 & 0 & 0 & 0.4 & 0 & 0 & 0 & 0 & 0 & 0 & 0 & 0 & 0 \\ 0 & 1 & 0 & 0 & 0 & 0 & 0 & 0 & 1 & 0 & 0 & 0 & 0 & 0 & 0 \\ 0.6 & 0.4 & 1 & 0.6 & 0 & 0 & 0.4 & 0.7 & 0 & 0 & 0 & 0 & 0.8 & 0 & 0 \\ 0 & 0 & 0 & 1 & 0 & 0 & 0 & 0 & 0 & 0 & 0 & 0.8 & 0 & 0 & 0 \\ 0 & 0 & 0 & 0 & 1 & 0 & 0 & 0 & 0 & 0 & 0 & 0 & 0 & 0 & 0 \\ 0 & 1 & 0 & 0 & 0 & 1 & 0.9 & 0 & 0 & 0 & 0 & 0 & 0 & 0 & 0 \\ 0 & 0 & 0 & 0 & 0 & 0 & 1 & 0 & 0 & 0 & 0 & 0 & 0 & 0 & 0 \\ 0 & 0 & 0 & 0.4 & 0.9 & 0 & 0 & 1 & 0 & 0 & 0 & 0 & 0 & 0 & 0 \\ 0 & 0.7 & 0 & 0.7 & 0 & 0 & 0 & 0 & 1 & 0 & 0 & 0 & 0 & 0.7 & 1 \\ 0 & 0 & 0 & 0 & 0.3 & 0 & 0 & 1 & 0.7 & 1 & 0.1 & 0.2 & 0.5 & 0 & 0 \\ 0 & 1 & 0 & 0 & 0 & 0 & 0.9 & 0 & 0 & 0 & 1 & 0 & 0 & 0 & 0 \\ 0 & 0 & 0.8 & 0 & 0.8 & 0 & 0 & 1 & 0 & 0 & 0 & 0 & 0.9 & 0 & 0 \\ 0 & 0 & 0 & 0 & 0.7 & 0 & 0 & 1 & 0 & 0 & 0 & 0 & 1 & 0 & 0 \\ 0 & 0 & 0 & 0 & 1 & 0 & 0 & 0 & 0 & 0 & 0 & 0 & 0 & 1 & 1 \\ 0.8 & 0 & 0.8 & 0.8 & 0.7 & 0.9 & 0.8 & 0 & 0 & 0 & 0.3 & 0.8 & 0.4 & 0 & 1 \end{bmatrix} \quad (14)$$

Now, matrix: $[{}^{+}_{1}\tilde{A}]$ is the allocation matrix on critical factors ST and IC:

$$\left[{}^{+}_{1}\tilde{A}\right] = \begin{bmatrix} 1 & 0 & 0 & 0 & 0 & 0 & 0 & 0 & 0 & 0 & 0 & 0.2 & 0.3 & 0 & 0 \\ 0 & 1 & 0 & 0 & 0 & 0 & 0 & 0 & 0 & 0 & 0 & 0.2 & 0.3 & 0 & 0 \\ 0 & 0 & 1 & 0 & 0 & 0 & 0 & 0 & 0 & 0 & 0 & 0.2 & 0.3 & 0 & 0 \\ 0 & 0 & 0 & 1 & 0 & 0 & 0 & 0 & 0 & 0 & 0 & 0.2 & 0.3 & 0 & 0 \\ 0 & 0 & 0 & 0 & 1 & 0 & 0 & 0 & 0 & 0 & 0 & 0.2 & 0.3 & 0 & 0 \\ 0 & 0 & 0 & 0 & 0 & 1 & 0 & 0 & 0 & 0 & 0 & 0.2 & 0.3 & 0 & 0 \\ 0 & 0 & 0 & 0 & 0 & 0 & 1 & 0 & 0 & 0 & 0 & 0.2 & 0.3 & 0 & 0 \\ 0 & 0 & 0 & 0 & 0 & 0 & 0 & 1 & 0 & 0 & 0 & 0.2 & 0.3 & 0 & 0 \\ 0 & 0 & 0 & 0 & 0 & 0 & 0 & 0 & 1 & 0 & 0 & 0.2 & 0.3 & 0 & 0 \\ 0 & 0 & 0 & 0 & 0 & 0 & 0 & 0 & 0 & 1 & 0 & 0.2 & 0.3 & 0 & 0 \\ 0 & 0 & 0 & 0 & 0 & 0 & 0 & 0 & 0 & 0 & 1 & 0.2 & 0.3 & 0 & 0 \\ 0 & 0 & 0 & 0 & 0 & 0 & 0 & 0 & 0 & 0 & 0 & 1 & 0.3 & 0 & 0 \\ 0 & 0 & 0 & 0 & 0 & 0 & 0 & 0 & 0 & 0 & 0 & 0.2 & 1 & 0 & 0 \\ 0 & 0 & 0 & 0 & 0 & 0 & 0 & 0 & 0 & 0 & 0 & 0.2 & 0.3 & 1 & 0 \\ 0 & 0 & 0 & 0 & 0 & 0 & 0 & 0 & 0 & 0 & 0 & 0.2 & 0.3 & 0 & 1 \end{bmatrix} \tag{15}$$

Since $\left[{}^{+}_{1}\tilde{A}\right] = \left[{}^{+}_{1}\tilde{B}\right]$, the results are obtained, as follows:
$[\left[{}^{-}_{1}\tilde{A}\right] \circ \left[{}^{+}_{1}\tilde{M}\right] \circ \left[{}^{+}_{1}\tilde{B}\right] = \left[{}^{+}_{1}\tilde{M}*\right]$ and $[\left[{}^{+}_{1}\tilde{O}\right] = \left[{}^{+}_{1}\tilde{M}*\right](-)\left[{}^{+}_{1}\tilde{M}\right]$:
A convolution of matrices (*max–min*) is then constructed, as shown below:
$\left[{}^{+}_{1}\tilde{A}\right] \circ \left[{}^{+}_{1}\tilde{M}\right] \circ \left[{}^{+}_{1}\tilde{B}\right] = \left[{}^{+}_{1}\tilde{M}*\right]$

$$\left[{}^{+}_{1}\tilde{M}\right] = \begin{bmatrix} 1 & 0 & 0.2 & 0 & 0.3 & 0.4 & 0 & 0.3 & 0.3 & 0 & 0 & 0 & 0.3 & 0 & 0 \\ 0 & 1 & 0.2 & 0 & 0.3 & 0 & 0 & 0.3 & 1 & 0 & 0 & 0 & 0.3 & 0 & 0 \\ 0.6 & 0.4 & 1 & 0.6 & 0.3 & 0 & 0.4 & 0.7 & 0.3 & 0 & 0 & 0 & 0.3 & 0 & 0 \\ 0 & 0 & 0.2 & 1 & 0.3 & 0 & 0 & 0.3 & 0.3 & 0 & 0 & 0.8 & 0.3 & 0 & 0 \\ 0 & 0 & 0.2 & 0 & 1 & 0 & 0 & 0.3 & 0.3 & 0 & 0 & 0 & 0.3 & 0 & 0 \\ 0 & 1 & 0.2 & 0 & 0.3 & 1 & 0.9 & 0.3 & 0.3 & 0 & 0 & 0 & 0.3 & 0 & 0 \\ 0 & 0 & 0.2 & 0 & 0.3 & 0 & 1 & 0.3 & 0.3 & 0 & 0 & 0 & 0.3 & 0 & 0 \\ 0 & 0 & 0.2 & 0.4 & 0.9 & 0 & 0 & 1 & 0.3 & 0 & 0 & 0 & 0.3 & 0 & 0 \\ 0 & 0.7 & 0.2 & 0.7 & 0.3 & 0 & 0 & 0.3 & 1 & 0 & 0 & 0 & 0.3 & 0.7 & 1 \\ 0 & 0 & 0.2 & 0 & 0.3 & 0 & 0 & 1 & 0.7 & 1 & 0.1 & 0.2 & 0.5 & 0 & 0 \\ 0 & 1 & 0.2 & 0 & 0.3 & 0 & 0.9 & 0.3 & 0.3 & 0 & 1 & 0 & 0.3 & 0 & 0 \\ 0 & 0 & 0.8 & 0 & 0.8 & 0 & 0 & 0.3 & 1 & 0 & 0 & 1 & 0.9 & 0 & 0 \\ 0 & 0 & 0.2 & 0 & 0.7 & 0 & 0 & 0.7 & 1 & 0 & 0 & 0 & 1 & 0 & 0 \\ 0 & 0 & 0.2 & 0 & 0.3 & 1 & 0 & 0.3 & 0.3 & 0 & 0 & 0 & 0.3 & 1 & 1 \\ 0.8 & 0 & 0.8 & 0.8 & 0.7 & 0.9 & 8 & 0.3 & 0.3 & 0 & 0.3 & 0.8 & 0.4 & 0 & 1 \end{bmatrix} \tag{16}$$

The Forgotten Effects matrix is subsequently obtained:
$[{}^{+}_{1}\tilde{O}] = [{}^{+}_{1}\tilde{M}*](\text{-})[{}^{+}_{1}\tilde{M}]$

$$\left[{}_{1}^{+}\tilde{O}\right] = \begin{bmatrix}
0&0&0.2&0&0.3&0&0&0.3&0.3&0&0&0.2&0.3&0&0\\
0&0&0.2&0&0&0&0&0.3&0&0&0&0.2&0.3&0&0\\
0&0&0&0&0.3&0&0&0&0.3&0&0&0.2&0&0&0\\
0&0&0.2&0&0.3&0&0&0.3&0.3&0&0&0&0.3&0&0\\
0&0&0.2&0&0&0&0&0.3&0.3&0&0&0.2&0.3&0&0\\
0&0&0.2&0&0.3&0&0&0.3&0.3&0&0&0.2&0.3&0&0\\
0&0&0.2&0&0.3&0&0&0.3&0.3&0&0&0.2&0.3&0&0\\
0&0&0.2&0&0&0&0&0&0.3&0&0&0.2&0.3&0&0\\
0&0&0.2&0&0.3&0&0&0.3&0&0&0&0.2&0.3&0&0\\
0&0&0.2&0&0&0&0&0&0&0&0&0&0&0&0\\
0&0&0.2&0&0.3&0&0&0.3&0.3&0&0&0.2&0.3&0&0\\
0&0&0&0&0&0&0&0.3&0&0&0&0.2&0&0&0\\
0&0&0.2&0&0&0&0&0&0&0&0&0.2&0&0&0\\
0&0&0.2&0&0.3&0&0&0.3&0.3&0&0&0.2&0.3&0&0\\
0&0&0&0&0&0&0&0.3&0.3&0&0&0&0&0&0
\end{bmatrix} \tag{17}$$

The absolute values of the negative elements of $\varphi = [|{}_{1}^{-}\tilde{M}|]$ are shown in Eq. 18.

$$[|{}_{1}^{-}\tilde{M}|] = \begin{bmatrix}
0&0&0&0&0&0&0&0&0&0&0&0&0&0&0\\
0&0&0&0&0&1&0&0&1&0&0&0&0&0&0\\
0&0&0&0&0&0&0&0&0&0&0&0&0&0&0\\
0&0&0&0&0&0&0&0&0&0&0&0&0&0&0\\
0&0&0&0&0&0&0&1&0&0&0&0&0&0&0\\
0&0&0&0&0&0&0&0&0&0&0&0&0&0&0\\
0&0&0&0&0&0&0&0&0.7&0&0&0&0&0.6&0\\
0&0&0&0&0&0&0&0&0&0&0&0&0.6&0.5&0\\
0&0&0&0&0&0&0&0&0&0.7&1&0&0&0&0\\
0&0&0&0&0&0&0.2&0&0&0&0&0&0&1&1\\
0&0&0&0&0&0&0&0&0&0&0&0&0&0&0\\
0&0&0&0&0&0&0&0&0&0&0&0&0&0&0\\
0&0&0&0&0&0&0&0&0&0&0&0.5&0&0&0\\
0&0&0&0&0&0&0&0&0&0&0&0&0&0&0\\
0&0&0&0&0&0&0&0&0&0&0&0&0&0.5&0
\end{bmatrix} \tag{18}$$

Then, the negative incidences of matrix $[{}_{1}^{-}\tilde{A}]$ on ST and IC variables are presented in Eq. 19:

$$[|_1^- \tilde{A}|] = \begin{bmatrix} 0&0&0&0&0&0&0&0&0&0&0&0.2&0.3&0&0 \\ 0&0&0&0&0&0&0&0&0&0&0&0.2&0.3&0&0 \\ 0&0&0&0&0&0&0&0&0&0&0&0.2&0.3&0&0 \\ 0&0&0&0&0&0&0&0&0&0&0&0.2&0.3&0&0 \\ 0&0&0&0&0&0&0&0&0&0&0&0.2&0.3&0&0 \\ 0&0&0&0&0&0&0&0&0&0&0&0.2&0.3&0&0 \\ 0&0&0&0&0&0&0&0&0&0&0&0.2&0.3&0&0 \\ 0&0&0&0&0&0&0&0&0&0&0&0.2&0.3&0&0 \\ 0&0&0&0&0&0&0&0&0&0&0&0.2&0.3&0&0 \\ 0&0&0&0&0&0&0&0&0&0&0&0.2&0.3&0&0 \\ 0&0&0&0&0&0&0&0&0&0&0&0.2&0.3&0&0 \\ 0&0&0&0&0&0&0&0&0&0&0&0.2&0.3&0&0 \\ 0&0&0&0&0&0&0&0&0&0&0&0.2&0.3&0&0 \\ 0&0&0&0&0&0&0&0&0&0&0&0.2&0.3&0&0 \\ 0&0&0&0&0&0&0&0&0&0&0&0.2&0.3&0&0 \end{bmatrix} \tag{19}$$

The causal matrix and the effects matrix are equal: $\left[\left|_1^- \tilde{A}\right|\right] = \left[\left|_1^- \tilde{B}\right|\right]$, thus resulting in $[|_1^- \tilde{A}|]^\circ[|_1^- \tilde{M}|]^\circ[|_1^- \tilde{B}|] = [|_1^- \tilde{M}*|]$ (Eq. 20) And $[|_1^- \tilde{O}|] = [|_1^- \tilde{M}*|](-)[|_1^- \tilde{M}|]$ (Eq. 21).

$$[|_1^- \tilde{M} * |] = \begin{bmatrix} 0&0&0&0&0&0&0&0&0&0&0&0.2&0.3&0&0 \\ 0&0&0&0&0&0&0&0&0&0&0&0.2&0.3&0&0 \\ 0&0&0&0&0&0&0&0&0&0&0&0.2&0.3&0&0 \\ 0&0&0&0&0&0&0&0&0&0&0&0.2&0.3&0&0 \\ 0&0&0&0&0&0&0&0&0&0&0&0.2&0.3&0&0 \\ 0&0&0&0&0&0&0&0&0&0&0&0.2&0.3&0&0 \\ 0&0&0&0&0&0&0&0&0&0&0&0.2&0.3&0&0 \\ 0&0&0&0&0&0&0&0&0&0&0&0.2&0.3&0&0 \\ 0&0&0&0&0&0&0&0&0&0&0&0.2&0.3&0&0 \\ 0&0&0&0&0&0&0&0&0&0&0&0.2&0.3&0&0 \\ 0&0&0&0&0&0&0&0&0&0&0&0.2&0.3&0&0 \\ 0&0&0&0&0&0&0&0&0&0&0&0.2&0.3&0&0 \\ 0&0&0&0&0&0&0&0&0&0&0&0.2&0.3&0&0 \\ 0&0&0&0&0&0&0&0&0&0&0&0.2&0.3&0&0 \\ 0&0&0&0&0&0&0&0&0&0&0&0.2&0.3&0&0 \end{bmatrix} \tag{20}$$

And $[|_1^- \tilde{O}|] = [|_1^- \tilde{M}*|](-)[|_1^- \tilde{M}|]$ (Eq. 21):

$$[\left|{}_{1}^{-}\tilde{O}\right|]=\begin{bmatrix}0&0&0&0&0&0&0&0&0&0&0&0.2&0.3&0&0\\0&0&0&0&0&0.6&0&0&0&0.8&0&0.2&0.3&0&0\\0&0&0&0&0&0&0&0&0&0&0&0.2&0.3&0&0\\0&0&0&0&0&0&0&0&0&0&0&0.2&0.3&0&0\\0&0&0&0&0&0&0&0.6&0&0&0&0.2&0.3&0&0\\0&0&0&0&0&0&0&0&0&0&0&0.2&0.3&0&0\\0&0&0&0&0&0&0&0&0.2&0&0&0.2&0.3&0.3&0\\0&0&0&0&0&0&0&0&0&0&0&0.2&0.3&0.4&0\\0&0&0&0&0&0&0&0&0&0.2&0.6&0.2&0.3&0&0\\0&0&0&0&0&0&0.7&0&0&0&0&0.2&0.3&0.8&0.6\\0&0&0&0&0&0&0&0&0&0&0&0.2&0.3&0&0\\0&0&0&0&0&0&0&0&0&0&0&0.2&0.3&0&0\\0&0&0&0&0&0&0&0&0&0&0&0.2&0.3&0&0\\0&0&0&0&0&0&0&0&0&0&0&0.2&0.3&0&0\\0&0&0&0&0&0&0&0&0&0&0&0.2&0.3&0.4&0\end{bmatrix} \tag{21}$$

Finally, the Forgotten Effects process is defuzzified, as the sum of the positive elements' matrices (Eq. 17) and the negative elements matrix (Eqs. 21). Where the Eq. 21 matrix (that of temporary positive elements) is multiplied by (-1) so as to obtain the Eq. 22 matrix, in order to apply the MCD model.

$$\varphi^{new}=\begin{bmatrix}1&0&1.2&0&0.3&0.4&0&0&0.3&0.3&0&0&0&0&0\\0&1&0.2&0&0.3&-1.8&0.3&0.2&0&0&0&0&0&0&0\\0.6&0.4&1&0.6&0.3&0&0.4&0.7&0.3&0&0&0&0.5&0&0\\0&0&0.2&1&0.3&0&0&0.3&0.3&0&0&0.6&0&0&0\\0&0&0.2&0&1&0&0&-0.5&0.3&0&0&0&0&0&0\\0&1&0.2&0&0.3&1&0.9&0.3&0.3&0&0&0&0&0&0\\0&0&0.2&0&0.3&0&1&0.3&-0.6&0&0&0&0&-0.6&0\\0&0&0.2&0.4&0.9&0&0&1&0.3&0&0&0&-0.9&-0.5&0\\0&0.7&0.2&0.7&0.3&0&0&0&1&-0.7&-1&-0.2&-0.3&-0.7&1\\0&0&0.2&0&0.3&0&0.9&1&0.7&1&0.1&0&0.2&-1.8&-1.8\\0&1&0.2&0&0.3&0&0.9&0&0&0&1&0&0&0&0\\0&0&0.8&0&0.8&0&0&1&1&0&0&0.6&0&0&0\\0&0&0.2&0&1.2&0&0&1&1.3&0&0&-0.5&1&0&0\\0&0&0.2&0&0.3&1&0&0.3&0.3&0&0&0&0&1&1\\0.8&0&0.8&0.8&0.7&0.9&0.8&0.3&0.3&0&0.3&0.6&1&-0.9&1\end{bmatrix} \tag{22}$$

If the MCD model is applied one more time, the results shown in Table 4, are obtained.

If the methodology is employed, a new matrix is obtained and the oscillations of variables get to be smoothed, as shown below (Fig. 3).

Table 4 MCD Results (initiators: ST y IC) with the New Matrix

t							C(t)									
0	0	0	0	0	0	0	0	0	0	0	0	1	1	0	0	
1	1	1	1	1	1	0	0	1	1	0	0	0	1	0	0	
$R = C''\varphi$																
0	0	1	0	2	−1.8	−0.8	0.5	0.4	−0.7	−1	−0.7	0.4	−4.5	−1.8	1	1
1.6	2.1	2.4	2.7	4.5	−0.4	−0.5	3.1	3.4	−0.7	−1	−0.1	0.3	−4.5	−1.8	1	1
$C(t+1) = f(R)$																
1	1	1	0	0	i	i	0	0	0	1	0	0				
1	1	1	0	0	i	i	0	0	0	1	0	0				

4.3 The Distance of Hamming

The distance between vector ***R*** and vector R^{new}, once an equilibrium is reached, is summarized as follows (Table 5).

The obtained differences, in descendent order, among cognitive elements are presented below:

$$H > IR > O > IC > E > SR > ST > A > EE, AR > PS > F > AC > RT$$

According to these results and in order to achieve a SOC in the four fat-acids-producing organizations some elements stand out, once obtained the Hamming distance between the two vectors ***R*** and R^{new}. Happiness (H) obtained the greatest distance at 6.4, followed by Interpersonal Relationships (IR) at 4.2, Optimism (O) at 2.8, Impulse Control (IC) at 2.2, Empathy (E) and Social Responsibility (SR), both at 1.7, followed by Stress Tolerance (ST) at 1.4 and so forth. Therefore, by establishing strategies aimed at improving work happiness and human relationships (as being the elements that showed the greatest distances) the other emotional-intelligence-related elements would have a tendency to reduce their negative impact and thus, significant changes in these organizations can be achieved. Remember that the study has been done from pessimistic approach. Thus, in the future lines research triangular fuzzy numbers (TFN): pessimistic, medium and optimistic values.

5 Conclusions

The proposed methodology in this work requires the division of positive and negative elements of the allocation matrix, in terms of absolute value, in order to allow their fuzzification so then the FE model may be applied. Subsequently, a defuzzification process is performed so as to apply the FCM model. So doing allows to conduct an

Table 5 Distance among cognitive elements: matrix R and R^{new}

	EE	A	AC	AR	I	E	IR	SR	PS	RT	F	ST	IC	H	O
R	2.4	3.1	2.6	3.5	4.1	1.3	3.7	1.4	3.7	−0.7	0.7	1.3	2.5	1.9	1
R^{new}	1.6	2.1	2.4	2.7	4.5	−0.4	−0.5	3.1	3.4	−0.7	−1	−0.1	0.3	−4.5	−1.8
Differences	0.8	1	0.2	0.8	−0.4	1.7	4.2	−1.7	0.3	0	−0.3	1.4	2.2	6.4	2.8

early diagnosis on the strategies that must be followed. In this case, the new strategies ought to be directed towards the tension associated to the following elements (cognitive factors): happiness, optimism, empathy, flexibility, reality test and stress tolerance to generate a SOC.

The determination of the cognitive factors through the Hamming absolute distance metrics allows formalizing the basis for the new strategies. Followed by evaluations conducted on the new-strategies-favoring staff (accumulated tension, $S_{(i+1)}$) and comparing them to those conducted on new-strategies-resisting staff (accumulated inertia, $I_{(i+1)}$). Therefore, if $S_{(i+1)} > I_{i+1}$,, then a SOC will take place, [14].

In future research works, the criteria employed for the strategic diagnosis can be expanded by dividing the initiating concept vector into fuzzy triplets (triangular fuzzy numbers): pessimistic, medium and optimistic. In the proposed study case, only the pessimistic assessment was applied, resulting in a 0.2 value for the Stress Tolerance factor and 0.3 for the Impulse Control factor, the application of triangular fuzzy numbers, will allow to broaden the spectrum in the diagnosis of strategies. On the other hand, the proposal can be extended to other types of projects so as to formalize the sensitive and subjective factors that involve the change of order in the organizational systems.

References

1. Bartunek et al (1987) The journal of applied behavioural. Sci J 23(4):496-503
2. Bar-On R, Parker JDA (eds) (2000) The handbook of emotional intelligence: Theory, development, assessment, and application at home, school, and in the workplace. Jossey-Bass
3. Carlsson C (1996) Knowledge formation in strategic management. HICSS-27. Proceedings. IEEE. Computer Society Press, Los Almitos
4. Curia L, Lavalle A (2011) Decisions strategies in dynamic systems using fuzzy cognitive maps. Application to a socio-economic example. Ed. Tecsifeausp, Jistem, Brazil, 8(3):672-674
5. Davenport et al (1999) Human capital, Ediciones Gesti6n 2000 S. A. Barcelona Spain
6. Echevarría R (2009) The observer and his world. Ed. Granica. Buenos Aires, Argentina
7. Extremera, Fernandez-Berrocal (2002) The importance of developing emotional intelligence. Review Iberoamericana de Educación
8. Extremera N, Fernández-Berrocal P (2004) The role of emotional intelligence in students: empirical evidence. Review Electronica de Investigación Educativa 6(2):4-8. Extracte: 23/04/2009. http://redil.uabc.mx/vol.6no2/contenido-extremera-htm/
9. Fit-enz J (2003) The ROI of human capital. Deusto, Barcelona, Spain
10. Gil A (2005) Elementos de una teoría de decisión en la incertidumbre. Ed. Milladoiro, España
11. Goleman D (1996) Emotional intelligence. Cairos, Madrid
12. Goleman D (1998) The Emotional Intelligence, 6ta edn. Zeta, Buenos Aires
13. Hiliera, Martínez (2000) Artificial neural networks fundamentals models and applications. RA-MA Ed. Madrid
14. Huff et al. (2002) The strategic change. Oxford University, impress in Mexico
15. Leyva et al (2012) Modelling and analysis of critical success factors of solfware projects using fuzzy cognitive maps. Magazine: Ciencias de la informacion. 43(2):41-46
16. Kaufmann, Gil (1988) Model for to Investigation of Forgotten Efects, 1a edn. Edition Milladoiro, Spain
17. Kaufmann G, Terceño (1994) Mathematics for economics and business management. 1a. ed. Edition Foro científico, Barcelona, Spain

18. Kosko HB (1986) Fuzzy Cognitive Maps. Int J Man Mach Stud 24:65–75
19. Kosko H, B (1997) Fuzzy Engineering Ed. Prentice-Hall New Jersey
20. Montoya R et al (2005) Magazine: school of business administration, vol 53. Bogotá, Colombia, p 84
21. Nelson R (1991) Why do firms differ and how does it matter? Strateg Manag J 12:61–74
22. Nonaka I, Takeuchi H (1995) The knowledge-creating company. Oxford University Press, Nueva York, How Japanese companies create the dynamics y innovation
23. Nonaka I, Takeuchi H (1997) The knowledge-creating Organization. Oxford University Press, México
24. Peláez CE, Bowles JB (1995) Applying fuzzy cognitive maps knowledge-representation to failure modes effects analysis. IEEE proceedings annual reability and maintainability symposium, pp 0149-144X/95
25. Piquera et al, (2009) Negative emotions and their impact on mental and physical health. Review Suma Psicológica 16(2):85-112
26. Porter M (2005) Competitive strategy. Ed. CECSA, Mexico
27. Probst et al (2001) Manage knowledge. Pearson Educación, Mexico
28. Tichy M (2003) Leaders in action. 1st. Press, CECSA Mexico
29. Ugarriza N (1997) The evaluation of emotional intelligence through of the Bar-On inventory (I-CE) in a sample from Metropolitan Lima. University of Lima, Lima
30. Ugarriza, Pajares (2001) Adaptation and standardization of the emotional intelligence inventory of Bar-On ICE: NA, in adolescent children (2nd edn). Amigo, Lima
31. Vargas J (2013) Emotional intelligence in education, 1st edn. Groppe Books, Mexico
32. Weakland H, Fisch R, Watzlawick P, Bodin A (1974) Brief therapy: focused problem resolution. Review Family Process 13(2):141–168
33. Winter S (1982) An evolutionary theory of economic change. Cambridge University, Press Cambridge
34. Zadeh J (1965) Fuzzy set. Rev Inf Control 8(3):338–353

Aggregating Fuzzy Sentiments with Customized QoS Parameters for Cloud Provider Selection Using Fuzzy Best Worst and Fuzzy TOPSIS

Walayat Hussain, José M. Merigó, Fethi Rabhi, and Honghao Gao

Abstract Consumers often get confused to select the best cloud providers from the huge marketplace. The hesitancy of consumers further escalates when multiple service providers offer the same type and quality of services. To deal with such an uncertainty, the decision-makers always combine multiple factors to make an informed choice. Sentiment mining is one of the key parameters to determine the service quality and get an insight into the business. It assists service providers in precisely deduce consumer's emotions regarding the product. The analysis helps providers fine-tune the product based on consumer's sentiment and accommodates the consumer's request to find an optimal service provider. Several existing literature for cloud service selection mainly focuses on the Quality of Service (QoS) of the offered services. However, very few of them have considered the user experience of a consumer in the decision-making process. Moreover, there is minimal literature that amalgamates Quality of Experience (QoE) with customized Quality of Service (QoS) requirements to decide on a complex framework. The paper addresses the issue by aggregating consumer's sentiments with customized QoS parameters to choose an optimal service provider. The paper uses the fuzzy Best Worst Method (BWM) to determine the weights of selection criteria and use the fuzzy TOPSIS to handle the uncertain linguistic preference. Analysis results demonstrate the applicability and effectiveness of the framework.

W. Hussain (✉)
Victoria University Business School, Victoria University, Melbourne, VIC 3011, Australia
e-mail: walayat.hussain@uts.edu.au

W. Hussain · J. M. Merigó
Faculty of Engineering and IT, University of Technology Sydney, Ultimo 2007, Australia
e-mail: jose.merigo@uts.edu.au

F. Rabhi
School of Computer Science, University of New South Wales, NSW Sydney, Australia
e-mail: f.rabhi@unsw.edu.au

H. Gao
School of Computer, Engineering and Science, Shanghai University, Shanghai 200444, China
e-mail: gaohonghao@shu.edu.cn

E. León-Castro et al. (eds.), *Soft Computing and Fuzzy Methodologies in Innovation Management and Sustainability*, Lecture Notes in Networks and Systems 337,
https://doi.org/10.1007/978-3-030-96150-3_6

Keywords Fuzzy sentiment · Quality of service (QoS) · Cloud provider selection · Fuzzy best worst method · Fuzzy TOPSIS · Quality of experience · Service level agreement (SLA)

1 Introduction

Cloud services have gained significant importance for organizations due to their flexible, cost-efficient, scalable and easy to manage features. Most organizations focus their cloud projects intending to attain the maximum of cloud benefits. However, selecting an optimal service provider to trust the organization's data and infrastructure is a complicated process. The lack of a common framework makes it difficult to choose the best provider among many service providers who offer the same services. Existing studies [1, 2] consider many factors to choose the service providers mostly based on QoS parameters. These parameters include—security, services, computer power, pricing and many others. Many existing approaches [3, 4] predict the QoS parameters for future interval and combine them with other factors such as reputation, size, and services offered by the service provider to determine the right provider [2]. Various QoS attributes have varying weightage or preferences for a consumer. Some consumer prioritizes certain parameter than the others which may impact the decision-making process. Although QoS parameters play a key role in service selection, most of the existing approaches have ignored the QoE or sentiment of previous consumers regarding each service level objectives (SLOs) and QoS parameters.

Quality of Experience (QoE) focuses on the actual user experience of offered services, while QoS focuses on the system's technical performance. The QoE is the main source to present the overall happiness or frustration of services, which directly impacts a potential consumer's choice to adopt a business [4]. Consumer reviews are the main source to portray the QoE of an existing consumer and the quality of offered services. The reviews are in general, and it is very difficult to identify a consumer sentiment for the custom individual QoS requirement of a requesting user. Unlike other products such as hotel booking or household product, there is no centralized system for cloud consumer reviews that show various aspects of cloud services. These reviews are mostly available at the cloud provider website. Other sources, including social networking websites and some reviewing websites, also provide a limited feedback. The lack of a centralized system to aggregate QoS and QoE creates many challenges for the decision making process. The challenges include latency and bandwidth issue [5, 6] for exchanging QoS data, lack of a mechanism to map QoE and QoS and offline or unresponsiveness of recommenders systems.

The paper presents a framework that combines individual sentiments in relation of offered QoS parameters to address above discuss issues. The approach obtains custom prioritized QoS requirement from a requesting consumer, identifies multiple related service providers, fetch the previous sentiment, and identifies an optimal service provider. To best handle the uncertainty and avoid losing any information, the paper uses a fuzzy system for criteria weightage and sentiment aggregation.

Relative weights for each criterion are calculated using fuzzy Best Worst Method (BWM) and then use the fuzzy TOPSIS method for decision making.

The rest of the paper is organized as follows. Section 2 presents related literature. Section 3 presents the mythology and working of the proposed approach. Section 4 presents the implementation and evaluation, and finally, Sect. 5 concludes the paper.

2 Related Literature

Selecting the best service provider from the myriad of service providers is a trivial job for a consumer in a vast cloud marketplace [1]. Many service providers offer the same type and same standard of services. However, the service selection not only depends on specific business needs but may depend on multiple factors. Existing literature [7, 8] mostly focus on the QoS of offered services to determine the right service provider. The service provider's wrong choice is the reason for service level agreement (SLA) violations and causes penalties both in terms of money and reputation or trust [9]. SLA is the key agreement between a consumer and the service provider for agreed QoS for a certain time frame. The service provider usually offers variety of services. For a viable SLA, it is imperative to assess individual required services, affiliated SLOs and QoS parameters. Alrashed and Hussain [10] proposed a fuzzy re-SchdNeg decision model. The proposed approach predict service violation and then necessary action to re-schedule or re-negotiate to avoid possible service violation. Rajavel and Thangarathanam [7] proposed an agent-based automated dynamic SLA negotiation model. The stochastic behavioral learning negotiation method integrates behavioral learning with the stochastic method to avoid any uncertainties in the decision-making process.

Alkalbani and Hussain [1] proposed a cloud service discovery method that uses domain-specific ontology to extract meaningful information from SaaS reviews. Precision, recall and F-score are used as a benchmark to evaluate the accuracy of the approach. In another approach, Alkalbani, Hussain and Kim [11] proposed a CCSR framework to find an optimal service provider. The framework use harvesting as a service (HaaS) harvester to extract real-time dataset. The dataset further combined and stored in a centralized repository to use for service selection. Hussain et al. [12–14] considered different factors such as the service provider's risk attitude, consumer reputation, transaction trends to select the right service provider and manage the SLA in a viable way. Alghamdi et al. [15] analyzed different QoS parameters to build a trusted relationship in the cloud. Hussain and Sohaib [16] used three QoS parameters—throughput, response time, and availability to analyze the prediction accuracy based on the prediction methods' freshness and control parameters.

Sentiment analysis is an important factor for service selection that gives a clear overview of consumer satisfaction for offered services. Alarifi et al. [17] used big data and machine learning techniques to evaluate sentiment to reduce noise and improve accuracy. The approach uses a greedy algorithm with cat swarm optimization based long short term memory neural networks. The approach gives better results with optimized features. Alharbi and Alhalabi [18] used a hybrid approach to analyze consumers sentiments using fuzzy method. The proposed approach combines a fuzzy inference process and dictionary-based method to categorise sentiment into five categories. Dang, Moreno-García and De la Prieta [19] analyzed various deep learning methods to identify sentiment polarity. The analysis results revealed that the CNN method offers the best trade-off between the processing time and the accuracy results.

Although the approaches mentioned above assist cloud stakeholders in managing SLA and building a sustainable trusted relationship, there are still many gaps due to decentralized approaches. Most of the existing approaches focused on individual QoS and QoE parameters. Such information spread across different nodes located at different places that a consumer or provider have interacted with earlier. However, no method combines fuzzy sentiment with customized QoS parameters in a complex cloud environment. The paper addresses the issue by proposing a centralized approach to combine fuzzy sentiment with custom prioritized QoS parameters. The proposed method is presented in Sect. 3.

3 Methodology

The section presents the proposed framework. The decision-making process comprises two modules—the data preparation module and the decision-making module. Data preparation module is responsible for obtaining customized QoS parameters from a consumer and then harvest related sentiment of the best-matched service providers. There are many existing methods for matrix comparison for decision making, such as—Euclidian space, Wasserstein distance [20], Weighted average, Ordered Weighted Averaging (OWA) aggregation operator [21–23]. However, most of these approaches are not suitable due to the unbalanced comparison process. The criteria below user requirements are averaged by those who are exceeding the user requirements, due to which they are unable to get an optimal result. Therefore, in this paper, we use the fuzzy BWM and fuzzy TOPSIS methods. In the second module, the system determines criteria weight by identifying the best and worst criteria. In real life, the quantitative assessment is not possible due to imprecise knowledge. Therefore, the decision-maker usually adopted qualitative assessment using linguistic variables. To handle the uncertainty without losing any meaningful information, the system uses the fuzzy approach. The system use criteria sentiment from the first module and apply fuzzy TOPSIS to execute the decision-making process. The working of each module are explained as follows:

3.1 Data Preparation and Sentiment Harvesting Module

In this module, the system takes the requested QoS parameter from a requesting consumer to form an SLA. The system finds related cloud providers that best match the requirements of a consumer. In this paper, we assume that there are four service providers that best match a consumer's requirements. All service providers offer the same type of services with the same quality level. For optimal decision-making, the system needs to use the quality of experience (QoE) of consumers who have executed their SLA with service providers. The system uses Harvesting as a Service (HaaS) dataset [11] for analyzing consumers sentiment. The dataset was harvested from SaaS providers that offer services with QoS parameters as requested by a consumer. We use the first 200 comments for each service provider for simplicity and consistency across all service providers and use SentiWordNet 3.0 [24] to extract consumer's fuzzy sentiment for each QoS parameters.

3.2 *Sentiment Evaluation and Decision Making Module*

The decision-making process comprise of two steps—criteria weightage and decision making. For a new SLA, there are many cases in which a consumer is requesting resources with a specific set of QoS parameters that are unique from existing SLAs. Some of the parameters are highly weighted than others that directly impact decision-making for choosing the service provider.

We address the problem by finding relative weights of each QoS parameters using the fuzzy BWM [25]. The method provides a higher comparison consistency that results in better accuracy [26]. In this paper, we have considered four QoS parameters to evaluate the service providers. Among those parameters, *price* or *cost of services* is highly desirable or best criterion, and reliability is the least favourite or worst criterion for a consumer. Using Eqs. 1–4, the system determines the relative weight for each QoS parameters.

The system assesses previous consumers sentiment and use the relative weights of each QoS parameters obtained from the best worst method to determine the best service provider. The paper uses the fuzzy TOPSIS method for the decision making process. The working of fuzzy best worst and fuzzy TOPSIS method is presented as follows:

a. **Fuzzy best worst method for criteria weights**

A consumer determines the most desirable or 'Best' and least desirable or 'Worst' criteria. The system then determines the fuzzy preference of the best and worst criterion over all other criteria using a triangular fuzzy number. The linguistic terms and membership function used in the paper is presented in Table 1.

The obtained fuzzy best vector $\widetilde{FP}_{best}$ is presented as follows:

$$\widetilde{FP}_{best} = \left(\widetilde{fp}_{best,1}, \widetilde{fp}_{best,2}, \ldots, \widetilde{fp}_{best,n}\right) \tag{1}$$

where $\widetilde{fp}_{best,j}$ is the fuzzy preference of the best criterion—'*best*' over the criterion—*j*.

Table 1 Linguistic variables of the decision-maker

Linguistic terms	Membership function
Equally important	(1, 1, 1)
Weakly important	(2/3, 1, 3/2)
Fair important	(3/2, 2, 5/2)
Very important	(5/2, 3, 7/2)
Absolute important	(7/2, 4, 9/2)

Similarly, the fuzzy preference for the worst criterion $\widetilde{FP}_{worst}$ overall the best criteria are calculated as follows:

$$\widetilde{FP}_{worst} = \left(\widetilde{fp}_{1,worst}, \widetilde{fp}_{2,worst}, \ldots, \widetilde{fp}_{n,worst}\right) \tag{2}$$

Considering the fuzzy preference $\widetilde{FP}_{best}$ and $\widetilde{FP}_{worst}$ we determine the criteria weight in such a way that $w_{best}/w_j \cong \widetilde{fp}_{best,j}$ and $w_j/w_{worst} \cong \widetilde{fp}_{j,worst}$ for all j. The weights $(\widetilde{w}_i^k)$ are obtained by minimizing and maximizing the absolute difference $\left|\frac{w_{best}}{w_j} - \widetilde{fp}_{best,j}\right|$ and $\left|\frac{w_j}{w_{worst}} - \widetilde{fp}_{j,worst}\right|$ for all j. Therefore, normalizing weights, the optimal weights can be achieved by solving the following nonlinear problem:

$$\text{minmax} = \left\{\left|\frac{w_{best}}{w_j} - \widetilde{fp}_{best,j}\right|, \left|\frac{w_j}{w_{worst}} - \widetilde{fp}_{j,worst}\right|\right\} \tag{3}$$

$$s.t.\left\{\begin{array}{c} \left|\frac{\left(w_{best}^{l,k}, w_{best}^{m,k}, w_{best}^{u,k}\right)}{\left(w_j^{l,k}, w_j^{m,k}, w_j^{u,k}\right)} - \left(\widetilde{fp}_{best,j}^{l,k}, \widetilde{fp}_{best,j}^{m,k}, \widetilde{fp}_{best,j}^{u,k}\right)\right| \leq \xi_L \\ \left|\frac{\left(w_j^{l,k}, w_j^{m,k}, w_j^{u,k}\right)}{\left(w_{worst}^{l,k}, w_{worst}^{m,k}, w_{worst}^{u,k}\right)} - \left(\widetilde{fp}_{j,worst}^{l,k}, \widetilde{fp}_{j,worst}^{m,k}, \widetilde{fp}_{j,worst}^{u,k}\right)\right| \leq \xi_L \\ \sum_{j=1}^{J} S\left(w_j^k\right) = 1 \\ w_j^{l,k} \leq w_j^{m,k} \leq w_j^{u,k} \\ w_j^{l,k} \geq 0 \\ j = 1, 2, \ldots, J \end{array}\right\} \tag{4}$$

b. **Fuzzy TOPSIS method**

The fuzzy Technique for Order Performance by Similarity to Ideal Solution (TOPSIS) is a widely used method for solving MCDM problems [27]. The working of TOPSIS and decision-making process is performed as follows:

- Assignment rating to criteria and alternatives: We have considered the first 200 comments to assess four customized QoS parameters, as discussed in the above section.
- The aggregated fuzzy sentiment $\widetilde{FS}_{i,j} = \left(a_{i,j}, b_{i,j}, c_{i,j}\right)$ of all commenting consumers is calculated as follows:

Table 2 Fuzzy linguistic scale

Linguistic terms for sentiment	Triangular fuzzy numbers
Strong positive	1, 1, 3
Positive	1, 3, 5
Neutral	3, 5, 7
Negative	5, 7, 9
Strong negative	7, 9, 9

$$a_{i,j} = \min_k\left\{a_{i,j}^k\right\}, b_{i,j} = \frac{1}{K}\sum_{k=1}^{K} b_{i,j}^k, c_{i,j} = \max_k\left\{c_{i,j}^k\right\} \tag{5}$$

where

i, *j* represents the number of alternatives and the number of criteria, respectively, $a_{i,j}, b_{i,j}, c_{i,j}$ are the lower, middle and upper bounds of the fuzzy numbers, respectively.

- Each criterion's linguistic weightage is substituted with an equivalent fuzzy number using a fuzzy linguistic scale [28] as presented in Table 2.
- The normalization process for beneficial and non-beneficial criteria is calculated using Eqs. 6–7 as presented below:

  For beneficial criteria:

$$\tilde{r}_{i,j} = \left(\frac{a_{i,j}}{c_j^*}, \frac{b_{i,j}}{c_j^*}, \frac{c_{i,j}}{c_j^*}\right), \text{where} c_j^* = \max\left(c_{i,j}\right) \tag{6}$$

  For cost criteria:

$$\tilde{r}_{i,j} = \left(\frac{\tilde{a}_j}{c_{i,j}}, \frac{\tilde{a}_j}{b_{i,j}}, \frac{\tilde{a}_j}{a_{i,j}}\right), \text{where} \tilde{a}_j = \min\left(a_{i,j}\right) \tag{7}$$

- The weighted normalized fuzzy decision matrix is computed as follows:

$$\tilde{v}_{i,j} = \tilde{r}_{i,j} \times \tilde{w}_i^k \tag{8}$$

where $\tilde{w}_i^k$ are weights obtained from the Fuzzy best–worst method.

- Compute the fuzzy positive ideal solution (FPIS) and fuzzy negative ideal solution (FNIS) using Eqs. 9–10.

$$\breve{A} = (\tilde{v}_1, \tilde{v}_2, \ldots, \tilde{v}_n) \tag{9}$$

$$\overline{\overline{A}} = \left(\overline{\overline{v_1}}, \overline{\overline{v_2}}, \ldots, \overline{\overline{v_n}}\right) \tag{10}$$

where $\tilde{v}_i = \max_i\{v_{ij3}\}$ and $\overline{\overline{v}}_1 = \min_i\{v_{ij1}\}$

- Compute each alternative's distance to the FPIS and FNIS using the vertex method [27] as presented below Eqs. 11, 12.

$$d\left(\tilde{x}, \tilde{y}\right) = \sqrt{\frac{1}{3}(a_1 - a_2)^2 + (b_1 - b_2)^2 + (c_1 - c_2)^2} \tag{11}$$

$$\breve{d} = \sum_{i=1}^{n} d\left(\tilde{v}_{i,j}, \tilde{v}_{\tilde{j}}\right), \overline{\overline{d}} = \sum_{i=1}^{n} d\left(\tilde{v}_{i,j}, \tilde{v}_{\overline{\overline{j}}}\right) \tag{12}$$

- Compute the closeness coefficient for each alternative as presented in Eq. 13.

$$CC_i = \frac{\overline{\overline{d}}_i}{\overline{\overline{d}}_i + \breve{d}_i} \tag{13}$$

- Finally, rank each alternative based on CC_i value. The alternative with the highest CC_i value is the best alternative.

4 Implementation

This section presents the implementation of the proposed approach. We consider a scenario in which a consumer search for an optimal service provider to form an SLA for its business. Among numerous service providers, a consumer has short-listed four service providers that best match consumer's requirements and business needs. We assume that all service providers offer the same type of services with the same requirements. For simplicity, we assume that a consumer is interested only in four QoS parameters—successability (SCY), cost (CST), reliability (REL) and throughput (THP). Therefore, the system assesses each service providers based on these four criteria. The first 200 comments of SaaS consumers reviews extracted by HaaS [11] are used to find consumers' QoE. The dataset has multiple columns—review date, reviewer name, service used, the title of a review, authentication of the reviewer and a review. SentiWordNet 3.0 [24] extracts consumer's fuzzy sentiment in linguistic terms for each QoS parameters [29–32]and then replaces them with fuzzy numbers using Table 2. The decision-making process is performed as follows:

4.1 Criteria Weight Determination

A consumer determines the best and the worst criteria (QoS parameter) from a requested set of QoS parameters. The relative fuzzy weights of each criterion are determined using Eqs. 1–4, as presented in Table 3.

Table 3 Relative QoS fuzzy weights using the best worst method

QoS parameters—criteria	Weighted symbol	Fuzzy weights	Crisp weights
Cost	CSTW	(0.428, 0.610, 0.859)	0.63
Successability	SCYW	(0.085, 0.131, 0.218)	0.14
Throughput	THPW	(0.117, 0.196, 0.309)	0.18
Reliability	RELW	(0.044, 0.063, 0.099)	0.05

4.2 Decision-Making Process

Aggregated fuzzy sentiment using from the first 200 feedbacks for all four service providers are collected and then converted to an equivalent fuzzy number as presented in Table 4.

After normalizing beneficial and non-beneficial criteria and applying QoS weights, the weighted normalized fuzzy decision matrix is achieved as presented in Tables 5 and 6.

The FPIS ($\breve{A}$)and FNIS ($\overline{\overline{A}}$) is calculated as presented in Table 7.

The distance from each alternative to the FPIS and FNIS is presented in Tables 8 and 9.

Finally ranking of each alternative is presented in Table 10.

The analysis result shows that service provider 2 (SPR-2) is the most suitable service provider that best satisfies prioritized QoS parameters. Thus, the proposed system helps the cloud consumer find a suitable service provider based on previous customer satisfaction level.

Table 4 The aggregated sentiment for each QoS parameters

	CST	SCY	THP	REL
SPR-1	3, 5.667, 9	5, 8.333, 9	5, 7, 9	1, 4.333, 7
SPR-2	5, 7, 9	1, 4.333, 7	1, 2.333, 5	3, 7, 9
SPR-3	1, 2.333, 7	5, 8.333, 9	3, 5.667, 7	1, 4.333, 9
SPR-4	3, 5, 9	1, 2.333, 5	1, 8.333, 9	1, 2.333, 5

Table 5 Criteria weights for QoS parameters with reference to previous consumers sentiment

Service provider	Criteria weights			
	CST (0.428, 0.610, 0.859)	SCY (0.085, 0.131, 0.218)	THP (0.117, 0.196, 0.309)	REL (0.044, 0.063, 0.099)
SPR-1	3, 5.667, 9	5, 8.333, 9	5, 7, 9	1, 4.333, 7
SPR-2	5, 7, 9	1, 4.333, 7	1, 2.333, 5	3, 7, 9
SPR-3	1, 2.333, 7	5, 8.333, 9	3, 5.667, 7	1, 4.333, 9
SPR-4	3, 5, 9	1, 2.333, 5	1, 8.333, 9	1, 2.333, 5

Table 6 Weighted normalized fuzzy decision matrix

	CST	SCY	THP	REL
SPR-1	1.284, 3.457, 7.731	0.425, 1.092, 1.962	0.585, 1.372, 2.781	0.044, 0.273, 0.693
SPR-2	2.14, 4.27, 7.731	0.085, 0.568, 1.526	0.117, 0.457, 1.545	0.132, 0.441, 0.891
SPR-3	0.428, 1.423, 6.013	0.425, 1.092, 1.962	0.351, 1.111, 2.163	0.044, 0.273, 0.891
SPR-4	1.284, 3.05, 0.859	0.085, 0.306, 1.09	0.117, 1.633, 2.781	0.044, 0.147, 0.495

Table 7 FPIS and FNIS

	CST	SCY	THP	REL
$\breve{A}$	2.14, 4.27, 7.731	0.425, 1.092, 1.962	0.117, 1.633, 2.781	0.132, 0.441, 0.891
$\bar{\bar{A}}$	0.428, 1.423, 6.013	0.085, 0.306, 1.09	0.117, 1.633, 2.781	0.044, 0.147, 0.495

Table 8 Distance of alternatives with FPIS

	CST	SCY	THP	REL	$\breve{d}$
SPR-1	0.682	0	0.309	0.158	1.149
SPR-2	0	0.439	0.985	0	1.424
SPR-3	2.159	0	0.486	0.109	2.754
SPR-4	1.381	0.705	0	0.289	2.375

Table 9 Distance of alternatives with FNIS

	CST	SCY	THP	REL	$\bar{\bar{d}}$
SPR-1	1.614	0.705	0.309	0.135	2.763
SPR-2	2.159	0.293	0.984	0.289	3.725
SPR-3	0	0.705	0.958	0.239	1.902
SPR-4	1.09	0	0	0	1.09

Table 10 Ranking of each alternative

	$\breve{d}$	$\bar{\bar{d}}$	CC_i	Rank
SPR-1	1.149	2.763	0.706	2
SPR-2	1.424	3.725	0.723	1
SPR-3	2.754	1.902	0.409	3
SPR-4	2.375	1.09	0.315	4

5 Conclusion

The growing cloud marketplace has emerged many challenges. One of the critical challenges for a consumer is to find the right service provider. A consumer often gets confused when choosing the service provider when multiple providers offer the same type and quality of services. Previous feedback can help a consumer to decide the best provider in an uncertain conditions. This paper proposed a framework that obtain custom prioritized QoS requirements and then aggregates them to prior consumer's fuzzy sentiment. The framework extracted cloud consumers sentiment in linguistic term and categorized them into five categories. Fuzzy relative weights were obtained using fuzzy BWM and used to rank each alternative using the fuzzy TOPSIS method. The analysis results demonstrate the applicability of the approach to handle uncertainty in decision making. In the future, we will evaluate the framework from a cloud provider perspective to manage its resources wisely. We will further analyze the applicability of the approach in IoT and mobile computing.

References

1. Alkalbani AM, Hussain W (2021)Cloud service discovery method: a framework for automatic derivation of cloud marketplace and cloud intelligence to assist consumers in finding cloud services. Int J Commun Syst 1–17
2. Papadakis-Vlachopapadopoulos K, González RS, Dimolitsas I, Dechouniotis D, Ferrer AJ, Papavassiliou S (2019) Collaborative SLA and reputation-based trust management in cloud federations. FutGener Comput Syst 100:498–512
3. Hussain W, Hussain FK, Hussain OK (2015)Comparative analysis of consumer profile-based methods to predict SLA violation. In: 2015 IEEE international conference on fuzzy systems (FUZZ-IEEE). IEEE, pp 1–8
4. Brunnström K et al (2013)Qualinet white paper on definitions of quality of experience
5. Adomavicius G, Sankaranarayanan R, Sen S, Tuzhilin A (2005) Incorporating contextual information in recommender systems using a multidimensional approach. ACM Trans Inf Syst (TOIS) 23(1):103–145
6. Schmitt AJ, Sun SA, Snyder LV, Shen Z-JM (2015) Centralization versus decentralization: risk pooling, risk diversification, and supply chain disruptions. Omega 52:201–212
7. Rajavel R, Thangarathanam M (2021) Agent-based automated dynamic SLA negotiation framework in the cloud using the stochastic optimization approach. Appl Soft Comput 101:107040
8. Hussain W, Hussain FK, Saberi M, Hussain OK, Chang E (2018) Comparing time series with machine learning-based prediction approaches for violation management in cloud SLAs. Fut Gener Comput Syst 89:464–477
9. Hussain W, Hussain FK, Hussain OK (2014)Maintaining trust in cloud computing through SLA monitoring. In: Neural information processing. Springer, pp 690–697
10. Alrashed BA, Hussain W (2020) Managing SLA violation in the cloud using Fuzzy re-SchdNeg decision model. In: 2020 15th IEEE conference on industrial electronics and applications (ICIEA). IEEE, pp 136–141
11. Alkalbani AM, Hussain W, Kim JY (2019) A centralised cloud services repository (CCSR) framework for optimal cloud service advertisement discovery from heterogenous web portals. IEEE Access 7(1):128213–128223

12. Hussain W, Hussain FK, Hussain OK, Damiani E, Chang E (2017) Formulating and managing viable SLAs in cloud computing from a small to medium service provider's viewpoint: a state-of-the-art review. Inf Syst 71:240–259
13. Hussain W, Hussain FK, Hussain O, Bagia R, Chang E (2018) Risk-based framework for SLA violation abatement from the cloud service provider's perspective. Comput J 61(9):1306–1322
14. Hussain W, Sohaib O, Naderpour M, Gao H (2020) Cloud marginal resource allocation: a decision support model. Mob Netw Appl 25:1418–1433
15. Alghamdi A, Hussain W, Alharthi A, Almusheqah AB (2017) The need of an optimal QoS repository and assessment framework in forming a trusted relationship in cloud: a systematic review. In: 2017 IEEE 14th international conference on e-business engineering (ICEBE). IEEE, pp 301–306
16. Hussain W, Sohaib O (2019) Analysing cloud QoS prediction approaches and its control parameters: considering overall accuracy and freshness of a dataset. IEEE Access 7:82649–82671
17. Alarifi A, Tolba A, Al-Makhadmeh Z, Said W (2020) A big data approach to sentiment analysis using greedy feature selection with cat swarm optimization-based long short-term memory neural networks. J Supercomput 76(6):4414–4429
18. Alharbi JR, Alhalabi WS (2020) Hybrid approach for sentiment analysis of twitter posts using a dictionary-based approach and fuzzy logic methods: study case on cloud service providers. Int J Semant Web Inf Syst (IJSWIS) 16(1):116–145
19. Dang NC, Moreno-García MN, De la Prieta F (2020) Sentiment analysis based on deep learning: a comparative study. Electronics 9(3):483
20. Santambrogio F (2017) {Euclidean, metric, and Wasserstein} gradient flows: an overview. Bull Math Sci 7(1):87–154
21. Yager RR (1988) On ordered weighted averaging aggregation operators in multicriteria decisionmaking. IEEE Trans Syst Man Cybern 18(1):183–190
22. Merigó JM, Gil-Lafuente AM (2009) The induced generalized OWA operator. Inf Sci 179(6):729–741
23. Merigo JM, Casanovas M (2011) Decision-making with distance measures and induced aggregation operators. Comput Ind Eng 60(1):66–76
24. Haque M (2014) Sentiment analysis by using fuzzy logic. arXiv:1403.3185
25. Guo S, Zhao H (2017) Fuzzy best-worst multi-criteria decision-making method and its applications. Knowl-Based Syst 121:23–31
26. Rezaei J (2015) Best-worst multi-criteria decision-making method. Omega 53:49–57
27. Chen C-T (2000) Extensions of the TOPSIS for group decision-making under fuzzy environment. Fuzzy Sets Syst 114(1):1–9
28. Nădăban S, Dzitac S, Dzitac I (2016) Fuzzy TOPSIS: a general view. Procedia Comput Sci 91:823–831
29. Hussain W, Merigó JM, Raza M, Gao H (2022) A new QoS prediction model using hybrid IOWA-ANFIS with fuzzy C-means, subtractive clustering and grid partitioning. Inf Sci 584:280–300
30. Hussain W, Merigo JM, Gao H, Alkalbani AM, Rabhi FA (2021) Integrated AHP-IOWA, POWA framework for Ideal cloud provider selection and optimum resource management. IEEE Trans Ser Comput. https://doi.org/10.1109/TSC.2021.3124885
31. Hussain W, Merigó JM, Raza MR (2022) Predictive intelligence using ANFIS-induced OWAWA for complex stock market prediction. Int J Intell Sys. https://doi.org/10.1002/int.22732
32. Hussain W, Merigó JM (2022) Centralised quality of experience and service framework Using PROMETHEE-II for cloud provider selection. In: intelligent processing practices and tools for E-commerce data, information, and knowledge. Springer, pp 79–94

Conditions of Technology Access for Remote Work in the Quaternary Sector in Mexico in Times of COVID-19

Ingrid Nineth Pinto López, Cynthia M. Montaudon Tomas, and Alicia L. Yáñez Moneda

Abstract This article analyzes the conditions of access to technology for teleworking that professionals in the quaternary sector in Mexico have experienced during the months of confinement derived from the health contingency caused by the COVID-19 Pandemic. Data collection was carried out between five and seven months after the confinement measures were decreed in Mexico, and the professional activities were migrated to telework mode, giving the participants time to adapt to the new normal. A validated scale was applied to 966 participants in 27 of the 32 states of the Mexican Republic. The results show relevant data about the conditions of access to technology, training, and tools required to work remotely.

Keywords Teleworking · Technology · Confinement · Quaternary sector · Mexico · COVID-19

1 Introduction

The outbreak of respiratory disease due to coronavirus COVID-19 declared by the World Health Organization as a global pandemic had an impact on people's health, but also on the social environment, the governments of the world adopted various measures to try to contain the disease. Given the disruption of these measures, many

I. N. P. López (✉)
Department of Business Intelligence, Business School, UPAEP University, Barrio de Santiago, 21 Sur 1103, Puebla, Mexico
e-mail: ingrid.pinto@upaep.mx

C. M. M. Tomas
Department of Administration, Business School, UPAEP University, Barrio de Santiago, 21 Sur 1103, Puebla, Mexico
e-mail: cynthiamaria.montaudon@upaep.mx

A. L. Y. Moneda
Department of Hospitality and Tourism, Business School, UPAEP University, Barrio de Santiago, 21 Sur 1103, Puebla, Mexico
e-mail: cynthiamaria.montaudon@upaep.mx

E. León-Castro et al. (eds.), *Soft Computing and Fuzzy Methodologies in Innovation Management and Sustainability*, Lecture Notes in Networks and Systems 337,
https://doi.org/10.1007/978-3-030-96150-3_7

organizations found it necessary to migrate their processes to telework mode as a necessary strategy to continue operating [1]. These changes made workers adapt to a new environment, with many adversities, because teleworking conditions were not ideal. They faced various situations, such as access to technology, time management, and tasks, as well as the unprecedented fusion of work and private and family life [2], in many cases resulting in effects on the health and personal well-being of workers.

This article aims to analyze the conditions of access to technology that workers in the quaternary sector Mexico experience considering the confinement measures and teleworking situation derived from the COVID-19 pandemic. The study analyzes variables such as internet access problems, service providers, work tools, training, communication, and digital problems. The study focuses on workers whose job-related activities are based on intellectual work or the knowledge economy; government, culture, scientific research, education, and information and communication technologies workers are considered. The data collection was carried out six to seven months after the workers transitioned into working from home, allowing a period of adaptation to the new normal. For the study, 966 surveys were applied to workers belonging to the quaternary sector. The results show the conditions of technology access that workers in this sector have been working during approximately six to seven months of confinement.

2 Theoretical Framework

2.1 Telework/Remote Work

Traditionally, work has been associated with a physical space, but jobs with flexible contracts and without a specific location or working hours are the new rule, allowing people to work anywhere and anytime [2]. Teleworking has been defined as one of the flexible modalities of the new ways of working [3], understanding by flexibility the benefit that employers provide to workers, allowing them a certain level of control over when and when where to work, instead of complying with the traditional working day [4]. Among the flexible forms of work, terms such as remote work [5, 6], work-from-home [7, 8], e-work [3], and mobile work are also recognized [9].

The benefits of teleworking include that it allows a balance between work, family, and personal life, an increase in flexibility, a reduction in travel times, increased productivity, and a cost reduction for the organization in terms of office space, light, electricity, heating, among others [10].

The International Labor Organization defines telework based on two different components (ILO 2020):

I. Work is performed-completely or partially-at an alternate location other than the predetermined workplace.

II. Job duties are executed using personal electronic devices, such as a computer, tablet, or phone (mobile or landline). The use of personal electronic devices should be a critical part of job performance.

2.2 COVID-19 Pandemic

The coronavirus (COVID-19) outbreak was first reported in Wuhan, China, on December 31st, 2019 [11, 2]. In January 2020, the World Health Organization (WHO) declared it a public health emergency of international impact, and on March 11th of the same year, the outbreak was declared a pandemic. According to data from Johns Hopkins University [12], by March 2021, around 114.9 million cases of coronavirus have been registered in the world; Latin America and the Caribbean accounts for approximately 20.7 million cases.

For governments worldwide, maintaining a controlled contagion rate and a low mortality rate became a priority. Seeking to keep the pandemic under control, the WHO urged leaders to face this problem and prepare to deal with it with a series of drastic measures, among which the closure of schools, closure of places work, travel restrictions, quarantine, social distancing, confinement, and others were included [13]. According to ILO data, worldwide, 93% of workers reside in countries where some type of workplace closure measures were applied; as these restrictions came into place, a large number of people had to stay at home and work remotely [1, 13].

Faced with the disruption these measures involved, a large number of organizations found it necessary to migrate their processes to teleworking mode as a strategy to continue operating (ILO 2010). These changes meant that workers had to adapt to a new environment without being prepared to do so [14], in many cases with many adversities because the conditions for teleworking were not ideal, and people faced different situations, such as access to technology and time and task management, as well as the unprecedented fusion of work and private and family life [2]. In many cases, this impacted health [14] and workers' personal well-being.

2.3 COVID-19 in Mexico

The first case of COVID-19 in Mexico was detected on February 27th, 2020. In March 2020, the government, in coordination with the Ministry of Health, implemented a series of measures to prevent and control infections in the country. According to the degree of transmission of the disease, a total of 3 epidemiological phases were set by the health authorities; on March 24th, phase 2 came into play, which primarily included the suspension of certain economic activities, the restriction of massive congregations, the recommendation of confinement for the general population, and as the suspension of face-to-face school activities at all educational levels [15, 16].

In Mexico, around 72% of the employed people were working from home after the confinement measures [16, 17]. In other countries, flexible forms of work are highly developed and legislated, however, in Mexico, there was no regulation at that time. Although flexible work under normal conditions brings many benefits, the speed with which the restrictions and confinement measures had to be adopted, as well as the disruptive way of carrying out work activities through teleworking, led to working conditions not suitable for all. The majority of workers faced very adverse situations related to conditions at home, access to technology, time and task management, job performance, digital skills, and the unprecedented convergence of work and private life in the same space [2]. In January 2021, a reform of the Federal Labor Law incorporated teleworking or home office, recognizing the performance of remunerated activities in places other than the establishment of the employer and establishing the following guidelines [18]:

- The use of information technologies to establish communication between the worker and the employer is required.
- Supplies, work equipment, and contact mechanisms must be provided by employers.
- Workers are entitled to the same salary, and they must be provided with computer equipment, internet connection, and the cost of electricity.
- Remote workers should be given the same treatment as those who attend the office.
- The right to disconnect is acknowledged.
- The need to establish training mechanisms to guarantee adaptation and learning is recognized.
- Video cameras and microphones should only be used to supervise workers in a supplementary manner.
- There will be labor inspectors to ensure compliance with the law.

2.4 Telework and Access to Technology

Teleworking is a working modality that is carried out using Information and Communication Technologies (ICT) [19, 3]. Workers require knowledge and intensive use of ICT in the development of their activities [20]. Access to technology is the main requirement to practice teleworking effectively, having digital tools and training for the use of technology are some of the most important elements. In this sense, having work equipment -such as a computer, tablet, or cell phone- is essential, as is having a reliable internet service without signal interruptions. In 2020, there were 80.6 million internet users in Mexico; however, there are still significant access gaps in some urban and rural areas; only 44.3% of Mexican households have at least one computer (CIRT 2020).

Due to the confinement measures derived from the health contingency, homes became the workplace, but they also became the place from where children attend school activities; in Mexico, for the 2020–2021 school year, more than 30 million students and 1.2 million teachers studied or worked remotely [15] generating a significant impact on access to technology in each of the households.

2.5 The Quaternary Sector

The economic sectors are the division of a State or territory's economic activity and are classified as primary, secondary, tertiary, quaternary, and quinary [21, 22]. The primary one groups mainly sectors that obtain products directly from nature. The secondary integrate sectors that transform raw materials into finished or semi-finished products. The tertiary is considered as a service sector. The quaternary includes economic activities based on knowledge and impossible to mechanize-such as the generation and exchange of information, technology, consulting, education, research and development, and financial planning-, mainly intellectual services or activities [23, 24]. The quinary sector groups non-profit services, including domestic activities such as those carried out by housewives or relatives who care for others in their own homes [21].

3 Method

3.1 Research Process

The collection of primary data that support this investigation was carried out in the months of August to November of the year 2020, between 5 and 8 months after the confinement measures were established by the central health authorities and abided by the 32 states of the Mexican Republic. It is important to mention that, at the time the data was collected, the workers were accustomed to teleworking since it had been their work modality for several months.

3.2 Research Design

The research is quantitative; the design is non-experimental with a descriptive scope [25]. Its purpose is to analyze the effects of access to technology in teleworking conditions from a set of variables and estimate its occurrence. It is cross-sectional because data collection is carried out in a single period in a representative sample of the population [25, 26]. The sampling strategy is non-probabilistic through the design of a scale sent electronically.

3.3 Scale and Sample

According to the National Survey of Occupation and Employment (ENOE), for the fourth quarter of 2020, there was an economically active population of approximately 55.8 million people in Mexico, of which around 30.7 million are concentrated in the tertiary sector million (60.6% of the total population), and around 7.02 million people in the quaternary sector, considering educational services (2.70 million), government activities and international organizations (2.40 million), professional, scientific and technical services (1.44 million) and leisure, cultural and sports services (0.48 million) [27, 28].

The variables analyzed in this research correspond to a subset of data from a scale developed for a larger study that assesses the conditions and effects of remote work in the life and productivity of workers that transitioned into teleworking as a result of the health crisis. The scale is made up of several dimensions, namely: effects on employment, income, and expenses; working conditions at home; access to technology and digital skills; time and task management; job performance; occupational health and wellness, family relationships, and the future of remote work.

The dimension access to technology and digital skills is the one used in this study, and only workers in the quaternary sector are considered. This dimension is composed of 30 variables, which include classification data, dichotomous questions, multiple-choice questions, and questions on a 7-point Likert scale.

3.4 Validation

The sample for the complete study in Mexico represents the study population at the national level; it consists of 2,136 observations with 113 variables.

The validation of the scale was carried out with Cronbach's Alpha [29], yielding a result of 0.9235, which shows that it is highly reliable. According to [30], values above 0.9 are excellent. For this analysis, 966 observations and 28 variables are considered. The Cronbach's Alpha of each of the variables used in this article is presented in Table 1, and the result for all variables is presented in Annex 1. The software used to perform the calculation was STATA.

4 Results and Analysis

The classification data shows that the average age of the participants is 46 years old, in a range of 18 to 79, 56.3% are female, 43.7% are male, and 0.31% did not specify. Regarding marital status, 56.9% are married, 29.29% single, 7.24% divorced, 4.95% common-law union and 1.76% are widowed.

Table 1 Cronbach's alpha results

Item	Indicator	Code	Cronbach's alpha
1	Age	edad	0.9237
2	Gender	genero	0.9241
3	Marital Status	civil	0.9243
4	Place of residence	país	0.9243
5	Economic sector in which you work	sector	0.9242
6	Number of people that reside in the home	personas_hog	0.9238
7	Number of people that study or work from home since the pandemic	trabajocasa_total	0.9240
8	Internet provider	internet	0.9242
9	Number of computers or other digital devices in the home that are used for study or work purposes	dispositivo_total	0.9244
10	Do you own a personal computer?	comp_propia	0.9241
11	When you moved to remote work, did you take the office computer with you?	comp_oficina	0.9240
12	Do you have a reliable internet connection?	conexión	0.9228
13	How many family members share the internet?	internet_comp	0.9238
14	You experience frequent internet failures	fallas	0.9231
15	Do you have internet problems at specific times?	horario_fallas	0.9232
16	You use your cellphone as a tool for work (WhatsApp, etc.)	celular	0.9241
17	What videoconferencing software do you use?	software1	0.9243
18	Total videoconferencing software you use	total_software	0.9243
19	You have received the necessary training to carry out your work from home	capacitación	0.9235
20	Your employer has provided you with the communication tools you need to do your job	herramientas_com	0.9230
21	You have received support from your family members to carry out your work using technology	apoyo_tecnología	0.9243
22	You have had problems accidentally leaving the camera or microphone open	micrófono-cámara	0.9231
23	You have forgotten to activate the microphone and started speaking without others being able to hear you	no-escucha	0.9235
24	Before the pandemic, you believed you had the digital skills necessary to carry out your work from home	habilidad_previa	0.9230

(continued)

Table 1 (continued)

Item	Indicator	Code	Cronbach's alpha
25	You have had difficulties using certain digital tools for your work	dificultad_digital	0.9225
26	You consider that since the beginning of the pandemic, you have developed the necessary digital skills	desarrollo_habilidad	0.9237
27	It has been difficult to adapt to the digital tools provided by your employer	adaptación	0.9222
28	Having to use different technologies has been difficult	dificultad	0.9217

Source Developed by the authors

The participants are located in 27 of the 32 states of the Mexican Republic; the percentage of participation from each state is shown in Fig. 1.

On average, 2.6 people work and/or study from home during the pandemic. The average number of devices in a household is 3.39, considering the use of different devices that include desktop computers, laptops, tablets, and cell phones (See Table 2). Additionally, 89.6% use the cell phone as a work tool.

As shown in Table 2, if only the number of people who use a device to work and/or study is considered, most of the interviewees meet their needs; however, some do not. Of those who have two devices, 14.8% require three or more; of those with three devices, 7.6% require four or more; of those with four devices, 5% require five or more, and of those with five devices, 1.3% require six or more.

If the number of people in the family who share the internet is considered, not only people who work and/or study from home use it, but also other family members, thus increasing the need for other devices. On average, 3.5 people per household share the internet. 68.3% of those surveyed report having between 2 and 4 devices at home; however, 77.4% need between 2 and 4 devices to cover their needs, so there is a negative gap of approximately 9.1%. The details can be observed in Table 3.

In Mexico, there are several companies that offer internet services. In the quaternary sector, 50.41% of the population uses Telmex, 15.22% Megacable, 12% Izzi, 11.18% TotalPlay, and 11.18% are with other companies (see Fig. 2).

Of the 966 people that were interviewed, 61.6% use Zoom as a work tool mainly for video conferencing, followed by Teams with 19.6%, Meet with 16.1%, Blackboard with 2.3%, and 0.4% use other platforms (see Fig. 3).

Regarding the total number of platforms they use, 32.4% use one, 33.3% use 2, 25.8% use 3, 7.4% use 4, and 1% use 5 or more. The average of different software that participants use is 2.11.

The results of the indicators formulated using a Likert scale are presented in Table 5. The values of the 7-point Likert scale are explained in Table 4. Table 5 also includes the mean, standard deviation, and level agreement or disagreement in every item. For the level of disagreement, the values from 1 to 3 are grouped, and for the level of

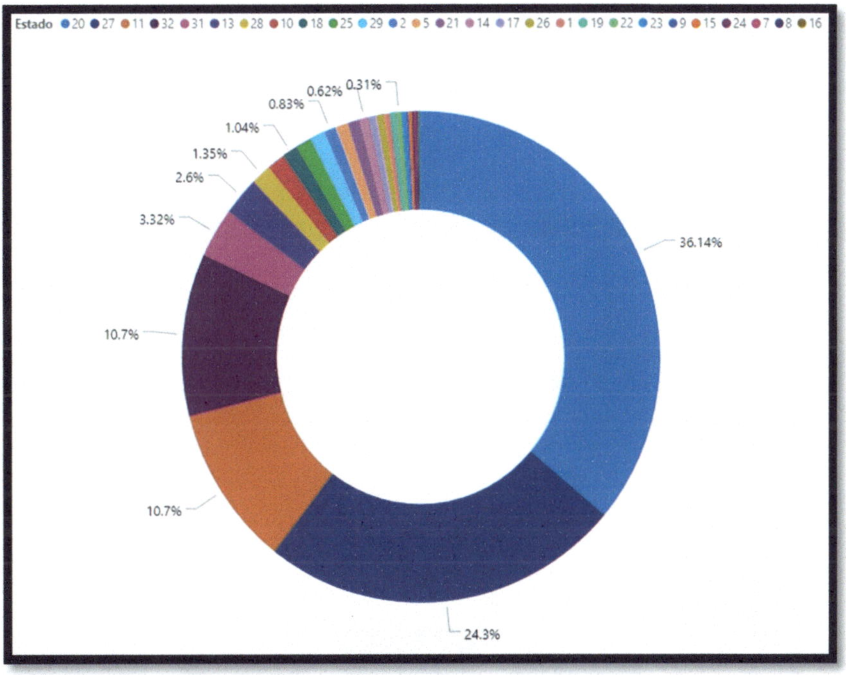

Fig. 1 Sample distribution by state. *Source* Elaborated by the authors. Abbreviations: 1. Aguascalientes, 2. Baja California, 5. Chiapas, 7. Coahuila de Zaragoza, 8. Colima, 9. Durango, 10. Estado de México, 11. Estado de México, 13. Hidalgo, 14. Jalisco, 15. Michoacán de Ocampo, 16. Morelos, 17. Nayarit, 18. Nuevo León, 19. Oaxaca, 20. Puebla, 21. Querétaro, 22. Quintana Roo, 23. San Luis Potosí, 24. Sinaloa, 25. Sonora, 26. Tabasco, 27. Tamaulipas, 28. Tlaxcala, 28, Veracruz de la Llave, 31. Zacatecas, 32. Mexico City (CDMX)

Table 2 Number of devices versus number of people who work or study from home

Number of devices	Number of people who work or study from home (%)					
	1	2	3	4	5	+6
1	61.9	22.6	8.3	5.9	1.2	0
2	39.8	45.1	8.8	3.9	1.7	0.4
3	20.4	28.5	43.4	4.68	2.12	0.8
4	6.1	27.5	23.9	37.2	3.5	1.5
5	5.3	20.0	32.0	12.0	29.3	1.3
6 +	7.6	10.4	29.8	31.2	13.8	6.9

agreement, the values from 5 to 7 are grouped, the value 4 is considered neutral. Table 6 groups the indicators formulated as dichotomous questions.

(a) **Technology/Connectivity**

Table 3 Number of people who share the internet

Number of people who share the internet in the household	Percentage (%)	Number of devices in the household	Percentage (%)	Gap
1	6.5	1	8.9	+
2	25.5	2	23.5	–
3	26.0	3	24.4	–
4	25.9	4	20.4	–
5	11.1	5	7.8	–
6 or more	5.0	6 or more	14.9	+

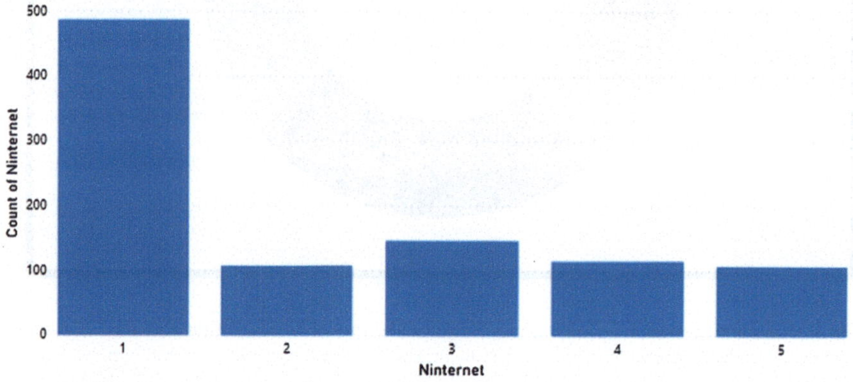

Fig. 2 Distribution of internet providers. *Source* Elaborated by the authors. Abbreviations: 1. Telmex, 2. Izzi, 3. Megacable, 4. TotalPlay, 5. Other companies

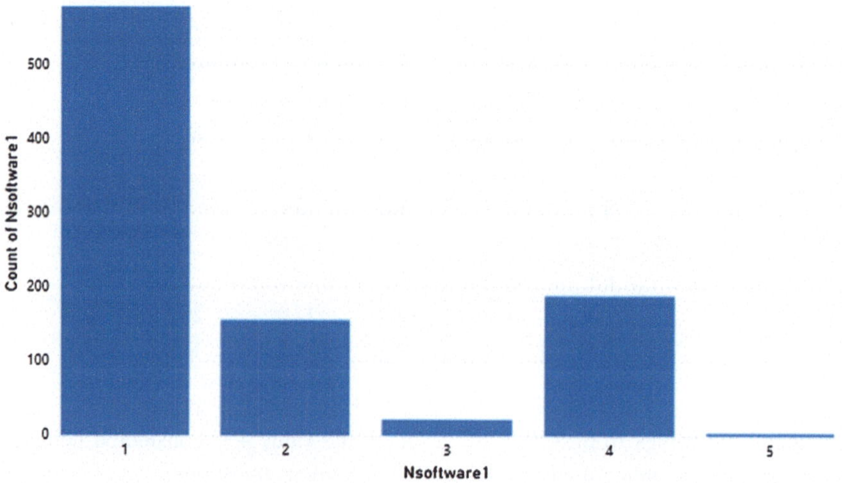

Fig. 3 Distribution of videoconferencing platforms. *Source* Elaborated by the authors. Abbreviations: 1: Zoom, 2. Meet, 3. Blackboard, 4. Teams, 5. Otros

Table 4 Likert scale values

Value	Interpretation
7	Completely in agreement
6	In agreement
5	Partially in agreement
4	Neither in agreement or in disagreement
3	Partially in disagreement
2	In disagreement
1	Completely in disagreement

According to the results from Table 5, in relation to technology and connectivity, the following results from the quaternary sector in Mexico stand out:

- 91.1% of those surveyed have their own computer, and 66.9% did not need to take their work computer home. 89.6% use the cell phone as a work tool. It is estimated that in Mexico, there are 80.6 million internet users and 86.5 million cell phone users [31], which may explain, in some cases, why they did not need to take the computer home from work.
- Regarding connectivity for remote work, 72.3% state that they have a reliable internet connection, although according to 47.1% of those surveyed, they experience frequent internet failures.
- 86.9% of those surveyed have not had problems for accidentally leaving the camera or microphone open; however, 61.9% have forgotten to activate the microphone at some point.

(b) **Tools for remote work**

With regard to teleworking tools, 68.8% stated that their company provided them with the communication tools necessary for the development of their work activities, 54% have not had difficulties in the use of digital tools, 87.4% have developed new digital skills, 60.6% state that they have no difficulties to adapt to the digital tools provided by employer and 54% of those surveyed have found it easy to use different technologies. Furthermore, 71.5% consider that before the pandemic they had the necessary skills to work from home.

(c) **Training**

Regarding training, 73.4% of the respondents stated that they had received the training they require for the development of their professional activities through teleworking, and 51.1% of the respondents have received technological support from some members of their family.

Table 5 Results for each indicator

Item	Indicator	Mean	Standard deviation	Level of agreement or disagreement		
				Disagreement (1–3)	Neutral (4)	Agreement (5–7)
Technology/connectivity						
12	You have a reliable internet connection	5.24	1.58	14.4	13.3	72.3
14	You experience frequent internet failures	4.08	1.95	40.9	11.9	47.1
16	You use your cellphone as a tool for work (WhatsApp, and others)	6.24	1.51	7.7	2.7	89.6
Tools for remote work						
20	Your employer has provided you with the communication tools you need to do your job	5.10	2.09	21.9	9.3	68.8
25	You have had difficulties using certain digital tools for your work	3.37	2.04	54.0	11.6	34.3
26	You consider that since the beginning of the pandemic, you have developed the necessary digital skills	6.04	1.32	5.0	7.6	87.4
27	It has been difficult to adapt to the digital tools provided by your employer	3.12	2.10	60.6	9.9	29.5
28	Having to use different technologies has been difficult	3.37	2.04	54.0	11.7	34.3

(continued)

Table 5 (continued)

Item	Indicator	Mean	Standard deviation	Level of agreement or disagreement		
				Disagreement (1–3)	Neutral (4)	Agreement (5–7)
24	Before the pandemic, you believed you had the digital skills necessary to carry out your work from home	5.34	1.69	15.3	13.1	71.5
Training						
19	You have received the necessary training to carry out your work from home	5.38	1.86	16.2	10.3	73.4
21	You have received support from your family members to carry out your work using technology	4.37	2.42	36.7	9.1	54.1

Table 6 Dichotomous questions

Item	Indicator	Yes	No
10	Do you own a personal computer?	91.1	8.9
11	When you moved to remote work, did you take the office computer with you?	33.1	66.9
15	Do you have internet problems at specific times?	43.4	56.4
22	You have had problems accidentally leaving the camera or microphone open	13.1	86.9
23	You have forgotten to activate the microphone and started speaking without others being able to hear you	61.9	38.1

5 Conclusion

According to the data analyzed, the conditions of access to technology have been good for this sector. In most of the indicators, there is a positive evaluation -above 50%- and in some other indicators, it exceeds 90%.

Considering that the analyzed sector refers to professionals who are engaged in economic activities based on knowledge, such as technology, consulting, education, research and development, and financial planning, among others, the results make sense.

However, even in the quaternary sector, there are some technological and training gaps that must be taken into account to ensure that 100% of the workers have the necessary conditions to work from home.

Working remotely during the pandemic was an abrupt change, and it has undoubtedly been a great challenge. Nevertheless, the quaternary sector seems to have overcome it well in terms of access to technology and the use of digital platforms. It would be interesting to analyze what has happened to family relationships and general health under these conditions.

Appendix

Cronbach's validity results

Item	Obs	Sign	Item-test correlation	Item-rest correlation	Average interitem correlation	Alpha
edad	2058	–	0.1903	0.1583	0.113	0.9237
genero	2058	–	0.1165	0.0852	0.1135	0.9241
civil	2058	+	0.0699	0.0392	0.1139	0.9243
pais	2048	+	0.0759	0.0451	0.1138	0.9243
mundo	2058	+	0.0518	0.0208	0.114	0.9244
sector	2058	+	0.0899	0.0576	0.1137	0.9242
personas_hog	2038	+	0.1686	0.1378	0.1131	0.9238
trabaja_casa	2034	–	0.1021	0.0683	0.1137	0.9241
trabajocas~l	2038	+	0.1222	0.0911	0.1135	0.924
internet	2012	–	0.0918	0.0606	0.1137	0.9242
disposit iv~l	2046	–	0.0521	0.0209	0.114	0.9244
casa_mts	1867	–	0.1495	0.1187	0.1132	0.9238
software1	2052	+	0.0757	0.0439	0.1138	0.9243
total_soft~e	2056	–	0.0756	0.0442	0.1139	0.9243
tv	2050	–	0.0769	0.0454	0.1138	0.9243
empleo	2029	+	0.1293	0.0972	0.1134	0.924
ingreso	2036	+	0.1875	0.1567	0.113	0.9237
prestaciones	2003	+	0.1502	0.1185	0.1133	0.9239
apoyo_hijos	2023	–	0.1224	0.091	0.1134	0.924
comp_propia	2051	–	0.103	0.0719	0.1137	0.9241

(continued)

(continued)

Item	Obs	Sign	Item-test correlation	Item-rest correlation	Average interitem correlation	Alpha
comp_oficina	2019	–	0.1252	0.0945	0.1135	0.924
conexion	2047	–	0.3615	0.3337	0.1117	0.9228
internet_c~p	2041	+	0.1706	0.1398	0.1131	0.9238
fallas	2050	+	0.2856	0.2563	0.1122	0.9231
horario_fa~s	2046	+	0.2766	0.2474	0.1123	0.9232
falla_diaria	2047	+	0.101	0.0707	0.1136	0.9241
habilidad ~a	2049	–	0.316	0.2873	0.112	0.923
dificultad~l	2035	+	0.4017	0.375	0.1114	0.9225
desar rollo~d	2030	–	0.1808	0.1503	0.113	0.9237
espacio_casa	2032	–	0.4344	0.409	0.1111	0.9223
trabaja_mas	2025	+	0.2289	0.1998	0.1127	0.9234
trabajo_li~a	2024	–	0.3725	0.3456	0.1116	0.9227
ropa_casa	2014	+	0.1423	0.1122	0.1133	0.9239
juntas_mas	2015	+	0.1289	0.0984	0.1134	0.924
mismo_hora~o	2015	–	0.3171	0.2892	0.112	0.923
Desempeño_~o	2007	+	0.534	0.5118	0.1104	0.9218
organizaci~a	2019	–	0.4982	0.4748	0.1106	0.922
gestion_ti~o	2022	+	0.5353	0.5129	0.1104	0.9218
trabajo_fin	2014	+	0.3853	0.3589	0.1115	0.9226
Motivacion	2022	+	0.643	0.6246	0.1095	0.9212
informacio~o	2014	–	0.4007	0.3748	0.1114	0.9225
integracion	2006	–	0.389	0.3627	0.1115	0.9226
contacto_j~e	1995	–	0.1889	0.1592	0.1129	0.9236
celular	2014	+	0.0998	0.0695	0.1136	0.9241
supervision	2011	+	0.3011	0.273	0.1121	0.9231
distraccion	2001	+	0.4571	0.4326	0.1109	0.9222
Adaptacion	1999	+	0.4586	0.4342	0.1109	0.9222
capacitacion	2000	–	0.2224	0.1934	0.1127	0.9235
herramient~m	1996	–	0.3172	0.2897	0.112	0.923
cansancio	2008	+	0.6218	0.6026	0.1097	0.9213
dolor_cabeza	2015	+	0.6223	0.6031	0.1097	0.9213
tension	2017	+	0.6019	0.5819	0.1099	0.9214
grita	2013	+	0.6516	0.6336	0.1095	0.9211
irritabili~d	2013	+	0.7185	0.7033	0.109	0.9208
conflicto	2016	+	0.7288	0.7141	0.1089	0.9207
sensible	2016	+	0.689	0.6725	0.1092	0.9209

(continued)

(continued)

Item	Obs	Sign	Item-test correlation	Item-rest correlation	Average interitem correlation	Alpha
enojo	2015	+	0.7241	0.7091	0.1089	0.9207
burnout	1992	+	0.6123	0.5927	0.1098	0.9214
ansiedad	2013	+	0.5446	0.5227	0.1103	0.9217
apoyo_medico	2010	+	0.4912	0.4677	0.1107	0.922
ojos	2015	+	0.5612	0.5398	0.1102	0.9216
articulaci~s	2011	+	0.5674	0.5463	0.1101	0.9216
estres	2011	+	0.7086	0.6929	0.1091	0.9208
dificultad	2014	+	0.5489	0.5272	0.1103	0.9217
microfono_~a	2012	+	0.3018	0.2739	0.1121	0.9231
privacidad	2015	+	0.5758	0.5551	0.1101	0.9216
asuntos_fa~s	2007	+	0.3909	0.3648	0.1114	0.9226
ruido	2007	+	0.5926	0.5725	0.1099	0.9215
mobiliario	2011	–	0.3295	0.302	0.1119	0.9229
relaciones~s	2002	+	0.629	0.6102	0.1097	0.9213
separacion~s	1998	+	0.4271	0.4019	0.1112	0.9224
tiempo_fam~a	1991	–	0.3208	0.2934	0.112	0.9229
separacion~o	1995	–	0.4676	0.4435	0.1109	0.9222
tiempo_libre	2002	+	0.494	0.4706	0.1107	0.922
habitacion~a	2000	–	0.3228	0.2954	0.112	0.9229
mas_estres~o	2000	+	0.7272	0.7124	0.1089	0.9207
mas_distra~n	1997	+	0.6026	0.5828	0.1099	0.9214
divorcio	1658	+	0.282	0.2543	0.1121	0.9231
Lapandemia~i	1989	–	0.211	0.1818	0.1128	0.9235
rivalidad	1985	+	0.5292	0.5069	0.1104	0.9218
tiempo_com~a	1987	–	0.3497	0.3227	0.1118	0.9228
conversaci~s	1987	+	0.5818	0.5612	0.11	0.9215
menos_paci~a	1992	+	0.6469	0.6287	0.1095	0.9212
invasion	1979	+	0.6017	0.5818	0.1099	0.9214
disfuta_fa~a	1981	–	0.3687	0.342	0.1116	0.9227
roces	1979	+	0.5804	0.5598	0.11	0.9216
mas_gastos	1988	+	0.3798	0.3535	0.1115	0.9226
apoyo tecn~a	1974	+	0.0573	0.0271	0.1139	0.9243
continuar	1997	–	0.38	0.3537	0.1115	0.9226
confianza_~o	1995	–	0.1636	0.1338	0.1131	0.9238
incertidum~e	1995	+	0.4009	0.3749	0.1114	0.9225
adaptar_re~o	1995	+	0.0648	0.0344	0.1139	0.9243

(continued)

(continued)

Item	Obs	Sign	Item-test correlation	Item-rest correlation	Average interitem correlation	Alpha
ma s capa ci~n	1998	+	0.3341	0.3068	0.1119	0.9229
no_escucha2	2007	+	0.2221	0.193	0.1127	0.9235
pa ntalla2	2001	+	0.1672	0.1376	0.1131	0.9238
ayuda_hijos2	1897	+	0.0863	0.0561	0.1136	0.9241
Test scale					0.1118	0.9235

References

1. Organización Internacional del Trabajo OIT (2020) COVID19: orientaciones para la recolección de estadísticas del trabajo. Recuperado de. https://www.ilo.org/global/statistics-and-databases/publications/WCMS_758333/lang--es/index.htm
2. Montaudon-Tomas CM, Pinto-López IN, Yáñez-Moneda AL, Amsler A (2021) The effects of remote work on family relationships. In: Chapter in future of work, work-family satisfaction, and employee well-being in the fourth industrial revolution. IGI Global
3. Ramos V, Ramos-Galarza C, Tejera E (2020) Teletrabajo en tiempos de COVID-19. Interam J Psychol 54(3):1–30
4. Gabini S (2016) Flexible work: conceptualization and state of the art. Investig Admin 46(118). http://orcid.org/0000-0002-0997-1423
5. Hill EJ, Ferris M, Martinson V (2003) Does it matter where you work? A comparison of how three work venues (traditional office, virtual office, and home office) influence aspects of work and personal/family life. J Vocat Behav 63(2):220–241
6. Messenger JC, Gschwind L (2016) Three generations of telework: new ICTs (R)evolution from homme office to virtual office. New Technol Work Employ 31(3):195–208
7. Aquije CP (2018) Home office as a strategy for organizational motivation and efficiency. Palermo Bus Rev 18:337–351
8. Vasconcelos P, Furtado E, Rogerio P (2015) An approach of multidisciplinary criteria for modeling alternatives of flexible working. Comput Hum Behav 51:1054–1060
9. Crawford JO, MacCalman L, Jackson CA (2011) The health and well-being of remote and mobile workers. Occup Med 61(6):385–394
10. Pérez C, Gálvez AM, Teletrabajo y vida cotidiana: Ventajas y dificultades para la conciliación de la vida laboral, personal y familiar. Athenea Dig Revista de Pensamiento e Investigación Soc 15:57–79
11. Peña-López BO, Rincón-Orozco B (2020) COVID-19 Pandemic overviewand its genetic association with SARS virus. Salud UIS 52(2):83–86. https://doi.org/10.18273/revsal.v52n2-2020001
12. John Hopkins University (2021) Coronavirus Resource Center. Recuperado de. https://coronavirus.jhu.edu/map.html
13. Organización Internacional del Trabajo OIT (2021) Observatorio de la OIT: La COVID-19 y el mundo del trabajo. Séptima edición. Recuperado de. https://www.ilo.org/wcmsp5/groups/public/---dgreports/---dcomm/documents/briefingnote/wcms_767045.pdf
14. Galeano N (2020) Teletrabajo y salud mental en tiempos de COVID-19. Gestión Práctica de los Riesgos Laborales 185:62–63
15. Secretaría de Educación Pública (2020) Recuperado de. https://www.gob.mx/sep

16. Secretaría de Salud (2020) Inicia Fase 2 por coronavirus COVID-19. Recuperado de. https://www.gob.mx/salud/prensa/095-inicia-fase-2-por-coronavirus-covid-19
17. STATISTA (2020) Porcentaje de empleados que trabajaron en casa durante el confinamiento en México en junio de 2020. Recuperado de. https://es.statista.com/estadisticas/1147384/porcentaje-empleados-home-office-mexico/
18. Diario Oficial de la Federación (DOF) (2021) Decreto por el que se reforma el artículo 311 y se adiciona el capítulo XII Bis de la Ley Federal del Trabajo en materia de Teletrabajo. Recuperado de. http://www.dof.gob.mx/nota_detalle.php?codigo=5609683&fecha=11/01/2021
19. Pérez M, Martínez A, de Luis P, Vela MJ (2004) La adopción del teletrabajo y las tecnologías de la información. Revista de Economía y Empresa 52–53:11–27
20. Osio L (2010) El teletrabajo: Una opción en la era digital. Observatorio Laboral Revista Venezolana 3(5):93–109
21. Hatt P, Foote N (1953) On the expansion of the tertiary, quaternary, and quinary sectors. Am Econ Rev
22. Kenessey Z (1987) The primary, secondary, tertiary and quaternary sectors of the economy. https://doi.org/10.1111/j.1475-4991.1987.tb00680.x
23. Rosenberg M (2020) The 5 sectors of the economy. Recuperado de. https://www.thoughtco.com/sectors-of-the-economy-1435795
24. Selstad T (1990) The rise of the quaternary sector. The regional dimension of knowledge-based services in Norway, 1970–1985. Nor J Geosy 44(1):21–37. https://doi.org/10.1080/00291959008552242
25. Hernández-Sampieri R, Mendoza CP (2018) Metodología de la Investigación: Las rutas cuantitativa, cualitativa y mixta. Mc Graw Hill
26. Martínez KI, Ávila OP, Pacheco A, Lira J (2011) Investigaciones longitudinales: Su importancia en el estudio del mantenimiento de cambio en las adicciones. Enseñanza e Investigación en Psicología 16(2):375–386
27. Instituto Nacional de Estadística Geografía e Informática INEGI (2020) Encuesta Nacional de Ocupación y Empleo ENOE. Recuperado de. https://www.inegi.org.mx/programas/enoe/15ymas/default.html#Tabulados
28. Flores E (2020) Un país de servicios y comercio: en esos dos sectores trabajan la gran mayoría de los mexicanos. Recuperado de. https://www.sinembargo.mx/24-02-2020/3733072
29. Quero M (2010) Confiabilidad y coeficiente Alpha de Cronbach. Telos 12(2):248–252
30. George D, Mallery P (2003) SPSS for windows step by step: a simple guide and reference, 4a edn. Ayllyn and Bacon, Boston
31. Instituto Federal de Comunicaciones IFT (2020) En México hay 80.6 millones de usuarios de internet y 86.5 millones de teléfonos celulares: ENDUTIH 2019. http://www.ift.org.mx/comunicacion-y-medios/comunicados-ift/es/en-mexico-hay-806-millones-de-usuarios-de-internet-y-865-millones-de-usuarios-de-telefonos-celulares

Economic Benefits to Conserve the Tourism Potential of Chifron Beach in the District of Capachica, Puno, Peru

Jhesus Wilson Panca Galindo, Blanca Roldán-Clarà, and Alfredo Pelayo Calatayud Mendoza

Abstract Chifron beach in the Capachica district, Puno, Peru, is a potential tourism resource in critical conservation due to pollution by solid wastes, visual contamination from graffiti painted on rock zones, and alteration of beaches by extraction of sand for construction. These perturbations threaten the quality of the beach and water, and have an impact on neighbouring households due to solid wastes carried by the wind to their homes. In consequence, in this paper we attempted to contribute to stimulate economic benefits for conserving Chifron beach through the identification of the socioeconomic profile of tourists affecting their willingness to pay an entrance fee for conservation, and the quantification of that fee through the application of a survey and the contingent valuation method. Results from our surveying of 252 visitors of the beach showed that 54% are male, 53% are married, 76% have higher education, 66% work in private or public institutions, 34% have their own business, 27% are residents of rural areas, and 73% live in urban areas of different cities. 94% of the tourists agreed that the beach should be cleaned and sanitary services be available. On the negative side, 66% of the surveyed tourists rated the cleanliness and hygiene of the sand as poor, and 64% considered the cleanliness and hygiene of the shoreline as poor. On the positive side, 59% of the tourists rated the landscape at the beach as very good. We found that 63% of the visitors would be willing to pay an entrance fee for conservation of the beach of PEN 2.40 (MXN 15.61) for adults and 50% of that amount for children for the conservation of the touristic resource. We concluded that, although the conservation condition of Chifron beach is critical, there is willingness to pay for visiting the beach among visitors, as long as it is in optimal conditions, that is, with clean sand and shoreline, and provided with sanitary facilities. Our analysis identified that monthly income and educational level of visitors was directly proportional to their willingness to pay for a conservation fee, and the hypothetical cost of visiting the beach was inversely proportional to that willingness to pay.

J. W. P. Galindo (✉) · B. Roldán-Clarà
Universidad Autónoma de Occidente, Unidad Mazatlán, Mexico

A. P. C. Mendoza
Universidad Nacional del Altiplano, Puno, Peru
e-mail: apcalatayud@unap.edu.pe

E. León-Castro et al. (eds.), *Soft Computing and Fuzzy Methodologies in Innovation Management and Sustainability*, Lecture Notes in Networks and Systems 337,
https://doi.org/10.1007/978-3-030-96150-3_8

Keywords Economic benefits · Conservation · Willingness to pay · Touristic resource · Contingent valuation

1 Introduction

This work was motivated by the current environmental issues at Chifron beach which are caused by the solid wastes left by visitors, a situation that causes the concern of local inhabitants about the deterioration of the touristic resource. Aware of the impact caused by this situation, we made an attempt to quantify tourists' willingness to pay an entrance fee for conservation of the beach. Chifron beach is located in the Capachica district, department of Puno, in southern Peru, on the shoreline of Lake Titicaca at an elevation of 3,820 m a.s.l. In the past few years, the beach has received an increasing number of national and international tourists. The Government of Peru declared the district of Capachica as a touristic zone of national interest through the June 3, 1998 Ministerial Resolution N°062-98-ITINCI. The district counts with varied touristic resources, Chifón beach among them, including beach viewpoints, natural landscapes, horseback riding, boating, and kayaking, and, in recent years the beach was also provided with quadricycle rides, restaurant, hosting, and convenience store services [5].

Chifron beach began its participation in touristic promotion events in 1999, when it attracted a large number of visitors with their automobiles. Other promotions and events—like the Miss. Beach, regional dance, and local gastronomy contests—attracted more visitors to the area looking for entertainment and camping on the Lake Titicaca shores to enjoy nature away from cities, especially during the weekends. An added advantage for tourism of the district of Capachica is its accessibility through a paved road. However, the massive arrival of visitors have generated negative consequences due to environmental and visual contamination, thus creating a bad image in the tourists' perception of the destination.

In order to change its bad image and preserve its touristic resource, Chifron beach requires to be cleaned, receive maintenance, and provide sanitary facilities. Conservation and service improvement of this important touristic resource are as essential to prevent future negative impacts from pollution as it is visitors to the beach being willing to pay an entrance fee for its conservation.

2 Objectives

The objectives of our research were to estimate the potential economic benefits needed for conservation of the Chifron beach by, first, identifying the socioeconomic characteristics of its visitors most determining their willingness to pay an entrance fee for visiting the beach if the sand and shore were clean and sanitary facilities were provided, and second, calculating the amount of that fee for conservation.

3 Methodology

Our methodology had two stages: First, determination of the sample size of the visitors' population, and application of a pilot questionnaire to 15 visitors to determine the scale of their willingness to pay, on which the application of the final survey was based; and second, gathering data based on the scale of the entrance fee and obtention of the estimates of the logit model using the STATA software.

For that, the latent variable of the logit model was defined as the willingness of visitors of Chifron beach to pay an entrance fee for its conservation using Eqs. 1 and 2.

$$AP = \beta^{'} X + \varepsilon \quad (1)$$

$$\begin{aligned} &DAP' = \{1, SiDAP > 0 \\ &0, SIDAP \leq 0 \end{aligned} \quad (2)$$

where *WTP* = Willingness to pay.

In the model, *WTP* is the latent (or hidden) variable, which in this case expresses the qualitative change in visiting a beach with clean sand and shore provided with sanitary facilities. If the change is positive, that is, if the beach was kept in optimal conditions, the visitors would be willing to pay a certain amount of money ($WTP >$ 0), but if the beach was contaminated, the visitors would be unwilling to pay any amount of money ($WTP < 0$).

In this context, we used the logit probability model—one of the techniques appropriate for this kind of studies—to detect the variables influencing visitors' willingness to pay.

3.1 Logit Model

The probability of being willing to pay for visiting Chifron beach in optimal condition is dependant on the socioeconomic characteristics of the visitors (income, hypothetical payment, education level, age, and place of residence, among others), as expressed in Eq. 3 by the vector X:

$$Prob\left(Y^{*} > 0\right) = \frac{e^{\beta^{'} X_i}}{1 + e^{\beta^{'} X_i}} = \Lambda\left(\beta^{'} X\right) \quad (3)$$

The logit model is estimated by the maximum likelihood method [4]. In consequence, the joint probability or likelihood function of a model with probability of willingness to pay $\Lambda(\beta^{'} X)$ and n independent observations will be:

$$Max \cdot L_{\text{Logit}} = \prod\nolimits_{i=1}^{n} \left[\Lambda(\beta' X_i)\right]^{Y_i} \left[1 - \Lambda\left(\beta' X_i\right)\right]^{1-Y_i} \quad (4)$$

$$\{\beta\}$$

$$Max \cdot LnL_{\text{Logit}} = \sum\nolimits_{i=1}^{n} \left[Y_i Ln\Lambda(\beta' X_i) + (1 - Y_i) Ln(1 - \Lambda(\beta' X_i)) \right] \quad (5)$$

$$\{\beta\}$$

$$Y = (WTP)$$

where $\Lambda(\beta' X)$ is the logistic distribution function.

Finally, the marginal consequence of factors having an influence on the probability of willingness to pay (X) is determined by:

$$\frac{\partial E[Y/X]}{\partial X} = \Lambda\left(\beta' X\right)\left[1 - \Lambda\left(\beta' X\right)\right]\beta \quad (6)$$

The unit of analysis was the 4,700 people that annually visit the Chifron beach, as recorded by the District Municipality of Capachica, and the dimensions of analysis were the following socioeconomic characteristics:

- Hypothetical entrance fee in Peruvian Soles (PEN)
- Monthly income in PEN
- Gender
- Educational level
- Age
- Marital status
- Place of origin
- Occupation.

The size of the sample was calculated by random sampling according to Eq. 7 as follows:

$$n = \frac{z^2 N.P.Q.}{[E^2(N-1)] + [z^2 P.Q]} \quad (7)$$

$$n = \frac{(1.96)(4,700)(0.5)(0.5)}{[(0.06)(4,699)] + [(1.96)(0.5)(0.5)} = 252$$

where n = sample size; Z = 1.96 at a 95%-confidence interval; N = 4,700 visitors/y; P = 50% probability of success; Q = 50% probability of failure (1 − P); and E = sampling error (6%).

4 Theoretical Approach: The Contingent Valuation Method

The contingent valuation method is applied to assign economic value of environmental commodities and services, an approach that has gained the approval of scholars and managers of environmental-policy because of its usefulness as a tool for knowing individual preferences in populations [6]. For that, as described by Farre [2], the method is based on a survey creating a hypothetical market in which the surveyed individuals represent the demand and the surveyor represents the offer. The surveyor offers a commodity or service to surveyants who can accept it or not. The contingent valuation method is an econometric instrument first introduced over 66 years ago, and that in 1963 became established for environmental economic studies of natural resources and conservation of protected areas [2, 3, 6]. Osorio and Correa ([6]: p. 16) list the recommendations from experts in the application of the method, among which the following apply:

- Person-to-person surveys are preferred over telephonic or mail surveys.
- Surveyed individuals must be reminded that they are being questioned about their willingness to pay for an improvement of the environmental commodity or service of interest.
- The questionnaire applied in the survey must only contain questions accepting a yes or a no for answer, each one revealing an upper limit (for a yes) of the measure of well-being.
- The interview must begin with describing to the interviewee the context of the survey so that they can understand the effects of the program or actions of interest.
- The questionnaire must include validation questions to verify that the interviewees understood and acknowledged the scenario of interest in order to detect socioeconomic and attitude variables to be included in the questionnaire that allows a better analysis of the results.

The contingent valuation method is applied in six steps, as described by Farre [2]: First, identify the object to be valued; second, define the population of interest for the survey; third, identify the well-being action to be assessed; fourth, define the modality of the survey (by mail, in person, by telephone, each one having their advantages and disadvantages); fifth, estimate the sample size according to the required level of confidence of the values to be estimated, afterwhich the questionnaire is designed; and sixth, apply the survey, make the statistical analysis of the obtained data.

Most evaluations using the contingency valuing method use the logit model for the statistical analysis because its results have a lower standard deviation value than those obtained with the probit model, making logit more adequate for this analysis. The tourists' willingness to pay was calculated using Eq. 8.

$$WTP = -\frac{\beta_0 + \beta_1 MI + \beta_2 A + \beta_3 G + \ldots + \beta_N \text{other socioeconomic variables}}{\beta_1} \tag{8}$$

where WTP = willingness to pay; $ß_0$ = intercept; $ß_{1\text{-}n}$ are the coefficients obtained for socioeconomic variables, where MI = monthly income; A = age; and G = gender.

5 Results

5.1 Socioeconomic Characterization of Visitors

The results from the 252 surveys applied to visitors of Chifron beach to know their socioeconomic profiles—from which the variables most influencing tourist's willingness to pay an entrance fee to visit the beach in optimal conditions were identified—were as follows.

Gender (G)—More males (54%) than females (46%) visited the beach.

Marital status (MS)—Of the 252 surveyed visitors, 53% answered they were married and 47% said they were single, which implies that married visitors are accompanied by other family members.

Educational level (EL)—76% of the surveyed visitors said they have college education and 24% lacked college education, which means the former could be students, traders, employees, or technicians having monthly incomes and being willing to pay an entrance fee, because of which, as stated by [1, 8], educational level is one of the factors of willingness to pay for a commodity or service.

Occupation (O)—66% of the surveyed visitors to Chifron beach stated they were hired workers in private or public institutions, and 34% said they were self-employed or owners of their own business; which implies that hired workers are more frequent visitors of the beach on weekends, holidays, and vacation periods than self-employed or independent business owners.

Place of origin (PO)—27% of the visitors said they were residents of rural areas and 73% answered that they lived in an urban zone—most of them inhabitants of Puno and Juliaca, the cities closest to Chifrón beach [5]. This implies that most of the visitors look for relaxing in a natural environment.

Agreement with sand and shore cleanliness and availability of sanitary services (SH)—94% of the surveyed visitors agreed that the sand and shore should be clean and sanitary services should be available to them. This means that most tourists value the beach and would like it to be clean and provided with sanitary services.

Willingness to pay an entrance fee (WP)—The answers to the question aimed at knowing if tourists would willingly pay an entrance fee to maintain the beach clean and provide sanitary services –an essential result for our study—showed that most (63%) of the surveyed visitors agreed to pay and 37% were not willing to pay an entrance fee regardless of their payment being used to keep the beach clean and provide sanitary facilities.

Cleanliness of beach sand (CS)—66% of the visitors rated the cleanliness of sand as poor and 34% as good, meaning most tourists have a negative perception of the cleanliness of the sand and think it should be cleaned.

Cleanliness and hygiene of the shoreline (SH)—Most surveyed tourists (64%) rated the shoreline as having poor cleanliness and hygiene and 36% said it was alright as it was, the majority thinking that the shoreline must be cleaned.

Beach landscape (L)—59% of the surveyed visitors of Chifron beach said the landscape was very good and 41% said it was good, all agreeing that they valued the landscape in the beach.

5.2 Descriptive Statistics of Socioeconomic Variables

Table 1 shows the results of the descriptive statistics analysis of the data obtained in the survey by variable.

Table 1 Descriptive statistics of the individual variables about the socioeconomic characteristics of the surveyed visitors to Chifron beach ($N = 252$)

Variable	Number of observations	Mean	Standard deviation	Minimum	Maximum
G	252	0.547619	0.4987178	0	1
A[a]	252	31.71429	9.274785	17	76
MS	252	0.531746	0.4999842	0	1
EL	252	0.7698413	0.421772	0	1
O	252	0.3492063	0.477668	0	1
PO	252	0.7380952	0.440546	0	1
SH	252	1.015873	0.6618112	0	3
SC	252	1.309524	0.6432523	0	3
L	252	2.876984	0.4147985	1	3
SF	252	0.9404762	0.2370733	0	1
WP	252	0.6309524	0.4835072	0	1
HP	252	1.75	0.8556119	0.5	3
MI[b]	252	1019.444	529.6572	100	5000
A^2	252	1091.476	775.9463	289	5776

Notes [a] The surveyed visitors' average age was 31 years. [b] The surveyed visitor's average monthly income was PN 1,019. *Abbreviations* G, gender; A, age; MS, marital status; EL; Educational level; O, occupation; PO, Place of origin; SH, sand cleanliness and hygiene; SC, shoreline cleanliness; L, landscape; SF, provision of sanitary facilities; WP, willingness to pay an entrance fee for conservation; HP, hypothetical payment for entrance fee for conservation; MI, monthly income; and A^2, square of age. *Source* Own elaboration with the data obtained from the application of the survey

5.3 Selection of the Better Logit Model for Estimating the Economic Benefits (Willingness to Pay an Entrance Fee) for Conservation of Chifron Beach

To estimate the economic benefits for conservation of Chifron beach, we needed to determine the tourists' willingness to pay an entrance fee for visiting the beach if its sand and shore were clean and sanitary facilities were implemented. For that, we applied the logit probability model through random sampling of visitors and explored the variables determining the probability of tourists' willingness to pay for a conserved, clean beach, and with sanitary services available to visitors. Five logit models were tested using as explanatory variables the socioeconomic characteristics of visitors to the beach that we obtained from the survey. In model 1 we included monthly income (MI), hypothetical payment (HP), educational level (EL), place of origin (PO), gender (G), age (A), and square of age (A2), and model 5 included monthly income (MI), hypothetical payment (HP), and having or not college education (EL). The best model was chosen by making a statistical analysis of the significance of each independent variable relative to the dependent variable, and analyzing the independence of variables and the goodness of fit for each model. Model 5 was chosen because it complied with the expected signs, had the highest statistical significance in individual and global coefficients, and provided the highest goodness of fit between the observed and expected results (Table 2).

Logit model 5 provided the best statistical fit because of which it offered the best relationship between willingness to pay (WP) for conserving the touristic resource, the hypothetical price (HP), and if visitors have or not college education. The following section describes the desirable features of logit model 5 that we observed.

5.4 Assessment of the Chosen Logit Model

5.4.1 Z-test

The level of significance of the coefficients associated to the individual independent variables was measured by a Z-test—similar to the *t*-test, in which the null hypothesis was that the coefficient is not statistically significant (Ho: $\beta_i = 0$) at a p value of 0.05). Logit model 5 showed coefficients different from zero (null hypothesis rejected) for variables MI, HP, and EL.

5.4.2 Likelihood Ratio Test

Using Eq. 9, we made a likelihood ratio (LR) test with the null hypothesis assuming the coefficients of the independent variables are equal to zero, or not statistically significant (Ho: $\beta_1 = \beta_2 = \beta_3 = \ldots.\beta_q = 0$. The LR test, similarly to the F-test, is chi-square distributed with q degrees of freedom.

Table 2 Logit probabilistic models tested

Independent variables	Model 1	Model 2	Model 3	Model 4	Model 5
Constant	0.1197	8.2778	7.8393	4.6792	0.8385
	0.056	0.054	0.062	0.203	0.233
MI	0.0009901	0.001136	−001,153	24,048	0.0014867
	0.209	0.147	0.134	0.001	0.036
HP	−1.2952	−1.2566	−1.2644	−1.1026	−1.43
	0.00	0.00	0.00	0.00	0.00
EL	0.39005				1.274415
	0.441				0.002
PO	1.74414	1.9224	1.80492		
	0.001	0.000	0.000		
G	−0.69275	−0.69629			
	0.039	0.038			
A	−0.54265	−0.56092	−0.54719	−0.3366	
	0.061	0.55	0.056	0.82	
A^2	0.0096949	0.0100737	0.0097803	0.0061	
	0.041	0.035	0.038	0.130	
LR	106.23 0.0000	105.64 0.0000	101.22 0.0000	82.34 0.0000	83.09
					0.0000
McFadden's R^2	0.3201	0.3183	0.3050	0.2481	0.2504
Sample size	252	252	252	252	252

Abbreviations MI, monthly income; HP, hypothetical price; EL; educational level; PO, place of origin; G, gender; A, age; A^2, square of age; LR, likelihood ratio. *Source* Own elaboration with the data obtained from the application of the survey

$$LR\text{Chi}^2(q) = -2(LnL_R - LnL_I) \tag{9}$$

where LnL_R = Log likelihood of the restricted model, and LnL_I = Log likelihood of the unrestricted model.

Solving for the coefficients from the logit model 5:

$$\text{LRChi}^2(3) \;=\; -2(-165.928 + 124.6835) \;=\; 83.09$$

Therefore, the null hypothesis was rejected at a 0.05 p value, meaning that the coefficients of the model were statistically significant.

Table 3 Marginal effects calculated for logit model 5

Socioeconomic variable	I	Dy/dx	Standard Error	z	P > IzI	95% confidence interval
MI	0,002,426	0.0001128	2.15	0.031	0.0000216	0.0004636
HP	−0.2337261	0.0212532	−11.00	0.000	−0.2753816	−0.1920705
EL	0.20795	0.064202	3.24	0.001	0.0821163	0.3337837

Footnote Abbreviations MI, monthly income; HP, hypothetical price; EL; educational level; PO, place of origin; G, gender; A, age; A^2, square of age; LR, likelihood ratio. *Source* Own elaboration with the data obtained from the application of the survey

5.4.3 McFadden's Pseudo R^2

The McFadden's pseudo R^2 statistic has values between 0 and 1, and is calculated as the ratio between the observed (unrestricted) log likelihood model and a simpler, less informative restricted model containing the constant as the only explanatory variable (Eq. 10).

$$R^2 = 1 - \frac{LnL_I}{LnL_R} \tag{10}$$

Solving for the logit model 5, I

$$R^2 = 1 - \frac{-124.6835}{-165.928} = 0.2504$$

It is important to state that the reading of the McFadden's pseudo R^2 is less demanding than that of the R^2 of linear models, and the value we obtained from the logit model 5 indicates its good classificatory performance.

5.4.4 Estimation of Marginal Effects in Logit Model 5

The marginal effects of non-linear models are not constant, because of which we estimated an average marginal effect for each variable in the logit model 5. Likewise, the marginal effects can be calculated for a specific value (Table 3).

6 Willingness to Pay an Entrance Fee for Conservation of Chifron Beach

The willingness to pay an entrance fee for conservation of the beach in optimal conditions was estimated by the maximum likelihood method based on the logit model 5, chosen because it provides the best fit:

$$Pr(WTP = 1) = \Lambda(0.8385 - 1.432383HP + 0.0014867MI + 1.274415EL$$

where $\Lambda(\ldots)$ is the cumulative logistic distribution function.

The mean (C') and median (C^*) of the willingness to pay were afterwards calculated using Eqs. 11 and 12 as follows:

$$C' = -\frac{Ln\lfloor 1 + e^{\propto} \rfloor}{\beta} \tag{11}$$

where C' = mean of willingness to pay; Ln = natural logarithm; $e = 2.718281$.

$$\alpha = \alpha_0 + \alpha_2 MI + \sum_{3}^{k} S_i \tag{12}$$

$$\beta = \alpha_1$$

Solving for model 5:

$$C' = -\frac{Ln(1 + e^{\alpha})}{\beta!} = -\frac{Ln\left(1 + e^{0.8385 + 0.0014867MI + 1.274415EL}\right)}{-1.432383} = 2.365 \approx 2.40$$

$$c^* = -\frac{\propto}{\beta} \tag{13}$$

Solving for model 5:

$$c^* = -\frac{\propto}{\beta} = -\frac{(0.8385 + 0.0014866MI + 1.274415EL)}{-1.432383} = 2.328431 \approx 2.40$$

We therefore estimated that the amount willing to be paid by tourists as an entrance fee for conservation of a clean and with sanitary facilities Chifron beach to be PEN 2.40 (MXN 15.61).

7 Conclusions

Chifron beach offers visitors nature viewpoints, horse rides, hosting, and boat rides, among other attractions. But the touristic potential of the destination is diminished because of the poor image it causes on visitors, as expressed by 94% of the surveyed visitors who agree that the beach must be cleaned and provided with sanitary services, which causes the concern of the local inhabitants.

Despite the critical situation of Chifron beach is critical, our survey showed that its visitors would be willing to pay an entrance fee for cleaning the beach and shore and providing sanitary services of PEN 2.40 (MXN 15.61) for adults and 50% of that price for children (PEN 1.20 or MXN 7.81). By applying logit model 5, which we chose because it offered the best statistical representation of the main factors involved, we found that the willingness to pay a conservation fee is directly proportional to the visitors' monthly income (MI) and educational level (EL), and inversely proportional to the hypothetical payment (HP).

References

1. Cayo N (2014) Valoración económica ambiental según la disponibilidad a pagar por el turismo rural vivencial en la isla de Taquile, Perú. *Comunicación* 5(2):25–34. Recuperado de. http://www.scielo.org.pe/scielo.php?script=sci_arttext&pid=S2219-71682014000200003
2. Farre M (2003) El valor de uso recreativo de los espacios naturales protegidos. Una aplicación de los métodos de valoración contingente y del costo de viaje. Estudios de Economía Aplicada 21(2), 279–320. Recuperado de. http://www.redalyc.org/articulo.oa?id=30121207
3. Garzón LP (2013) Revisión de método de valoración contingente: experiencias de la aplicación en áreas protegidas de América Latina y el Caribe. Espacio y Desarrollo 25:65–78. Recuperado de. http://revistas.pucp.edu.pe/index.php/espacioydesarrollo/article/view/10623
4. Green W (2002) Econometric analysis, 5ta edición. Editorial Prentice-Hall, New York, EE UU
5. Jara AF, Arana GV, Butrón JE, Medrano RA (2018) Plan de negocios: servicio turístico de aventura en la ciudad de Puno (tesis de maestría). Universidad San Ignacio de Loyola Lima, Perú
6. Osorio JD, y Correa FJ (2009) Un análisis de la aplicación empírica del método de valoración contingente. Semestre económico 12(25):11–30. Recuperado de. http://www.redalyc.org/articulo.oa?id=165013651001
7. Sánchez JM (2008) Valoración Contingente y Costo de Viaje Aplicados al área recreativa Laguna de Mucubají. Economía 26(XXXIII):119–150. Recuperado. http://revencyt.ula.ve/storage/repo/ArchivoDocumento/econo/n26/articulo5.pdf
8. Tudela JW, Martínez MA, Valdivia R, Romo JL, Portillo M, y Ventura R (2011) Valoración económica de los beneficios de un Programa de recuperación y conservación en el Parque Nacional Molino de Flores, México. RCHSFA 17(2):231- 244. doi: 1 0.5154/r.rchscfa.201 0.05.033

Monitoring Videocolonoscopy Examinations in Real-Time via Internet

Feng C. Wu, Huei D. Lee, Newton Spolaôr, Moacir F. Junior, Weber S. R. Takaki, Claudio S. R. Coy, João J. Fagundes, Raquel F. Leal, Renato B. Machado, and Maria de L. S. Ayrizono

Abstract Computational technologies have promoted several benefits for medicine, such as improvements in medical data storage, transmission, and analysis. They also have the potential to support the establishment of collaboration networks among professionals in medical centers, clinics, and hospitals, notably after the advent and expansion of the Internet. In particular, these networks could allow experts in different locations to monitor and discuss medical procedures in real time, as well as to collect, store, and share multimedia data associated with these procedures. This chapter presents an innovative method in telemedicine and a computational system that allows experts to interact in real time by text, audio, and video. The proposal also allows experts to perform real-time gathering and sharing of videocolonoscopic images and videos with authorized users connected to the system through the Internet. These multimedia data are stored and published into a centralized server implemented with security techniques. The presented method was experimentally evaluated in a simulation of a real environment in which the system could be applied. As a result, the system met the requirements of the telemedicine method and was able to collect, store, and share videocolonoscopic multimedia data securely in real time. This finding confirms the proposal as an alternative to support collaboration among experts and institutions that offer videocolonoscopy or other types of video-based examinations and medical procedures to patients.

F. C. Wu · H. D. Lee (✉) · N. Spolaôr · M. F. Junior · W. S. R. Takaki · R. B. Machado
Laboratory of Bioinformatics, Graduate Program in Electrical Engineering and Computing, Western Paraná State University, Presidente Tancredo Neves Avenue 6731, Foz do Iguaçu 85867-900, Brazil
e-mail: huei.lee@unioeste.br

C. S. R. Coy · J. J. Fagundes · R. F. Leal · M. de L. S. Ayrizono
Coloproctology Service, University of Campinas, Tessália Vieira de Camargo Street 126, Campinas 13083-887, Brazil

E. León-Castro et al. (eds.), *Soft Computing and Fuzzy Methodologies in Innovation Management and Sustainability*, Lecture Notes in Networks and Systems 337,
https://doi.org/10.1007/978-3-030-96150-3_9

1 Introduction

The development of the technological field and, especially, Information and Communications Technology (ICT), has promoted computational applications with different purposes to assist several areas of knowledge [13, 14, 40, 42]. In particular, the use of such technologies in the medical area has brought important contributions and stimulated the development of research in a multidisciplinary manner [13, 14, 40].

Due to the application of ICTs to different medicine fields, it was necessary to categorize different types of computational systems [42], such as Hospital Information Systems (HIS), Radiology Information Systems (RIS), and Picture Archiving and Communication Systems (PACS). In order to integrate these computational approaches, standards were developed, such as the Digital Imaging Communications in Medicine (DICOM) protocol [37].

An important class of healthcare application is telemedicine, which has spurred the development of various computational methods. Among these, the classic work that demonstrated the possibility to transmit electrocardiogram data over existing telephone lines stands out [11]. That work continued through the era of radio and television broadcast, finally reaching the present day, in which it uses Internet transmission [9, 10, 18].

Accordingly, different benefits such as a reduction in the transportation and communication costs, the dissemination of specialized medicine to remote areas and the sharing of knowledge among medical specialists, became possible by using integrated and collaborative tools, which support long-distance medicine [23–26, 28–30, 47–49].

In this scenario, there are many technological resources that can be applied to telemedical tasks, making it possible, for example, to transfer digital images, audio, video, and multimedia information. Undoubtedly, the Internet architecture is one of the most consistent models associated with multimedia and cryptographic session management techniques. Through this model, it is possible to transmit data in real-time and to allow researchers to interact through video, voice and text messaging communication [1, 8, 15, 20, 21, 32, 46].

The Laboratory of Bioinformatics (LABI—Laboratório de Bioinformática) at the Western Paraná State University (UNIOESTE—Universidade Estadual do Oeste do Paraná), in partnership with the Coloproctology Service of the Faculty of Medical Sciences (Serviço de Coloproctologia da Faculdade de Ciências Médicas) at University of Campinas (UNICAMP—Universidade Estadual de Campinas), has been developing research in telemedicine to support the monitoring of patients and the remote transmission of medical procedures [23–26, 28, 29, 47, 48].

Related work on telemedicine includes [6], which proposes the idea of using a medical kiosk to allow, for example, patients to interact with physicians. Some pieces of work, such as [17], provide tools and methods for remote monitoring and interaction between patients and physicians without taking into account images and videos collected from medical equipments. Another reference is found in [43], which proposes a method to allow users in a specific place to follow surgeries and share media collected from medical equipments.

In this chapter, we present an original method in telemedicine for remote real-time monitoring and interaction between professionals in the health field, while conducting endoscopies, in particular videocolonoscopies.[1] This method, implemented in a computational system, was evaluated in terms of Frames Per Second (FPS) considering different settings. It should be emphasized that, different from related work [6, 17, 43], our method makes possible:

1. The real-time monitoring of medical procedures and the discussion before, during and after the procedures by experts and authorized personnel located anywhere with an Internet connection;
2. The communication among medical equipments based on videos and images;
3. Shooting and sharing of images collected from medical equipments.

An additional contribution of the described method involves the centralization of medical videos, images, audio, and data from exams in a databases server. These databases can be the starting point for further research and development. For example, by applying Artificial Intelligence approaches, such as the Data Mining process [19], it is possible to extract knowledge from the data. As part of this process, different data descriptors, illustrated in [4, 31, 33], are considered as input to learn patterns. In the end, valid knowledge extracted from the telemedicine databases can be useful to provide second opinion for experts' Decision Making.

This chapter is organized as follows. Section 2 describes the method and the implemented computational system, as well as the experimental setting considered. In Sect. 3 the results are presented and in Sect. 4 we discuss the performance evaluation of the results. Sect. 5 presents the concluding remarks and future work.

2 Materials and Methods

The system developed from the proposed telemedicine method was carried out in four phases: specification, method definition, computational implementation and experimental evaluation.

All steps were performed by using a software development process based on prototyping [36] and the Unified Modeling Language (UML).

2.1 Requirements Specification

In this step, the principal concepts and features identified as fundamental for the computational solution were based on the characteristics of HIS and PACS [3, 42].

[1] In this chapter, we use the terms videocolonoscopy and endoscopy indistinctly.

First, the problem domain was analyzed and discussed during meetings with medical experts. These meetings involved different topics, such as understanding the protocol for videocolonoscopy execution [12], the equipment, the communication mechanisms available, brands of components and models, video resolutions, quality of images generated and input and output video technologies, among others.

The main functional requirements defined in this phase were:

- Restrict access to patient data and examinations by means of authentication of authorized users;
- Capturing and storing images and videos during the examinations;
- Real-time monitoring of video colonoscopy exams using a web browser, at the location of the endoscopic procedure itself, or remotely, via text messaging and voice, video and image sharing, through the Internet;
- Implementation of features that allow the users to analyze endoscopic examinations;
- Development of strict security features.

2.2 *Method Definition*

The telemedicine method for real-time monitoring of endoscopic procedures has been designed using the architectural, the data and the security models presented as follows.

2.2.1 Architectural Model

From the study of the specific domain and of the definition of functional requirements, we built an architectural model [24, 27, 28, 47] consisting of the following components (Fig. 1):

- Endoscopic Procedures Room (EPR): location where colonoscopic examinations are performed. The EPR is composed of a Hospital Equipment (HE) and a Local Execution Unit (LEU), i.e., a system to acquire and forward the exam's video to the Application Server (AS), located at the Management Center (MC);
- Hospital Equipment (HE): equipment used for the endoscopic examination;
- Local Execution Unit (LEU): communication component between the computer and the videocolonoscopy equipment. The LEU is responsible for the acquisition and forwarding of the video to the Application Server (AS) in the Management Center (MC);
- Remote Monitoring and Interaction Units (RMIU): units where authorized users are able to monitor examinations being performed with videocolonoscopes. Through the RMIU it is possible the interaction with other professionals at the EPR and/or at other RMIUs, provided that there is an Internet connection available;

Fig. 1 Architectural model [24]

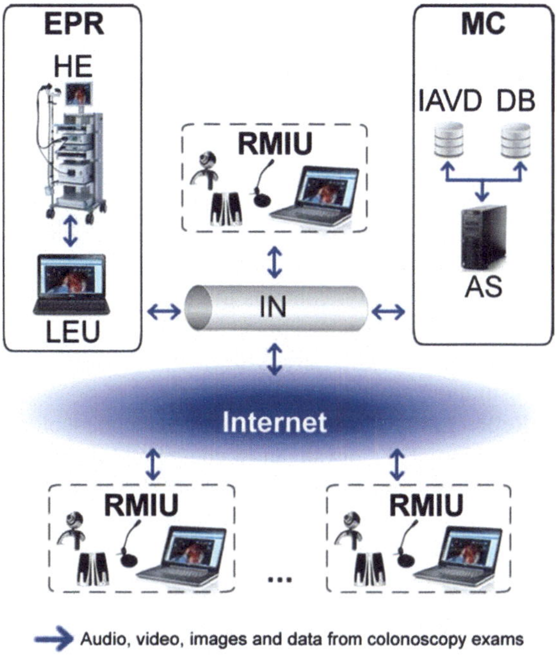

- Management Center (MC): management location for the computational resources used for examination conduction, monitoring, and interaction between local and remote participants at both the EPR and the RMIU. The MC is composed of the AS, the Database (DB) and the Images, Audio, and Video Database (IAVD);
- Application Server (AS): computer responsible for access control, reception, management and distribution of text, audio, and video streams between the LEU and the RMIUs;
- Database (DB): computational resource where data from exams and medical reports are physically stored;
- Images, Audio, and Video Database (IAVD): computational resource that organizes and stores images, audio and videos of colonoscopy exams. This database will allow, in the future, the application of techniques for intelligent data analysis;
- Internal Network (IN): hardware and software used for communication and transmission of data, video, images and audio within the institution;
- Internet: data communication networks external to the institution through which the RMIUs get connected to the AS.

2.2.2 Data Model

The data model was designed using the Entity–Relationship model (ER model). The entities defined to build this model are:

- Professional Entity: stores information about the health professionals who have access to the system with general identification data, such as the professional's name, individual taxpayer registration number, general registry number, birth date, gender, telephone number, address, institutional affiliation and, authentication information (user name and password);
- Patient Entity: holds the patient's data, such as name, taxpayer registration number, general registry number, birth date, e-mail, gender, telephone number, and address;
- Institution Entity: stores information about address, contact, and medical specialty of the institution;
- Equipment Entity: contains the identification of the equipment used for videocolonoscopies, such as brand and model;
- Examination Entity: stores the interrelationship between the professional, patient, institution, and equipment entities;
- Report Entity: maintains data related to videocolonoscopy reports, including identification of the physician in charge and the examination, test results, indicated treatment, observations, and date of the medical consultation;
- Auxiliary Entities: composed of databases for cities, states, countries, institutional associations of professionals, and health insurance plans.

2.2.3 Security Model

The Security Model (SM) is responsible for the interface between the internal network and the Internet.

To this end, firewall rules defining what is open and what is closed to external access, as well as lists of trusted and banned sites were implemented.

To complement this strategy, tools for intrusion detection, storage, and log analysis, as well as proxy systems, were defined within the servers that connect the internal network to the Internet.

Publication of audio and video streams are made using a method of Key Generation Based on Genetic Algorithms (LABI-PUBLISH) proposed by our research group. This process consists in the use of concepts of the Theory of Evolution of Species and computing resources to generate secret keys, as described in detail in [24, 29].

2.3 Computational Implementation

The Method in Telemedicine for Remote Monitoring and Real-Time Medical Procedures [24, 28] was computationally implemented into the Real-Time Telemedical System (S2TR) [24, 47]. For the construction of the S2TR, the following tools were considered: the Model-View-Controller (MVC) design pattern [36], Java, and Flex programming languages, the JBOSS Application Server, the Red5[2] Streaming Server,

[2] https://github.com/Red5/red5-server

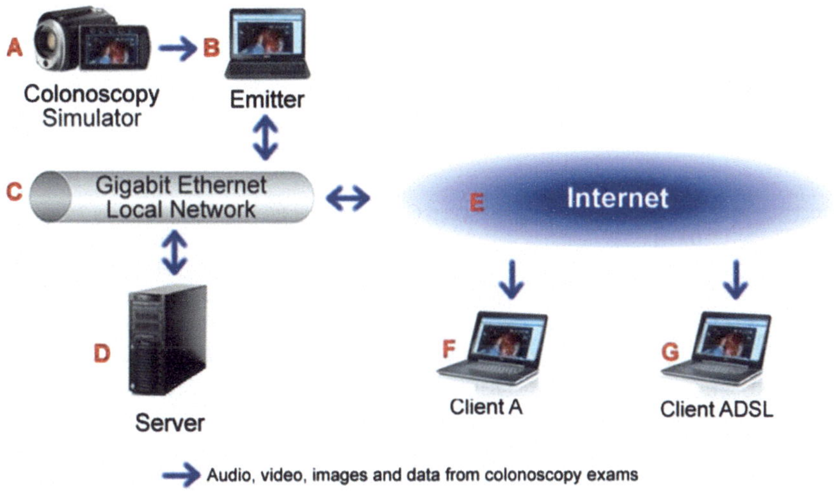

Fig. 2 Components of the experimental evaluation [24]

the JBOSS Seam development framework, the MySQL Database Management System and the Flamingo[3] framework.

2.4 *Experimental Evaluation Setting*

The experimental evaluation was performed using a local network, located at the Laboratory of Bioinformatics (LABI) at UNIOESTE Campus in Foz do Iguaçu, connected to the Internet through a 100 Mbps link provided by the National Research Network (RNP). Figure 2 shows the components in the experimental evaluation [24].

The videocolonoscope was simulated by a JVC Everio GZ HD 520 digital camcorder (Fig. 2, component A). The Emitter computer (Fig. 2, component B) was responsible for capturing the video from the JVC Everio GZ HD 520 digital camcorder and sending the audio and video stream to the Server computer (Fig. 2, component D). The Server computer then forwarded these flows to Client A (Fig. 2, component F) and Client ADSL (Fig. 2, component G) through the Internet (Fig. 2, component E).

The Client A (Fig. 2, component F) was situated in an Institution that shares an Internet connection of 1 Gbps provided by the RNP.

The Client ADSL (Fig. 2, component G) was connected through a residential Asymmetric Digital Subscriber Line (ADSL) Internet connection of 15 Mbps provided by the Global Village Telecom (GVT) telephone company.

[3] https://code.google.com/archive/p/exadel-flamingo/downloads

Table 1 Experimental model configuration of hardware and software components [24]

	CPU	RAM	Network Interface	OS
Server	Intel Core 2 Duo, 2.2 GHz	DDR 2, 4 GB	Broadcom NetXtreme BCM5755 Gigabit Ethernet	Linux Debian Kernel 2.6.32, 64 bits
Emitter	Intel Core i7, 2.0 GHz	DDR 2, 6 GB	JMicron PCI Express Gigabit Ethernet	Windows 7 Professional SP1, 64 bits
Client A	Intel Core i5-3210 M, 2.5 GHz	DDR 2, 4 GB	JMicron PCI Express Gigabit Ethernet	Windows 7 Home SP1, 64 bits
Client ADSL	Intel Core i5-2410 M, 2.3 GHz	DDR 2, 6 GB	Realtek PCIe FE Family Ethernet 1702 b/g/n	Windows 7 Home SP1, 64 bits

The evaluation consisted in collecting and analyzing the sending and receiving frame rate, measured in frames per second, of the video stream broadcasted over the Internet, using a video representing a colonoscopy examination. FPS values were collected from different computers connected to the local network and the Internet.

The experiments were conducted during working days in one week, according to the following schedule: 10:00 AM, 11:00 AM, 03:00 PM, and 04:00 PM. Each experiment lasted 30 min, and the data related to FPS was collected once every second, yielding 1,800 values per experiment.

Table 1 displays the hardware configuration (CPU, RAM, and network card) and operating system (OS) of the computers used in the experiments.

The computational system was accessed, on all computers, using the Mozilla Firefox browser version 18.0.2, with Java 7 U13 virtual machine and the Adobe Flash Player 11.5 R502 plug-in.

The video used to perform the experiments presented the following properties:

- Resolution of 480 × 360 pixels;
- Frame rate of 25 FPS;
- Bitrate of 6000 Kbps;
- Lossless compression;
- Sorenson Spark video codec (H.263).

As for the audio, the following features were adopted in the experiments:

- Speex[4] audio codec;
- Bitrate of 34.2 Kbps.

The frame rate of the Emitter computer was considered the control group. Resultant FPS values for the Emitter, Clients A and ADSL were analyzed using Friedman's nonparametric statistical test and Dunn's post test [41]. The percentage of samples

[4] http://www.speex.org

with FPS values lower than 10, i.e., the ones with video crashes perceptible to humans [7, 34, 39, 45] were also analyzed. For all analyses performed, rejection of the null hypothesis was set for *p*-value 0.05.

3 Results

In what follows, the computational system Real-Time Telemedical System (S2TR) and the experimental results are presented.

3.1 S2TR Construction

The first result was the proposal of the method in telemedicine for real-time monitoring of endoscopic procedures [24, 28].

Another important result was the construction of the computational system S2TR [47], which implements the proposed method. The building process of S2TR consisted in three main aspects: database creation, security policies definition and graphical interface implementation, i.e., the user-friendly layout of the computational solution.

The available interfaces followed a standard that comprises a screen for management (Fig. 3) and others for insertion, editing and visibility (Fig. 4) [24]. These screens were created for all entities defined in the database. In Fig. 5, the system interface that allows monitoring the video and visualizing the videocolonoscopy examinations is presented.

3.2 Experimental Performance Results

Tables 2 and 3 describe the average values ($\overline{x}$) of the transmission frame rates and the corresponding standard deviations (SD) achieved by the Emitter and the Clients A and ADSL, during the experiments performed [24]. In both tables, the last column averages the values achieved by the Emitter or a client across all samples for mornings ($\overline{x}$ *(M)*) and afternoons ($\overline{x}$ *(A)*).

Figure 6 displays the average FPS rate achieved by the Emitter and Clients A and ADSL, and their respective average SD between brackets, across all executions for each of them.

Table 4 shows the percentage of seconds that the FPS value was either below or above or equal to 10, which was considered in this work as the threshold representing the limit of human perception [7, 34, 39, 45].

Table 5 shows the results of the statistical analysis for morning, afternoon and both periods [24]. In particular, the table describes the existence of Statistically Significant

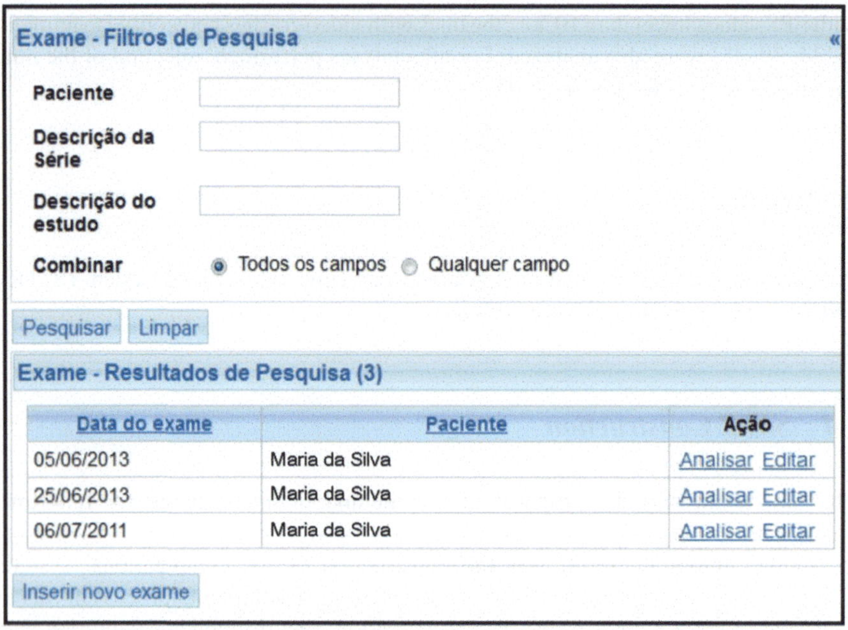

Fig. 3 Colonoscopy examinations management screen [24]

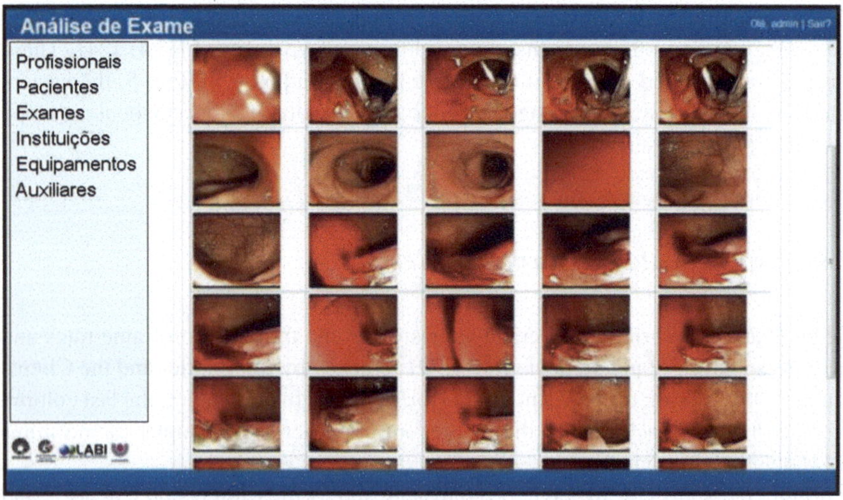

Fig. 4 S2TR examination analysis screen [24]

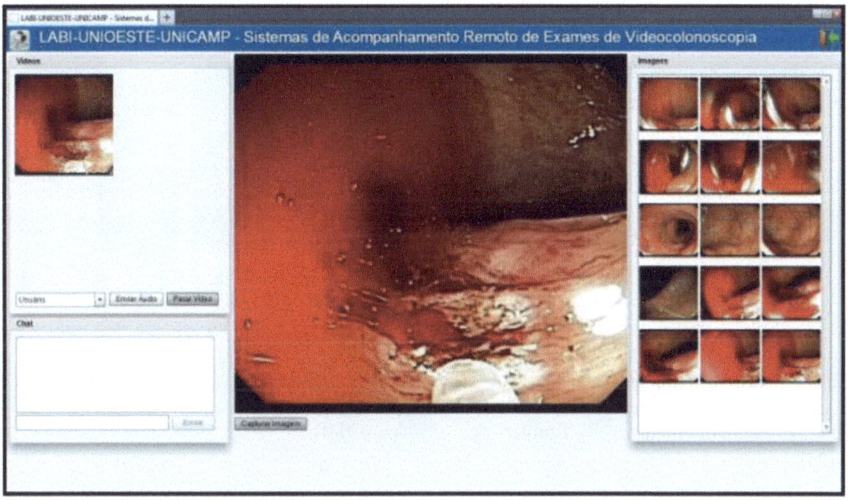

Fig. 5 Monitoring and remote interaction during videocolonoscopic exams in S2TR [24]

Table 2 Daily FPS average and standard deviation of morning (M1–M5) samples for Emitter, Client A and Client ADSL [24]

FPS		M1	M2	M3	M4	M5	$\bar{x}(M)$
Emitter	$\bar{x}$	23.97	23.92	23.83	23.90	23.94	**23.91**
	SD	0.52	0.58	0.67	0.62	0.66	**0.61**
A	$\bar{x}$	21.34	21.31	21.36	20.97	21.12	**21.22**
	SD	1.83	1.85	2.49	2.75	2.42	**2.30**
ADSL	$\bar{x}$	13.44	14.29	14.77	13.48	14.09	**14.01**
	SD	6.29	6.20	6.10	6.31	6.35	**6.27**

Table 3 Daily FPS average and standard deviation of afternoon (A1–A5) samples for Emitter, Client A and Client ADSL [24]

FPS		A1	A2	A3	A4	A5	$\bar{x}(A)$
Emitter	$\bar{x}$	23.95	23.85	23.83	23.94	23.94	**23.90**
	SD	0.59	0.71	0.79	0.70	0.55	**0.67**
A	$\bar{x}$	21.21	20.92	20.61	20.89	20.92	**20.91**
	SD	2.12	2.66	3.18	2.93	2.53	**2.72**
ADSL	$\bar{x}$	13.74	14.94	14.69	14.81	14.71	**14.58**
	SD	6.26	6.28	6.34	6.31	6.28	**6.31**

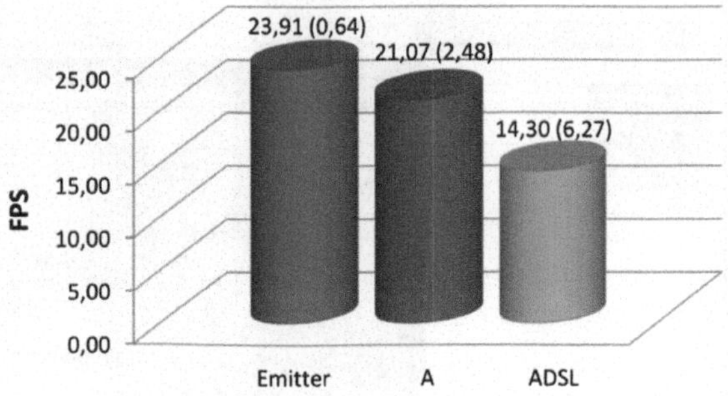

Fig. 6 Average FPS rate (average SD between brackets) achieved by the Emitter, Client A, and Client ADSL

Table 4 Relative frequency of occurrence of FPS lower and greater than or equals to 10 [24]

FPS value	<10 (%)	10 (%)
Emitter	0.01	99.99
A	1.30	98.70
ADSL	37.17	62.83

Table 5 *P*-Values regarding the comparative analysis between the FPS distributions for the Emitter, Client A and Client ADSL [24]

Computers	Morning comparisons	Afternoon comparisons	Morning and afternoon comparisons
Emitter versus A	<0.0001	<0.0001	<0.0001
Emitter versus ADSL	<0.0001	<0.0001	<0.0001
A versus ADSL	<0.0001	<0.0001	<0.0001

Difference (Diff) of the *p*-values (*P*-value) obtained in comparisons between the Emitter, Client A, and Client ADSL, considering all samples.

4 Discussion

Although there are several papers in the area of telemedicine that aim to develop technologies to assist in medical activity [2, 9, 10, 13, 14, 16, 18, 22, 38, 40], there is still a lack of computational methods that enable real-time monitoring and discussion on medical procedures along with communication and endoscopic diagnostic equipment. Another key aspect regarding the topic of medical data transmission is maintenance, with maximum focus on privacy, confidentiality and protection of the stored information.

This reinforces the need for a solution built using open source tools. The solution we constructed consists of a telemedical method for remote and real-time monitoring of medical procedures. This approach was defined in order to promote interaction during the endoscopic examination through different media such as video, audio, text and images. By using this process, medical professionals in different places are able to participate actively in the examination, capturing and sharing images [23–26, 28, 29, 47–49].

The computational implementation of the method, S2TR [47], was developed for web environment, thus running on web browsers. This feature allows greater flexibility and access from health care professionals in remote areas, contributing to minimize the impact of medical doctors' shortage on remote and isolated areas [24]. Furthermore, the architecture of the solution is independent of hardware platform and operating system, facilitating the addition of new features as well as the extension of the solution to other fields of medicine.

S2TR implements strong guidelines for building user-friendly interfaces.

4.1 Performance Analysis

Videos with frame rates lower than 10 FPS cause a feeling of discontinuity because their faults are perceptible to human vision. Given this scenario, the proposed method aims to employ the highest quality supported by a videocolonoscopy equipment while keeping the resolutions suitable to the characteristics of the defined network environments, even if it entailed a reduction in FPS rates [7, 39, 45].

Thus, in order to obtain efficiency and accuracy in the FPS transmission rates and to ensure consistent visual perception of the data, in the present work we set the transmission rate at the Emitter to 25 FPS, i.e., 150% above of that referred by most papers in the literature on the subject [7, 39, 45].

Another key aspect was the definition of the period in which we would perform the tests, i.e., in the morning at 10:00 AM and 11:00 AM and in the afternoon at 03:00 PM and 04:00 PM. They represent periods of massive Internet usage, when a large number of users are using some kind of service on the network, which influences the speed and transmission quality of data. Thus, it was possible to perform the experiments in a scenario with maximum use, as stated in a survey regarding the use of Internet in Brazil [5].

In the experiments, we used a video resolution of 480×360 pixels for experimental evaluation, which is high compared to other applications that provide videos on the Internet, such as YouTube, which employs a default resolution of 320×240 pixels.

The Emitter computer, responsible for sending the video, achieved an average of 23.91 FPS (SD of 0.61) and 23.90 FPS (SD of 0.67) during the morning and afternoon periods, respectively. The values of these rates showed that there were losses of frames during the capture and transmission of images. This data loss may be due to a number of factors, such as the processing mechanisms of the Central Processing Unit (CPU) of the Emitter and of the Server; synchronization when capturing audio and video; and the switch in the data communication network. Besides these reasons, other elements also interfere in the transmission rate, such as electrical problems, damage to mechanical components and hard drives, processor overheating, hardware incompatibility, application corruption, the presence of viruses or a lack of cleaning of the computer.

Client A received the video at an average rate of 21.22 FPS (SD of 2.30) in the experiments performed in the morning and 20.91 FPS (SD of 2.72) in the afternoon. Among the clients, A was the one with best performance. The factors that contributed to this performance were the 1 Gbps connection provided by the RNP and the low number of intermediate devices (only four hops) between the source (Emitter) and the destination (Client A).

Hop count influences lag time. In other words, for each intermediate device through which a Transmission Control Protocol (TCP) package travels, there is a processing cost related to the characteristics of the hardware and software of such device, such as protocol conversion, execution of routing algorithms, input, and output queue management and traffic management policy use, among others. Thus, the number of intermediate devices directly influences the latency of TCP and hence affects the performance of the video frame transmission.

Regarding the Client ADSL, we obtained video streaming reception rates of 14.01 FPS (SD of 6.27%) in the experiments in the morning on average. For experiments conducted in the afternoon, reception rates of 14.58 FPS (SD of 6.31%) were obtained. The connection channel used by this client presents the lowest cost and is widely available. In fact, it is used primarily in residential and enterprise environments and represents approximately 69% of the Internet connections currently used in Brazil. For this client, the connection speed contracted was 15 Mbps, and the number of hops between the source and the destination was 10.

These results indicated that the method can be used with ADSL connections, which are common, widespread and very cheap, presenting the most affordable prices.

This alternative can be implemented by clinics and institutions that do not have other demands that consume large amounts of bandwidth. The current 15 Mbps plans have an approximate cost of U$ 30.00 monthly. Telecommunication companies are now offering connections with greater bandwidth, such as 100 Mbps, at an average monthly cost of U$ 150.00 by the time this study was carried out. These are encouraging data for clients using ADSL channels.

It is also necessary to emphasize that ADSL connections, unlike the other alternatives dealt with here, use only 10% of their bandwidth for uploading. Thus, it would be required a 100 Mbps link to use the method in an institution for performing medical procedures and not only for remote monitoring as a client.

We considered the value of 10 FPS as a reference to analyze video streaming transmission rates sent by the Emitter and received by the clients. As can be observed in Table 4, the Emitter sent data at rates below 10 FPS just 0.01% of the time, whereas the clients exhibited reception rates lower than 10 FPS in the following percentages of time: 1.30% for Client A and 37.17% for Client ADSL.

These data reinforce the performance results of the Emitter and of each client, as well as the factors that influenced this performance. Considering the influence of the

FPS rate on the feeling of continuity of the video by human perception [7, 39, 45], the discontinuity would be felt by the 15 Mbps ADSL client.

Although the frame rate measured at this client was lower than 10 FPS during 37.17% of the time, the average frame rate was approximately 40% higher than the minimum required. This was due to the reasonable variability in performance yielded by the ADSL connection, which can be noticed by observing its average standard deviation (6.27) associated with an average frame rate of 14.30 FPS. This level of variability in the Client ADSL performance could be mitigated with QoS strategies like resource reservation, traffic shaping and/or differentiated services. However, since the connection established used the Internet, no QoS strategy could actually be guaranteed, as the route and the ability to recognize QoS settings of each router may vary [44].

Even in these conditions, monitoring of examinations is still possible, but with a loss in video performance, remote participation and interactivity. It is noteworthy that this transmission restriction is not due to the method proposed in the present work, but rather to external factors of the technological infrastructure.

An alternative to QoS strategies is adaptive streaming, which relies on the application's ability to detect and switch to the best suited media stream available according to the client's processing and network capacities, instead of on router's settings. Nevertheless, this strategy requires an additional component, which is capable of generating multiple media streams in real-time, with varying characteristics suitable for clients with different capacities [35].

Table 5 showed the statistical comparison results considering the Emitter and Clients A and ADSL, based on mornings, afternoons, and mornings and afternoons together. In all comparisons, there was a statistically significant performance difference (p-value < 0.0001).

This result was expected considering the processing time for video capture and transmission, as well as the different technologies of the Internet links used and the number of hop counts involved.

However, it was possible to obtain the performance and FPS rates designed while maintaining the highest quality of images transmitted, which is central to the discussion of examinations by specialists and for performing diagnoses.

5 Conclusion

The method implemented in the computational system and presented in this chapter meets the requirements for data transmission, information security and real-time interaction between users. This is an innovative solution and can be used for remote monitoring and interaction, through the Internet, while performing endoscopic examinations.

The main contributions expected from applications of this method include: reducing costs with transport and communication, dissemination of specialized medicine to remote areas and the strengthening of interactions and knowledge sharing among general practitioners, specialists, and researchers from higher education and research institutions, congregating them from their specific action sites.

These networks and computer applications can also assist in quality improvement processes and continuing education of health professionals.

Future work include the application of intelligent techniques for knowledge discovery [19] from the colonoscopy examination database, in order to aid the prediction and diagnosis of diseases in colonoscopy exams. In particular, the Data Mining process can be applied to the database to provide a second opinion potentially useful for medical Decision Making.

References

1. Adhikari S, Blaivas M, Lyon M, Shiver S (2014) Transfer of real-time ultrasound video of FAST examinations from a simulated disaster scene via a mobile phone. Prehosp Disaster Med 29(3):290–293. https://doi.org/10.1017/S1049023X14000375
2. Antoniou SA, Antoniou GA, Franzen J, Bollmann S, Koch OO, Pointner R, Granderath FA (2012) A comprehensive review of telementoring applications in laparoscopic general surgery. Surg Endosc 26(8):2111–2116. https://doi.org/10.1007/s00464-012-2175-x
3. Bakker AR (1991) HIS, RIS, and PACS. Comput Med Imaging Graph 15(3):157–160
4. Baniya B, Lee J (2016) Importance of audio feature reduction in automatic music genre classification. Multimed Tools Appl 75(6):3013–3026. https://doi.org/10.1007/s11042-014-2418-z
5. Brazilian Institute of Public Opinion and Statistics (IBOPE): Survey on Internet use in Brazil (in Portuguese) (2013). http://www.ibope.com.br/pt-br/noticias/Paginas/Numero-de-usuarios-ativos-na-internet-cresce-4.aspx
6. Cashman S, Spirk JW, Tillk JG, Nottingham JR, Kalman J (2015) Medical kiosk and method of use. https://www.google.com/patents/US9043217. US Patent US,904,32,17B,2
7. Chen JYC, Thropp JE (2007) Review of low frame rate effects on human performance. IEEE Trans Syst Man Cybern Part A Syst Hum 37(6):1063–1076
8. Colin P (2003) RTP audio/video transport for the Internet, 1st edn. SAMS
9. Costa FJ (1924) The family homeopathist physician (in Portuguese), 4th edn. Almeida Cardoso
10. Dwyer TF (1973) Telepsychiatry: psychiatric consultation by interactive television. Am J Psychiatry 130(8):865–869
11. Einthoven W (1906) The telecardiogram (in French). Arch Int Physiol 4(1):132–164
12. Eusebio EB (1989) A practical aid in colonoscopy. Dis Colon Rectum 32(11):996–997
13. Eysenbach G (2000) Consumer health informatics. BMJ 7251:1713–1716
14. Eysenbach G (2008) Medicine 2.0: social networking, collaboration, participation, apomediation, and openness. J Med Internet Res 10:1–22
15. Fang JL, Carey WA, Lang TR, Lohse CM, Colby CE (2014) Real-time video comunication improves provider performance in a simulated neonatal resuscitation. Resuscitation 85(11):1518–1522. https://doi.org/10.1016/j.resuscitation.2014.07.019
16. Ganeshalingam A, Pritchett S, Tam T, Cafazzo JA, Rossos PG (2010) Effectiveness of asynchronous tele-endoscopy. Gastrointest Endosc 71(3):461-467.e2. https://doi.org/10.1016/j.gie.2009.10.020
17. Gazula K (2011) Novel one integrated system for real-time virtual face-to-face encounters. https://www.google.com/patents/US20110106557. US Patent US,201,10,106,557,A1
18. Gershon-Cohen J, Cooley AG (1950) Telegnosis. Radiology 55(1):582–587
19. Han J, Kamber M (2011) Data mining: concepts and techniques, 3rd edn. Morgan Kaufmann
20. Jurik A, Weaver A (2009) Body sensors: wireless access to physiological data. IEEE Softw 26(1):71–73. https://doi.org/10.1109/MS.2009.5

21. Kang J, Yoo S, Oh D (2013) Development of a portable embedded patient monitoring system. Int J Multimed Ubiquitous Eng 8(6):141–150. https://doi.org/10.14257/ijmue.2013.8.6.14
22. Kiah MLM, Al-Bakri SH, Zaidan AA, Zaidan BB, Hussain M (2014) Design and develop a video conferencing framework for real-time telemedicine applications using secure group-based communication architecture. J Med Syst 38(10):133. https://doi.org/10.1007/s10916-014-0133-y
23. Lee HD, Jung W, Silva ACd, Costa LHDd, Espindola B, Coy CsR, Fagundes JJ, Wu FC (2011) Prototype system to manage data on coloproctology surgery. J Coloproctol (Rio J) 31(1):351–361
24. Machado RB (2013) Computational method for real time monitoring and remote interaction for videocolonoscopy (in Portuguese). Ph.D. thesis, University of Campinas
25. Machado RB, Lee HD, Ayrizono MdLS, Leal RF, Coy CSR, Fagundes JJ, Cherman EA, Wu FC (2012) Telemedicine computational system to monitor and remotely interact during the conduction of videocolonoscopic exams based on nonproprietary software (in Portuguese). In: Brazilian congress of coloproctology, pp 31–31
26. Machado RB, Lee HD, Ayrizono MdLS, Leal RF, Coy CSR, Fagundes JJ, Wu FC (2012) Prototype of a computer system for managing data and video colonoscopy exams. J Coloproctol (Rio J) 32(1):50–59
27. Machado, RB, Maciel JN, Voltolini RF, Maletzke AG, Coy CSR, Fagundes JJ, Wu FC (2013) Method and system for real-time audio and video transmission of colonoscopy exams: A study case on a local network. In: International workshop on knowledge discovery, knowledge management and decision support, pp 330–338
28. Machado RB, Wu FC, Lee HD, Coy CSR, Fagundes JJ, Nunes MJ, Voltolini RF, Maletzke AG, Leal RF, Ayrizono MdLS (2012) Telemedicine method to the remote and real time monitoring of medical procedures (in Portuguese). http://www.google.com/patents/WO2014094095A1?cl=pt&hl=pt-BR. Brazilian Patent BR1020120331250
29. Machado, RB, Wu FC, Lee HD, Coy CSR, Fagundes JJ, Nunes MJ, Voltolini RF, Maletzke AG, Leal RF, Ayrizono MdLS, Cherman EA (2012) Key generation method based on genetic algorithms (in Portuguese). http://projetos.unioeste.br/campi/nit/images/DPI/patentes/chaves.pdf. Brazilian Patent BR1020120331284
30. Mariani AW, Pego-Fernandes PM (2012) Telemedicine: a technological revolution. São Paulo Medical Journal 130(5):277–278. https://doi.org/10.1590/S1516-31802012000500001
31. Markatopoulou F, Pittaras N, Papadopoulou O, Mezaris V, Patras I (2015) A study on the use of a binary local descriptor and color extensions of local descriptors for video concept detection. Lect Notes Comput Sci (Incl Subser Lect Notes Artif Intell Lect Notes Bioinform) 8935:282–293
32. Minh CD, Shimizu S, Antoku Y, Torata N, Kudo K, Okamura K, Nakashima N, Tanaka M (2012) Emerging technologies for telemedicine. Korean J Radiol 13(1):21–30
33. Oliva JT, Lee HD, Spolaôr N, Coy CSR, Wu FC (2016) Prototype system for feature extraction, classification and study of medical images. Expert Syst Appl 63:267–283. https://doi.org/10.1016/j.eswa.2016.07.008
34. Ou YF, Liu T, Zhao Z, Ma Z, Wang Y (2008) Modeling the impact of frame rate on perceptual quality of video. In: 15th IEEE international conference on image processing (ICIP). IEEE, San Diego, pp 689–692. https://doi.org/10.1109/ICIP.2008.4711848
35. Ozer J (2011) Video compression for flash, apple devices and HTML5. Doceo Publishing
36. Paetsch FF, Eberlein A, Maurer F (2003) Telemedicine computational system to monitor and remotely interact during the conduction of videocolonoscopic exams based on nonproprietary software (in Portuguese). In: IEEE international workshops on enabling technologies: infrastructure for collaborative enterprises, pp 308–313
37. Pianykh OS (2008) Digital imaging and communications in medicine (DICOM): a practical introduction and survival guide, 1st edn. Springer
38. Raison N, Khan MS, Challacombe B (2015) Telemedicine in surgery: What are the opportunities and hurdles to realising the potential? Curr Urol Rep 16(7). https://doi.org/10.1007/s11934–015–0522-x

39. Read P, Meyer MP (2000) Restoration of motion picture film, 1st edn. Elsevier Butterworth-Heinemann
40. Sanjay PA, Sindhu M, Jesus Z (2012) A survey of the state of cloud computing in healthcare. Netw Commun Technol 1:12–19
41. Sheskin DJ (2011) Handbook of parametric and nonparametric statistical procedure, 5th edn. CRC Press
42. Shortliffe EH, Cimino JJ (2006) Biomedical informatics: computer applications in health care and biomedicine, 3rd edn. Springer Science Business Media
43. Silva JCF, Vieira E, Passos M, Melo EA, Mota GHMB, Tavares TA, Souza Filho GLd (2011) A tool for video collaboration in health (in Portuguese). In: Argentinian congress of informatics and health, pp 1–13
44. Tanenbaum AS, Wetherall DJ (2011) Computer networks, 5th edn. Prentice Hall
45. Wang Y, Claypool M (2005) Realtracer—Tools for measuring the performance of realvideo on the Internet. Kluwer Acad Publ 27(3):411–430
46. World Health Organization: Telemedicine: Opportunities and developments in member states (2011) Global observatory for eHealth, vol 2. http://www.who.int/goe/publications/ehealth
47. Wu FC, Coy CSR, Lee HD, Machado RB, Fagundes JJ, Ayrizono MdLS, Leal RF, Voltolini RF, Spolaôr N (2014) RT2S (in Portuguese). Brazilian registered computer program, BR512014000324-8
48. Wu FC, Coy CSR, Lee HD, Machado RB, Fagundes JJ, Ayrizono, MdLS, Leal RF, Voltolini RF, Spolaôr N (2014) SMDE/HI (in Portuguese). Brazilian registered computer program, BR512014000326-4
49. Yu L, Jihong W (2011) PACS and digital medicine: essential principles and modern practice, 1st edn. Taylor and Francis Group

Economic Growth and Universities: Empirical Evidence for Mexico

Antonio Kido-Cruz and F. A. Luna-Campos

Abstract In this paper, the relationship between the economic growth in the states of Mexico (EGM) and the number of colleges and universities (NU) was estimated. In order to prove the existence of long-term association among the study variables, yearly data from 1995 to 2015 was used and a fixed-effects data panel model was calculated. Results show that a 100% increase in the number of public universities is positively and significantly related to a growth of 38.2% in the GDP per capita. Our evidence can potentially be useful for the design of public policies of higher education, especially in some countries as Mexico while there are not many studies in this regard.

Keywords Measurement of economic growth · Public policy and regulation · Human development · Financing of education

1 Introduction

Higher education has been highlighted as one of the fundamental factors for the development of societies around the world, its favorable repercussions on variables such as human development, innovation, technological growth, per capita income, foreign direct investment, and exports are well known for countries with high investment in new universities.

The essential premise is that universities are producers of human capital; and specialized workers are more productive than unskilled workers. There are some channels through which universities may affect economic growth including more innovation, promote democracy, and a greater supply of human capital [16].

A. Kido-Cruz (✉)
Facultad de Contaduría Y Ciencias Administrativas, Morelia, Mexico
e-mail: antonio.kido@umich.mx

F. A. Luna-Campos
Universidad Michoacana de San Nicolás de Hidalgo, Morelia, Mexico
e-mail: fluna@umich.mx

E. León-Castro et al. (eds.), *Soft Computing and Fuzzy Methodologies in Innovation Management and Sustainability*, Lecture Notes in Networks and Systems 337,
https://doi.org/10.1007/978-3-030-96150-3_10

According to [14], the economic rationale of establishing a greater number of universities that affect the development or growth of a country or a region through public financing is found in the fact that education, especially that of superior level, can be understood as a good of partial public character that exhibits the property of non-rivalry.[1] Furthermore these authors signalize that education generates positive social externalities, such as the fact that the formation of human capital contributes significantly to economic growth and that university graduates have, on average, a higher income than cohorts generational with lower studies or without studies.

Additionally, higher education allows social mobility and forms individuals committed to highly desirable social values in a community, such as attachment to democracy, altruism, and family values that are referents of the need to finance this type of funding with government resources. The education in other words, in a benefit-social cost analysis, the marginal social benefit exceeds the marginal social cost [14].

Under this logic, several models of economic growth can be contemplated to explain, mathematically, the relationship between education and growth, such as the model of human capital that points out the educational factor as one more factor of production whose accumulation explains the economic growth of a country or region (impact on transitory dynamics for exogenous growth models, permanent impact on dynamics). It is also possible to point out the case of endogenous models of accumulation as in [11] or the stock of human capital as a facilitator of the imitation, adoption, diffusion, or even innovation of new technologies [5].

However, works as [1] assesses whether the returns to university education are due to the increase in productivity or should they respond to the fact that when attending a university indicates a greater capacity from the point of view of employers (theory of credentialing).

This research analyzes the reform to the Universidad de los Andes (Colombia) that in the year 2006 reduced the number of courses required to obtain a degree in economics and business, the number of students per class, and the time of completion of studies.

The evidence showed that before these reforms, neither the grade point average nor the graduation rate were affected, that is, the quantity and quality of the students remained unchanged. Using administrative data on salaries and college attendance, the results showed that salaries fell by approximately 16% in the economic related areas and 12% in the business fields. These results suggest that the theory of human capital plays an important role in the determination of wages and rejects a pure signaling model.

In contrast to these results, it is important to highlight the study by [4], where it is found that the education is eighty percent signaling and only twenty percent human capital development. In the same vein, the idea that education might not have

[1] Non-rivalrous goods are public goods that are consumed by people but whose supply is not affected by people's consumption. The education and the knowledge are non-rivalrous goods; once something is known it is in principle available to all at very low cost and should be organized so that it is [19].

the same impact on growth as a function of the country's distance to the technology frontier gained some credentials. First suggested by empirical studies as [3], and then it received considerable attention from pioneering researchers in the philosophy of neo-Schumpeterian growth models [14].

In addition, it is possible to mention the research of [17] which reveals that countries towards the technological frontier, higher education will contribute to a higher growth than those countries with lower levels of education, since innovation (high-skilled qualified workers) is the key factor for these countries (imitation is no longer an option). Confirmed by an econometric analysis in the OECD countries, the implicit message for the most advanced economies is that they must develop their higher education systems especially at their highest levels.

In the case of Mexico, little has been written about the relationship between economic growth and higher education. We highlight the work of [10, 12, 13, 15]. Where evidence is found suggesting a positive and significant relationship among innovation, higher education, and economic growth.

Based on the aforementioned motivation, the strategy aim goal of this research focuses on the incidence of the number of universities, taken as the set that generates all the positive externalities described in the previous paragraphs and also as a demand from the community, in Mexico's economic growth. Concatenated with the previous idea, the main objective of this study is to measure the long-term relationship between the EGM (measured by the gross domestic product per capita) and the NU.

This research paper is structured as follows. After this introduction, Sect. 2 contains a Literature Review on the theoretical and methodological proposals on the relationship between economic growth and variables such as human capital, education, innovation, and average schooling. Sect. 3 shows the methodology used in this research. The results of the research are shown in Sect. 4 and finally, Sect. 5 shows the corresponding conclusions of the study.

2 Literature Review

At the international level, the paper by [16] stands out, they point out that using data from 15,000 universities in 1,500 regions in about 78 countries, they find out that the increasing number of universities is positively and strongly related with a greater economic growth in a country or region and that this effect extends to the neighboring regions within the same country through the increase of human capital and the number of innovations. They also point out that the economic benefits of university expansion can outweigh the costs of generating more universities.

Hausman [8] points out that university innovation has a positive and direct effect on the economic growth in the counties of the United States of America (USA) and estimated that salaries increase in sectors closely linked to the innovative strength of a local university.

On the other hand, Torres et al. [15] study assesses the impact of the University of Lleida (Spain) on the regional economy through an input–output analysis. Their results indicate that the direct impact on the demand for goods and services in the city of Lleida comes, in its great majority, from the expenditure made by the students of the local University.

Regarding the human capital incidence in the economic growth of Mexico, human capital models are often used to support this thesis. Examples of these studies in Mexico are those of [7, 12]. About the fact that professionals have a higher income than those generations that only have a high school, middle school, or primary education, the Mincer wage equation is usually used for this purpose.

For Mexico, several studies such as [2, 12, 13, 18] have analyzed the relationship among factors such as human capital, innovation, income levels and education with economic growth, however, no specific evidence has been found that demonstrates a solid relationship between the variation in the number of universities and economic performance.

Once the need for this work has been theoretically grounded, the basic methodology of the research is presented. The panel data model is defined, as well as the data collection and processing to obtain the results.

3 Methodology

3.1 Panel Data Model

One of the first criticisms for the concept of human capital is that it was inherently an elusive concept that lacked a satisfactory measure. This idea begins to change when in a simple but elegant model, presents the idea of investing in education individually.

This model indicates that the level of education is representative of the individuals´ level overall skills, therefore, it could be thought that the time and money they invest to attend school should result in a level of future income that compensate for the skills acquired when people are inserted into the labor market.

Due in part to the power of Mincer's analysis and part to the immediate availability of data, the level of schooling became virtually synonymous with the measurement of human capital. In this way and due to the need for neoclassical economic growth models to have a measure of human capital, it became natural to think about measures of academic and/or academic achievement as the next variables of human capital.

The empirical form of the economic growth model is represented as follows:

$$\text{Economic Development} = \alpha + \beta \text{ Human Capital} + \beta \text{ Other Factors} + \varepsilon \quad (1)$$

This model indicates that the growth rate of a country can be considered as a function of the workers´ educational skills over time along with other systemic factors such as economic institutions, initial income levels, and technology. Thus, Eq. (1)

could be estimated substituting schooling for human capital and estimating the relationship of economic growth and schooling directly (it is pertinent to emphasize here that the modeling of economic growth rates based on the level of human capital is the general form of the models of endogenous growth, while the models of economic growth rates based on changes in human capital over time is the general form of neoclassical growth models).

This idea of human capital was widely accepted and has dictated the empirical approach of a great majority of empirical studies in the field of labor economics until today.

It turns to be inevitable to mention that the surveys and the censuses that provide information about the level of schooling were equally important in the empirical development of this field of specialization. However, globally, the differences in the level of schooling measurements have presented some uniformity problems in the information in human capital and a second problem with this measurement of human capital is that it presupposes schooling as the only source of human capital and abilities.

In this context, a variety of policies promoted by national governments and international development agencies emphasize not only the quality of the school, but also the role of families and the importance of improving health and nutrition as a way to develop human capital.

Besides, there are likely to be cultural factors that the model does not capture, but we assume are constant for all the entities of the Mexican Republic. It has been also taken into account that the time effect can be allowed if we believe that the function of interest changes over time due to other factors, such as technological advances, variations in government regulation policies, as well as other effects. Such temporary effects are easily taken into account, if we temporary introduce dichotomous variables into the modelo [7].

The initial econometric specification of the panel information model is as follows:

$$\mathrm{GDP}_{\mathrm{it}} = \beta \sum X_{\mathrm{it}} + \alpha_i + \lambda_t + \varepsilon_{\mathrm{it}} \tag{2}$$

where GDP pc represents the gross domestic product per capita in real terms, $\sum X_{\mathrm{it}}$ are the explanatory variables, α_i represents the fixed effect of each federative entity and λ_t represents the fixed effect over time. This model can also be represented by the inclusion of fictitious variables that simulate *n*−*1* binary indicators for each state and *T*−*1* binary indicators for periods such as:

$$GDP_{it} = \beta_0 + \beta_1 \sum X_{it} + \gamma_n D_{ni} + \delta_n T_{ti} + \varepsilon_{it} \tag{3}$$

However, it is important to point out the following considerations in the use of the dichotomous model of fixed effects, such as:

1. If too many dichotomous variables are introduced, the problem of degrees of freedom may arise. That is, there will not be enough observations for a meaningful statistical analysis.
2. With so many dichotomous variables in the model, both individual and interactive or multiplicative, the possibility of multicollinearity is always present, which can prevent doing a precise estimation of one or more parameters.
3. In some situations, it is possible that the model identifies the effect of variables that do not change over time.

Meanwhile, for the purposes of this research, the formulation presented by [16] is used, the empirical estimation model to be used contains the following specification:

$$\ln \mathrm{EGM}_{\mathrm{it}} = \beta \ln NU_{\mathrm{it}} + \varepsilon \tag{4}$$

where $\ln\mathrm{EGM}_{\mathrm{it}}$ represents the gross domestic product per capita in the states i in period t; NU is the reference for the number of universities in state i for time t.

3.2 Data Collection and Processing

According to [6], by the year 2015, a total of 275 public higher education institutions were registered and distributed in the states of Mexico, as shown in Table 1. Of this total, 159 belong to the system of technological universities representing (57.82%); 64 are located in the polytechnic system with (23.27%) and; 36 are state universities that participate with (13.09%). Finally, 16 are within the scheme of federal universities that correspond to (5.82%).[2]

The largest number of these public universities is located in Mexico City with a total of 66 and the Estado de México with a total of 21. They are followed by the state of Hidalgo with 16, Chihuahua and Guanajuato have 12 each one of them, and Puebla and Coahuila have 11 institutions per state.

About the evolution in the appearance of new universities in Mexico, there are three distinguishable moments in time. The first one covers from the beginning of the first University in Mexico (1581) to the year 1978 where there was only a total of 64 public higher education institutions. From 1979 to 1989, there is a record of only three new universities, one in Campeche, one in Yucatán, and one in Quintana Roo. Since 1990, there has been a boom in the creation of new institutions of higher

[2] For more information, visit: http://www.execum.unam.mx/.

Execum is a data explorer of the comparative study of the Mexican universities of the UNAM (2018).

National Autonomous University of Mexico (UNAM, by its acronym in Spanish).

Table 1 Distribution of public universities in Mexico, 2015

State	Federal	State	Technological	Polytechnic	Total
Aguascalientes	0	1	3	1	5
Baja California	0	1	1	1	3
Baja California Sur	0	1	1	0	2
Campeche	0	2	2	0	4
Chiapas	0	1	1	2	4
Chihuahua	0	2	9	1	12
Mexico City	9	2	54	1	66
Coahuila de Zaragoza	1	1	5	4	11
Colima	0	1	1	0	2
Durango	0	1	4	3	8
Estado de México	4	1	7	9	21
Guanajuato	0	1	7	4	12
Guerrero	0	1	3	2	6
Hidalgo	0	1	9	6	16
Jalisco	0	1	2	1	4
Michoacán de Ocampo	2	1	1	2	6
Morelos	0	1	1	2	4
Nayarit	0	1	4	1	6
Nuevo León	0	1	4	2	7
Oaxaca	0	1	2	0	3
Puebla	0	1	7	3	11
Querétaro	0	1	3	2	6
Quintana Roo	0	1	3	1	5
San Luis Potosí	0	1	1	1	3
Sinaloa	0	1	2	3	6
Sonora	0	2	7	0	9
Tabasco	0	1	2	3	6
Tamaulipas	0	1	4	3	8
Tlaxcala	0	1	1	2	4
Veracruz	0	1	3	1	5
Yucatán	0	1	4	1	6
Zacatecas	0	1	1	2	4
Total	16	36	159	64	275

Source Own elaboration with data from [6]

education, with positive growth rates practically reaching all the states of the Mexican Republic.

In this research, the following databases are used to construct the variables of the empirical model:

1. Gross Domestic Product series by state, 1993–2006. Contained in the INEGI[3] System of National Accounts.
2. Gross Domestic Product series by state, 2003–2017. Contained in the INEGI System of National Accounts.
3. Data Explorer of the Comparative Study of Mexican Universities (EXECUM) UNAM.
4. Population and Housing Censuses by state, 1895 to 2010. INEGI.
5. Main results of the 2015 Intercensal Survey of the United Mexican States INEGI.

In relation to the independent variable NU, it was measured taking the cumulative number of new universities per year and for each of the 32 states of Mexico.

To measure the EGM variable, the Gross Domestic Product series was used by state 1993–2006 (in thousands of pesos at current prices) and the Gross Domestic Product series by state 2003–2017 (in millions of pesos at current prices). This variable was calculated by deflating the GDP values (starting with five-year periods 1990, 1995, 2000, 2005, 2010 and 2015) with the historical series of the national consumer price index (INPC[4]) based on the second half of July 2018 = 100 published by INEGI on September 21, 2018. Subsequently, the data obtained from the GDP at constant prices based in 2018 was divided by the population registered in each state for the corresponding five-year periods of the series.

Regarding the population variable (POP), it was measured as the number of inhabitants registered in the population and housing censuses by state from 1895 to 2015 at the INEGI. It is noted that they are figures corresponding to the following census dates: March 12th (1990), November 5th (1995), February 14th (2000), October 19th (2005), June 12th (2010) and March 15th (2015).

4 Results

We present the results of the data model panel established in Eq. (4), for a fixed-effects model Table 2.

---

[3] The National Institute of Statistics and Geography (INEGI, by its acronym in Spanish). It is one of Mexico's autonomous constitutional institutions with its own management, legal personality and assets, responsible for regulating and coordinating the National System of Statistical Information and Geography [9].

For further information on the databases, visit: https://www.inegi.org.mx/.

[4] The INPC, by its acronym in Spanish, is a global economic indicator whose purpose is to measure, over time, the variation in prices of a basket of goods and services representative of the consumption of the country's households. The INPC has become one of the main indicators of the country's economic performance; its applications are numerous and of great importance in the economic, legal and social fields.

Table 2 Results of data panel

Variable	Coefficient	Std. Error	t-Statistic	Prob
C	10.95308	0.053077	206.3617	0.0000***
Log (NU)	0.382092	0.026390	14.47880	0.0000***

Notes *** indicates significance at the 1% level. *Source* Own elaboration based on data panel model

According to the results obtained for this data panel model, it can be seen that the estimator for the Universities (NU) variable is significant, given that the p-value is lower than any of the acceptable significance levels that are 0.1, 0.05, and 0.01. Regarding the sign of the estimator of the variable universities, it is observed that it coincides with the theoretical preconception. The value of the coefficient of the university variable indicates that a 100% increase in the number of universities in Mexico would generate an increase of approximately 38.20% of the GDP per capita in the states of the Mexican Republic.

In terms of public education policy, the evidence suggests that investment in the creation of new universities would generate an increase in average per capita income for the states of the Mexican Republic.

5 Conclusions

The objective of this study is to measure the relationship between the economic growth in the states of Mexico and the number of colleges and universities. A positive and significant relationship was found over time according to the evidence of the panel data model.

Our results coincide with the evidence found in the work of Mungaray Lagarda and Torres Preciado [12], where it was found that higher education and primary education have a similar effect in the long term, while in the short term higher education has a relatively modest effect. However, given the comparatively lower public funding received by the higher education system, the results imply a more efficient application of the resource by this educational segment.

In this regard, the results are consistent with the empirical evidence found in [16], which establishes that increases in the number of universities are positively associated with future per capita GDP growth.

Finally, it can be said that the relationship between economic growth and education has no simple answers, the mere idea of being able to formulate this relationship in different logical structures (empirical models) describes the complexity of its study. Even when this relationship is possible to identify, as we have done in this research, the message is still quite diffuse. Due to the above, it is necessary to continue asking, just as [5] did about the specific skills and competencies to develop among students and about the risks of short-term education when economic structures are not sufficiently developed to hire these recent graduate students. An additional aspect to consider is represented by the fact of a non-evident viability of educational policies (long-term), given the short-term political times that do not contribute to the development of public education policies that have an impact on the economic growth of the countries.

References

1. Arteaga C (2018) The effect of human capital on earnings: evidence from a reform at Colombia's top university. J Public Econ 157:212–225. https://doi.org/10.1016/j.jpubeco.2017.10.007
2. Barceinas Paredes F (2001) Capital Humano y Rendimientos de la Educación en México. Universitat Autònoma de Barcelona. https://www.tesisenred.net/handle/10803/3983#page=17
3. Benhabib J, Spiegel MM (1994) The role of human capital in economic development evidence from aggregate cross-country data. J Monet Econ 34(2):143–173. https://doi.org/10.1016/0304-3932(94)90047-7
4. Caplan B (2018) The case against education why the education system is a waste of time and money. Radical Teacher Princeton University Press, New Jersey
5. Demeulemeester JL, Diebolt C (2011) Education and growth: what links for which policy? Hist Soc Res 36(4):323–346. https://doi.org/10.12759/hsr.36.2011.4.323-346
6. Execum (2021) Execum: Estudio Comparativo de Universidades Mexicanas. http://www.execum.unam.mx/
7. Gujarati DN, Porter DC (2013) Econometría. In: Zúñiga Gutiérrez EC (ed) Quinta. Ciudad de México, McGraw-Hill
8. Hausman N (2012) University innovation, local economic growth, and entrepreneurship. US Census Bur Cent Econ Stud 12(10):48. https://doi.org/10.2139/ssrn.2097842
9. INEGI (2021) Instituto Nacional de Estadística y Geografía
10. Lomelí Vanegas L (2020) Educación superior y desarrollo: los desafíos de México. Economía UNAM 16(47):3–11. http://www.scielo.org.mx/scielo.php?script=sci_arttext&pid=S1665-952X2019000200003&lang=pt%0Ahttp://www.scielo.org.mx/pdf/eunam/v16n47/1665-952X-eunam-16-47-3.pdf
11. Lucas RE (1988) On the mechanics of economic development. J Monet Econ 3–42. https://www.parisschoolofeconomics.eu/docs/darcillon-thibault/lucasmechanicseconomicgrowth.pdf
12. Mungaray Lagarda A, Torres Preciado V (2010) Actividad económica y educación superior en México. Revista de La Educación Superior 39(156):7–18
13. Ríos Flores JA, Ocegueda Hernández JM (2017) Capacidad innovadora y crecimiento regional en México: un enfoque espacial. Econ Soc Y Territ xvii(1):743. https://doi.org/10.22136/est2017705
14. Sanyal BC, Martin M (2006) La financiación de la educación superior: perspectivas internacionales. In: Mundiprensa (ed) Informe: La educación superior en el mundo 2006 : La financiación de las universidades. Barcelona: Universidad Politécnica de Cataluña, pp 3–17. https://upcommons.upc.edu/handle/2099/5823
15. Torres T, Enciso P, Farré M, Sala M (2010) El impacto de la universidad en el ámbito económico y del conocimiento. El caso de la universidad de lleida. RegNal Sect Econ Stud 10(3):174–199. http://www.usc.es/economet/reviews/eers10310.pdf
16. Valero A, Van Reenen J (2018) The economic impact of universities: evidence from across the globe. Econ Educ Rev 68:53–67. doi:https://doi.org/10.1016/j.econedurev.2018.09.001

17. Vandenbussche J, Aghion P, Meghir C (2006) Growth, distance to frontier and composition of human capital. J Econ Growth 11(2):97–127. https://doi.org/10.1007/s10887-006-9002-y
18. Villarreal Peralta EM (2018) Endogeneidad de los rendimientos educativos en México. PerfEs Latinoam 26(51):265–299. https://doi.org/10.18504/pl2651-011-2018
19. Williams G (2016) Higher education: public good or private commodity? Lond Rev Educ 14(1):131–142. https://doi.org/10.18546/LRE.14.1.12

ABC Costing System Applied to an SME Specializing in Dairy Production in Colombia

Juan Carlos Ruíz-T and Gina P. Fonseca-Cifuentes

Abstract This paper aims to develop a methodology for implementation of management accounting associated with the ABC costing, in an SME specializing related to milk production in Colombia. For this, the methodologies proposed by Kaplan and Cooper, were taken as a reference, which were consolidated through nine stages for the construction of the cost system. Among the results, is the determination of the cost of the agricultural product as well as the measurement of biological assets according to IAS 41. It is concluded that this system costs allows appropriate decisions to this kind of agricultural SMEs, optimizing processes and increasing profitability indicators.

Keywords ABC costing system · Livestock · Milk · Management accounting

1 Introduction

The agricultural sector maintains a high participation in the Colombian economy, this being significant in the generation of employment and income for the economically active population, however, it is found that this has been affected by factors associated with the economic, political, environmental and social [12].

In the particular case of the dairy sector, milk has become one of the main commodities in the same conditions as beef and eggs and whose production comes from specialized systems or dual-purpose livestock (meat and milk production) [20]. In this way, dairy activity contributes to the Colombian economy about 9.1% of agricultural GDP, 24.3% of livestock GDP and 3.5% of industrial GDP, with an average

J. C. Ruíz-T (✉)
Facultad Seccional Chiquinquirá, Universidad Pedagógica Y Tecnológica de Colombia, Carrera 14A No 2-37, Chiquinquirá, Colombia
e-mail: juan.ruiz@uptc.edu.co

G. P. Fonseca-Cifuentes
Facultad de Ciencias Económicas Y Administrativas, Universidad Pedagógica Y Tecnológica de Colombia, Av. Central del Norte 39-115, Tunja, Colombia
e-mail: gina.fonseca01@uptc.edu.co

E. León-Castro et al. (eds.), *Soft Computing and Fuzzy Methodologies in Innovation Management and Sustainability*, Lecture Notes in Networks and Systems 337,
https://doi.org/10.1007/978-3-030-96150-3_11

participation of 45% of the production of the specialized livestock. Informal milk collection is the one that maintains the greatest relevance with 46% of production at the national level [21].

In this way, a high concentration of informality is evidenced, which results in a low inclusion of administrative processes and a lower estimate of profitability, allowing in general terms the value destruction. Thus, this constitutes an economic sector with high production costs, as a consequence of higher prices of inputs and lack of timely infrastructure for storage and production processes [11].

Therefore, the sector maintains related barriers in the financial and accounting fields, which hinder decision-making, as well as the evaluation of productivity. This situation generates the consolidation of management systems, among others, associated with management accounting, which allows the compilation of information, from a holistic conception to carry out processes of diagnosis, assessment, control and potentiation of activities specific to the value addition [26]. Thus, implicit elements are involved in the appropriate production process with the measurement of biological assets, from a fiscal-accounting perspective, referring to IAS 41. In the case of this standard, the differentiation of financial and tax accounting allows a broad reflection of the economic facts of the organization, making it possible to update the historical value towards the constitution of a fair value, increasing the levels of confidence in the information financial disclosed [25].

Thus, management accounting, from the application of ABC costing, recognizes causal relationships that occur between the consumption of resources associated with the activities and according to the transformation or service provision processes and the causal relationships between the activities and resources according to business objectives [8].

Therefore, this costing has to involve understanding the development of the productive cycle and the development of drivers in the costing [30, 31], which means the identification of the processes associated with the value chain, and therefore, the activities that represent benefits to the company. For Hicks [13], the philosophical foundation involves three elements (a) adapt the model to company size, (b) implement erroneously system costs affect the growth of the organization and (c) a need for rigor in the study of adaptability of the system cost to the company.

According to [29], the activity-based costing involves the grouping of indirect costs per activity, so that the grouping of these costs is assigned to the products from the multiplication rates for activity with the amount that is consumed by the product, so that the inclusion for activities rates become a mechanism more precisely.

In this way, this document aims to develop a methodological proposal for the application of management accounting associated with ABC costing, in an SME related to specialized milk production in Colombia. For this, the chapter is organized as follows, in numeral 2 the theoretical framework is presented, in numeral 3 the methodology developed is shown, later in numeral 4 are the results, finally in numeral 5 the conclusions of the study.

2 Theoretical Framework

ABC Costing has emerged with the purpose of generating changes in the conception of standard costing, for which, this cost has formulated the interaction between resources, activities and cost objects. Johnson and kalpan [14], Cooper [4–7] and Berrío Guzmán and Castrillón Cifuentes [1] consider that, "costing by activities was developed to assign, control and rationalize indirect costs. Above all, you want to have better information on the cost of the products, their marketing and the provision of services ". In this way it is possible to assign the resources to the activities and subsequently these to the cost objects. Cuevas Villegas [10] sets ABC costing as "a methodology that measures the cost and performance of activities, resources and cost objects. Resources are assigned to activities first; then the activity costs are assigned to the cost objects according to their use."

Therefore, ABC costing has conceived a level of hierarchy of activities that bases their assignment, this situation has generated that cost management in decision-making is compatible with other management systems such as management based on activities, activity-based budgeting among others [22]. Is so,

> …activity analysis serves to decompose a large and complex organization into elementary processes (activities) and the outputs of such activities that turn out to be understandable and manageable. Thus, to carry out the analysis of the activities, the foundation is in the observation of a global system, being too big to be managed, while its elementary pieces are not [2].

Likewise, Pérez Falco [23], p. 28, identifies ABC costing, as a company management system from which the cost for business management is integrated, providing substantial elements to the management for decision making. Therefore, the cost is seen from different perspectives, within which are business activities, processes, value chain, products and/or services, purchasing decisions, as well as the reduction or elimination of activities.

Thus, for the development of ABC Costing, the concept of the value chain is accentuated. Under this premise, Porter [24], explains the existence of cost-generating activities and value-generating activities. Cost-generating activities are its structural drivers, characterized among others by economies of scale, while, the generators of value are exclusively and inescapably linked to the organization, among other examples, business policies.

In this way, for the implementation of ABC costing, it is necessary to carry out a series of stages, within which are: 1. Set the activities 2. Assign costs according to activities and according to the resource controller. 3. Identify the controller of the activity. 4. Determine the activity controller. 5. Multiply the rate of the activity controller with the activity consumption controller [3].

3 Methodology

3.1 Data Sources

For the development of the cost system, financial information for the month of July 2020 was taken as a reference, from the SME Hacienda Agrícola Casa de Lata Ltda. This SME is specialized in dairy production. Likewise, this company is considered a leader in the sector for sustainability in farm management, in the livestock and environmental field, reasons for which it has been awarded by the National Federation of Cattlemen of Colombia "FEDEGAN". This company is located in the municipality of Chiquinquirá (zone of high dairy impact) in the department of Boyacá in Colombia.

3.2 Cost System Methodology

In the particular case of methodologies on the cost and management system in agricultural companies, the information in databases is scarce [16]. For this reason, it was necessary to resort to conceptual contributions from Kaplan et al. [13], Hicks [15] and Cuervo Tafur et al. [9]. According to Kaplan and Cooper [15], the application of this system promotes a comprehensive allocation of the organization's costs and expenses, also it involves other allusive information systems, operating systems and strategic learning and organizational improvement.

In the case of the Hicks [13] cost system methodology, the fundamental objective is to develop a more effective mechanism for determining the costs of goods and / or services individually, so as time passes, the new data input of the model takes into account the same procedures as defined in the original model. For [9], the methodology that involves each step has operationalized the expected results, being an opportunity for the evolution of cost management.

In this way and according to the conceptual contributions, the methodology of the cost system considered the following stages, (a) establishment of resources (resources refer to costs and expenses that consume activities; for example, salaries), (b) identification and assignment of activities (the activities are homogeneous actions that in practice consume resources), for example, salaries), (c) determination of the main cost drivers (cost distribution criterion to make an objective cost allocation), (d) establishment of cost objects (the object of the cost is what pays for example, products and/or services and/or processes), e) development of the cost diagram, (f) characterization of drivers related to resource allocation and activities, (g) allocation of resources to activities, for the peso value of each activity is used the following equation [27],

$$C/A = HD + M + E + S \tag{1}$$

where, C/A is estimated cost per activity, H is the number of hours of labor required to perform the activity once, D is salary per hour, M is cost of the material necessary to carry out the activity once, E is cost of the equipment to carry out the activity once, S are manufacturing indirect costs allocated for the activity once.

(h) designation of activities to cost objects, the total cost of the product and / or service, results from adding all the recurring activities in its process, expressed from the following equation [27],

$$CG = \sum_{i=1}^{n} C/Aij \tag{2}$$

where, CG is cost of the good (or service), C/Aij is cost of activity i for good (or service) j, i indicates the number of activity used in the production of the good (or service) j and n Total number of activities necessary for the production (or service) of good j.

(i) setting the total cost according to the cost objects, the direct costs are added to the cost of the activities [19],

$$V_i = q_i * P_i \tag{3}$$

where, V_i is total cost of direct element i, q_i is total amount of direct element i and $P_i$is unit price of direct element i.

4 Results

a. Establishment of resources. In the first instance, the resources are established according to the economic activity, which consists of a specialized milk production system, which are contained in Table 1. According to the activity of the company, significant expenses and costs were determined.

Table 1 Establishment of resources

Resources
Salaries, benefits and security contributions
Technical and livestock advice
Leases
Depreciation and amortization
Maintenance and repairs
Public services- energy, aqueduct and sewerage, internet
Fuels and lubricants
Other production resources

Table 2 Identification of basic activities

General activity	Specific activities
Production and raising of livestock (P)	Milking routine (P1)
	Supply of fodder, water and supplement (P2)
	Transfer of livestock between lots (P3)
	Cooling, measuring, milk dispensing and equipment washing (P4)
	Hato records (milk weightings and others) (P5)
	Calf rearing 0 to 6 months (P6)
	Care of heifers from 6 to 18 months (P7)
	Care of 18–20 month old heifers (P8)
	Care of heifers 20–27 months (P9)
	Caring for a non-pregnant cow (P10)
Reproduction and animal health (R)	Vaccination programs (R1)
	Mastitis control (R2)
	Insemination (R3)
	Control of matings (R4)
	Veterinary check-up (R5)
	Registration of livestock (R6)
	Body condition (R7)
	Deworming of livestock (R8)

b. Identification and assignment of activities. Later, the activities are identified hierarchically in general and specific, and according to the value chain [24], as seen in Tables 2 and 3. Thus in Table 2 are the basic activities, production an raising of livestock (P) and reproduction and animal health (R) with its specific activities. In the Table 3 are the support activities, management (M), forages and biological control (F) and infrastructure maintenance with its specific activities.
c. Determination of the main cost drivers.

   To determine the cost drivers, that is, the parameters for cost distribution, it has been pertinent to take as a reference elements associated with the simplicity in establishing the data, as well as the correlation between real consumption and its implicit consumption. Thus, three simple indicators have been determined, these are spent time activity, spent time weighted activity and weighting of the area used. A combined indicator has also been determined, this is weighting of the area used + Spent time weighted activity (see Table 4). Likewise, the percentage of time devoted to activities have been determined.
d. Establishment of cost objects. Then the cost objects were established according to two fundamental concepts, contemplated in the International Financial Reporting Standards (IAS 41 Agriculture) the agricultural product and the biological assets. Within the first categorization, raw milk is stipulated, which will be designated as cost object 1 (Obj1) and the biological assets were thus

Table 3 Identification of support activities

General activity	Specific activities
Management (M)	Production control (M1)
	Control of livestock inventories (M2)
	Planning of activities and tasks (M3)
	Materials and resources management (M4)
	Order processing, material control (M5)
Forages and biological control (F)	Manuring of pastures (F1)
	Sowing paddocks (F2)
	Manufacture of haylage and silos (F3)
	Pest control (F4)
	Soil analysis (F5)
	Soil preparation (F6)
	Irrigation adequacy (F7)
	Manure spreading (F8)
Infrastructure maintenance (I)	Fence arrangement (I1)
	Machinery and equipment maintenance (I2)
	Road and path maintenance (I3)
	General services (I4)

Table 4 Main cost driver identified

Cost driver class	Cost driver name	Initials	Description
Simple	Spent time activity	STA	It consists of the percentage determination of the time dedicated to each activity for each of the positions held
	Spent time weighted activity	STWA	It consists of the weighting of the time dedicated to each activity for each of the positions held
	Weighting of the area used	% AU	This driver is constituted by the weighting of the hectares used in the development of the livestock activity
Combined	Weighting of the area used + Spent time weighted activity	% AU + STWA	This combined driver is made up of the weighting of the area used plus the power of the time dedicated to each activity for each of the positions held

determined, calf of 0–6 months cost object 2 (Obj2), heifers of 6–18 months cost object 3 (Obj3), heifers of 18–20 months cost object 4 (Obj4) and heifers 20–27 months cost object 5 (Obj5).

e. Development of the cost diagram. The cost diagram is an outline in which resources, activities, and cost objects are interrelated, as shown in Fig. 1. In this figure, the resources (salaries, benefits, and security contributions; technical and livestock advice; leases; depreciation and amortization; maintenance and repairs; public services-energy, aqueduct and sewerage, internet; fuels and lubricants and other resources) are relationed with basic and support activities. Then, the basic and support activities are relationed with the cost object (raw milk, calf 0–6 months, heifers 6–18 months, heifers 18–10 months and heifers 20–27 months). Finally, the direct inputs (animal feed) are assigned.
f. Characterization of drivers related to resource allocation and activities. At this stage, the interrelationships of the inducers with the resources and activities are established.
g. Allocation of resources to activities. "In strategic cost management, it is recognized that cost is caused or driven by many factors that are interrelated in complex ways. Understanding cost behavior means understanding the complex interaction of the set of cost factors in any situation" [28]. In this way, we proceed to carry out the allocation of resources to activities as evidenced in the Tables 5 and 6. In Table 5 the cost of each resource (salaries, benefits, and security contributions; technical and livestock advice; leases; depreciation and amortization; maintenance and repairs; public services-energy, aqueduct and sewerage, internet; fuels and lubricants and other resources) has been assigned to each one basic activities (production and raising of livestock and reproduction and animal health). Then in the Table 6, the resources have been assigned to each one of support activities (management; forages and biological control and infrastructure maintenance).
h. Designation of activities to cost objects. After assigning resources to activities, the process continues with assigning activities to cost objects, taking the used area driver as a reference, as shown in Table 7. In this way, it is found that the cost objects that have generated the greatest demand for activities, according to the used area cost driver and therefore the highest cost are raw milk with 70%, then there are heifers of 18 to 20 months with 11% and finally Heifers 20–27 months with 10%. The most pertinent cost driver to the activities of the farm is Area used (Ha). In this way, it has been assigned for the corresponding distribution of costs.
i. Setting the total cost according to the cost objects. Finally, the total cost is obtained with the direct assignment of the direct material cost objects, in the case of activities these were obtained by the previously mentioned driver cost, which generated as a result the total cost for different costs objects, according to Table 8. In this way the allocation of the cost of raw materials and the cost of activities, again shows that the object of the most significant cost to the company is the raw milk with 74% of the total cost. Then, there are the heifers 6–18 months with 8% and finally heifers 20–27 months with 7%. This changes what has been

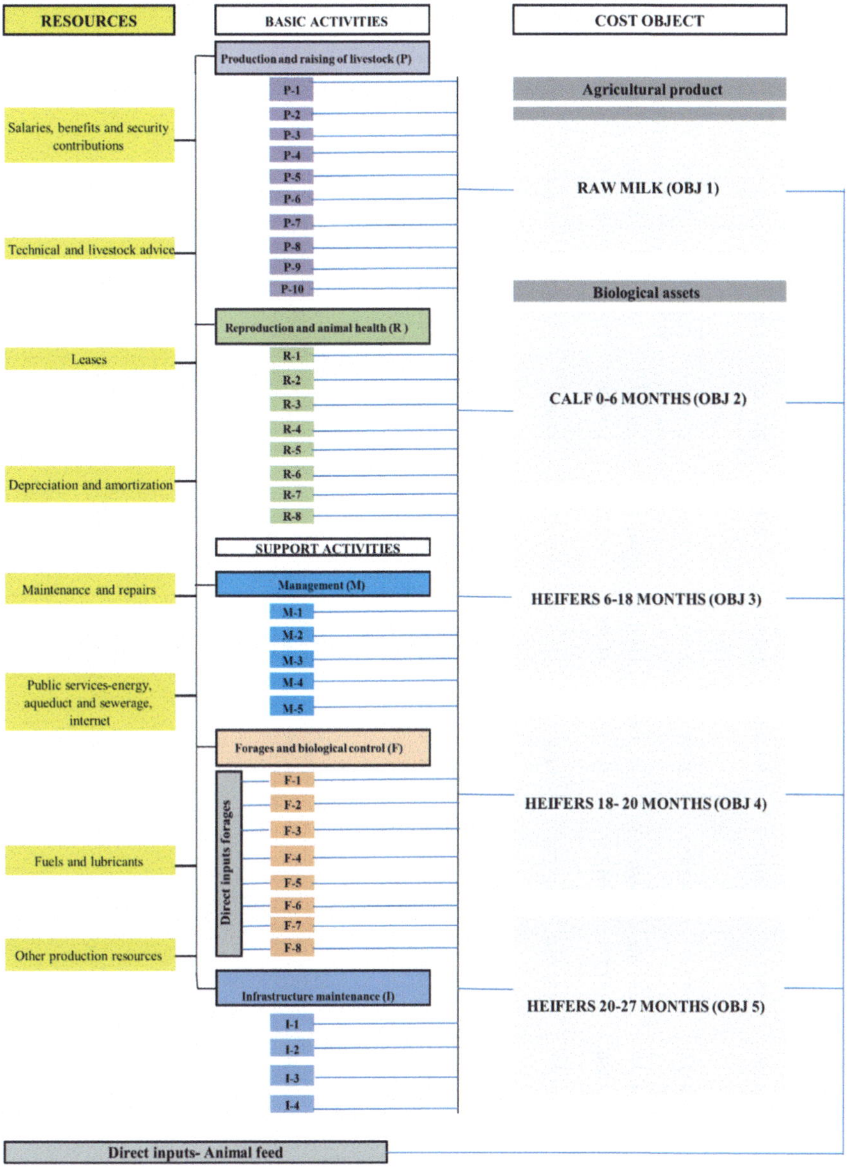

Fig. 1 Cost diagram

Table 5 Allocation of resources to basic activities

Resources/Basic activities		Production and raising of livestock (P)									
		P-1	P-2	P-3	P-4	P-5	P-6	P-7	P-8	P-9	P-10
1	Salaries, benefits and security contributions	7.454.189	558.165	201.263	508.396	191.617	373.542	287.465	287.465	648.987	115.312
2	Technical and livestock advice	312.857	37.143	12.857	22.857	8.571	30.000	22.857	22.857	52.857	8.571
3	Leases	4.124.850	489.708	169.514	301.359	113.010	395.534	301.359	301.359	696.892	113.010
4	Depreciation and amortization	1.623.750	192.774	66.729	118.630	44.486	155.702	118.630	118.630	274.332	44.486
5	Maintenance and repairs	1.162.921	138.064	47.791	84.962	31.861	111.513	84.962	84.962	196.475	31.861
6	Public services-energy, aqueduct and sewerage, internet	1.508.125	179.047	61.978	110.183	41.318	144.615	110.183	110.183	254.797	41.318
7	Fuels and lubricants	606.202	71.969	24.912	44.289	16.608	58.129	44.289	44.289	102.418	16.608
8	Other production resources	439.166	52.138	18.048	32.085	12.032	42.112	32.085	32.085	74.197	12.032
9	Total direct cost inputs forages	0	0	0	0	0	0	0	0	0	0
Total allocation to activities		27.235.387									

(continued)

Table 5 (continued)

Resources/Basic activities		Reproduction and animal health (R)							
		R-1	R-2	R-3	R-4	R-5	R-6	R-7	R-8
1	Salaries, benefits and security contributions	312.833	458.284	100.326	100.326	172.183	79.016	99.561	141.416
2	Technical and livestock advice	18.571	18.571	5.714	5.714	10.000	4.286	5.714	8.571
3	Leases	9.922	9.922	3.053	3.053	5.343	2.290	3.053	4.579
4	Depreciation and amortization	96.387	96.387	29.658	29.658	51.901	22.243	29.658	44.486
5	Maintenance and repairs	69.032	69.032	21.241	21.241	37.171	15.930	21.241	31.861
6	Public services- energy, aqueduct and sewerage, internet	89.523	89.523	27.546	27.546	48.205	20.659	27.546	41.318
7	Fuels and lubricants	35.985	35.985	11.072	11.072	19.376	8.304	11.072	16.608
8	Other production resources	26.069	26.069	8.021	8.021	14.037	6.016	8.021	12.032
9	Total direct cost inputs forages	0	0	0	0	0	0	0	0
Total allocation to activities		2.899.055							

Table 6 Allocation of resources to support activities

Resources/Support activities		Management (M)				
		M-1	M-2	M-3	M-4	M-5
1	Salaries, benefits and security contributions	648.401	79.016	165.064	522.246	292.358
2	Technical and livestock advice	31.429	4.286	8.571	27.143	14.286
3	Leases	30.225	4.122	8.243	26.103	13.738
4	Depreciation and amortization	163.116	22.243	44.486	140.873	74.144
5	Maintenance and repairs	116.823	15.930	31.861	100.893	53.101
6	Public services- energy, aqueduct and sewerage, internet	151.501	20.659	41.318	130.842	68.864
7	Fuels and lubricants	60.897	8.304	16.608	52.593	27.680
8	Other production resources	44.117	6.016	12.032	38.101	20.053
9	Total direct cost inputs forages	0	0	0	0	0
Total allocation to activities		3.338.289				

Resources/Support activities		Forages and biological control (F)							
		F-1	F-2	F-3	F-4	F-5	F-6	F-7	F-8
1	Salaries, benefits and security contributions	568.379	501.373	372.098	123.663	199.205	113.313	142.549	111.270
2	Technical and livestock advice	41.429	37.143	18.571	8.571	14.286	8.571	10.000	4.286
3	Leases	11.952	10.716	5.358	2.473	4.122	2.473	2.885	1.236
4	Depreciation and amortization	215.017	192.774	96.387	44.486	74.144	44.486	51.901	22.243
5	Maintenance and repairs	153.994	138.064	69.032	31.861	53.101	31.861	37.171	15.930
6	Public services- energy, aqueduct and sewerage, internet	199.706	179.047	89.523	41.318	68.864	41.318	48.205	20.659
7	Fuels and lubricants	80.273	71.969	35.985	16.608	27.680	16.608	19.376	8.304
8	Other production resources	58.154	52.138	26.069	12.032	20.053	12.032	14.037	6.016
9	Total direct cost inputs forages	5.250.378	4.707.235	2.353.618	1.086.285	1.810.475	1.086.285	1.267.333	543.143
Total allocation to activities		22.859.102							

Resources/Support activities		Infrastructure maintenance (I)			
		I-1	I-2	I-3	I-4
1	Salaries, benefits and security contributions	5.174.137	1.655.269	417.657	249.732
2	Technical and livestock advice	101.429	41.429	14.286	5.714
3	Leases	51.338	20.969	7.231	2.892
4	Depreciation and amortization	526.421	215.017	74.144	29.658
5	Maintenance and repairs	377.020	153.994	53.101	21.241
6	Public services- energy, aqueduct and sewerage, internet	488.935	199.706	68.864	27.546
7	Fuels and lubricants	196.531	80.273	27.680	11.072
8	Other production resources	142.378	58.154	20.053	8.021
9	Total direct cost inputs forages	0	0	0	0
Total allocation to activities		10.521.894			

Table 7 Designation of activities to cost objects

Cost object		Area used (Ha)	Driver weighting	Management -M-	Forages and biological control -F-	Infrastructure maintenance -I-	Production and raising livestock -P-	Reproduction y animal health -R-	Total
			(%AU)	3.338.289	22.859.102	10.521.894	27.235.387	2.899.055	
Raw milk	OBJ 1	60,70	69,8	2.329.128	15.948.822	7.341.137	19.002.161	2.022.674	46.643.922
Calfs 0–6 months	OBJ 2	1,50	1,7	57.557	394.122	181.412	469.576	49.984	1.152.650
Heifers 6–18 months	OBJ 3	6,52	7,5	250.180	1.713.119	788.537	2.041.089	217.263	5.010.187
Heifers 18–20 months	OBJ 4	9,48	10,9	363.758	2.490.854	1.146.524	2.967.718	315.897	7.284.751
Heifers 20–27 months	OBJ 5	8,80	10,1	337.666	2.312.185	1.064.283	2.754.844	293.238	6.762.216
		87,00	100,0	3.338.289	22.859.102	10.521.894	27.235.387	2.899.055	66.853.727

Table 8 Assigning cost to cost objects

Cost object		Cost of raw material	Cost of activities	Total cost
Raw milk	OBJ 1	58.667.843	46.643.922	105.311.765
Calfs 0–6 months	OBJ 2	5.121.121	1.152.650	6.273.771
Heifers 6–18 months	OBJ 3	6.780.930	5.010.187	11.791.118
Heifers 18–20 months	OBJ 4	1.160.825	7.284.751	8.445.576
Heifers 20–27 months	OBJ 5	3.448.967	6.762.216	10.211.183
Total		75.179.686	66.853.727	142.033.412

evidenced in Table 7, in which heifers 18–20 months had occupied the second place in the overall cost and Heifers 20–27 months had occupied the third place.

5 Conclusions

Agricultural companies in Colombia have high degrees of informality, particularly in the case of organizations in the livestock sector in the milk production chain (productive systems). Likewise, it evidenced in older formalized businesses concentration levels in financial accounting for the establishment of tax burdens, therefore the assessment of returns, the administrative process and generally adding value is not considered a major issue for managerial decision making. Thus, this study focused his attention on a proposal consistent with the needs of SMEs through the implementation of a system of activity-based costing.

Activity-based costing is understood as a process through which the purpose of adding value is intended, in which the consumption of resources is carried out by activities through a cause-effect relationship for the imputation of costs to both products as services. For this reason, the analysis and determination of the relevance of indirect costs at the operational and support levels is pertinent for making managerial decisions [17, 18].

For this was taken as reference for the proposed methodologies [13, 15] and [9]. In this, nine stages have been considered as follows (a) establishment of resources, (b) identification and assignment of activities, (c) determination of the main cost drivers, (d) establishment of cost objects, (e) development of the cost diagram, (f) characterization of drivers related to resource allocation and activities, (g) allocation of resources to activities, (h) designation of activities to cost objects and, (i) setting the total cost according to the cost objects. For which, financial information from July 2020 was considered, from a Colombian SME, Hacienda Agrícola Casa de Lata Ltda. This company is engaged in the field of dairy production and a leader in its sector for the agricultural and environmental farm management.

Thus the Activity Based Costing becomes a tool of management accounting that encourages proper decision-making, allowing accurately know the cost of cost objects related to milk production, and the activities to measurement of biological assets contemplated in IAS 41. The analysis of activities allows management of such agricultural enterprises eliminate those that do not create value and thus improve the determination of production costs, streamline processes and increase profitability ratios.

Within the limitations of the study is the absent empirical evidence about the cost systems in agricultural enterprises. Therefore, for the development of the cost system methodology, it was necessary to resort to general conceptual contributions and make adjustments according to the size and needs of the organization. Due to this, it is necessary to include research lines around cost systems and business management in this sector. This sector contributes 24.3% to livestock GDP in Colombia. In this way, this study becomes a contribution to the conception and formulation of cost systems in this sector of the economy, also, this study contributes significantly to the business and academic sector.

References

1. Berrío Guzmán D, Castrillón Cifuentes J (2011) Costos para gerenciar organizaciones manufactureras, comerciales y de servicios. Barranquilla, Ediciones Uninorte
2. Brimson JA (1991) Activity accounting: an activity-based costing approach
3. Bruggeman W, Everaert P, Anderson SR, Levant Y (2005) Modeling logistics costs using time-driven ABC: a case in a distribution company
4. Cooper R (1988a) The rise of activity-based costing: part one: what is an activity-based cost system. J Cost Manag 45–54
5. Cooper R (1988b) The rise of activity-based costing: part two: when do i need an activity-based cost system. J Cost Manag 41–48
6. Cooper R (1989a) The rise of activity-based costing: Part three: how many cost drivers do you need? and how should you select them. J Cost Manag 34–46
7. Cooper R (1989b) The rise of activity-based costing-part four: what do activity-based cost system look like? J Cost Manag 34–46.
8. Costa Marques M da C da (2012) Contribución del modelo ABC en la toma de decisiones: el caso universidades Cuadernos de Contabilidad. Cuadernos de Contabilidad, 13(33):527–544
9. Cuervo Tafur J, Osorio Agudelo JA, & Duque Roldán, M. I. (2013). Costeo basado en actividades ABC-Gestión basada en Actividades ABM. Ecoe Editores
10. Cuevas Villegas CF (2001) Contabilidad de costos. Enfoque gerencial y de gestión. Bogotá DC, Pearson Educación
11. Fedegan (2014) Foro Ganadería Regional Visión 2014 -2018. Cundinamarca
12. Fonseca-Cifuentes G, León-Castro E, Blanco-Mesa F (2020) Predicting the future price of a commodity using the OWMA operator: an approximation of the interest rate and inflation in the brown pastusa potato price. J Intell Fuzzy Syst, Preprint(Preprint), 1–11
13. Hicks DT (1998) El sistema de costos basado en las actividades (ABC) : Guía para su implantación en pequeñas y medianas empresas. México, Alfaomega
14. Johnson HT, Kaplan RS (1987) Relevance lost: the rise and fall of management accounting. Harvard Business School Press
15. Kaplan RS, Cooper R (1998) Cost and effect: Using integrated cost systems to drive profitability and performance. Harvard Business Press, Boston
16. Kuster Nieves C (2018) Contribución al estudio crítico de los sistemas de costos para los procesos agrícolas ovinos. Costos y Gestión 94:8–30
17. Lu T-Y, Wang S-L, Wu M-F, Cheng F-T (2017) Competitive Price Strategy with Activity-Based Costing-Case Study of Bicycle Part Company. Procedia CIRP 63:14–20

18. Mallo Rodríguez C, Jiménez Montañés MÁ (2009) Contabilidad de costes
19. Millán-Solarte JC, Sánchez-Mayorga X (2014) Modelo matricial para la asignación del costo utilizando activity basing cost. Entramado 10(2):144–155
20. Ministerio de Agricultura y Desarrollo Rural (2005) La competitividad de las cadenas agroproductivas en Colombia. Análisis de su estructura y dinámica (1991–2004)
21. Ministerio de Agricultura y Desarrollo Rural, M. de C. I. y T. (2017). Implementación política para mejorar la competitividad del sector lácteo Nacional.
22. Pérez Barral O, Alarcón Quinapanta M del R, Peñaloza López VL, Mora Rivera J beatriz (2017) Implementación de un procedimiento de gestión y costo basado en actividades a partir de la formación del capital intelectual contable para instalaciones turísticas. II Congreso: Ciencia, Sociedad e Investigación Universitaria.
23. Pérez Falco, G. (2007). *Modelo para la gestión de costos de la generación de energía en Cuba.*
24. Porter, M. E. (1995). *Ventaja competitiva: Creación y sostenimiento de un desempeño superior.* Grupo Editorial Patria.
25. Rodríguez García DM, Ruiz Torres JC (2016) Comparación del tratamiento contable y financiero de la NIC 41 agricultura-NIIF para pymes, sección 34: actividades especiales y el decreto 2649 de 1993. In Vestigium Ire 10(1):180–197
26. Ruiz-Torres JC, Fonseca-Cifuentes GP, García-Montaño CA (2020) Contabilidad de gestión en la ganadería bovina de leche en Colombia: caso Hacienda Agrícola Casa de Lata Ltda Management accounting in dairy cattle in Colombia :case Hacienda Agrícola Casa de Lata Ltda. Revista Espacios 41(15):24
27. Sánchez-Mayorga X, Millán-Solarte JC (2010) Propuesta para la implementación del costeo ABC en microempresas. Libre Empresa 7(1):107–119
28. Wang, M., & Han, D. (2013). Chapter 33. Measure Strategic Cost Based on Activity-Based Costing Method. In E. Qi, S. Jiang, & D. Runliang (Eds.), *The 19th International Conference on Industrial Engineering and Engineering Management. Engineering Management* (pp. 302–311).
29. Warren, C., Reeve, J., & Duchac, J. (2010). *Contabilidad administrativa.*
30. Zeuner, P. (2007). *Activity-based costing: Introducing process thinking into cost management.*
31. Zhang BX, Xie XT, Liang HR (2007) Activity cost based manufacturer cost and profit optimization. Proceedings of the Sixth International Conference on Machine Learning and Cybernetics, ICMLC 2007(4):2306–2310

Did Mexico Lose the Fight of Tomato Exports in the Times of USA Tariff, Facing the UMSCA?

María Eugenia Estrada-Chavira, Sylja Viridiana Guerrero-García, Maribel Rocío Hernández-Velázquez, and Guillermo Arredondo-García

Abstract The U.S.A. placed in 2019, a tariff of 17.5% in all the tomato imports that the U.S.A. made from Mexico. This tariff affected all the tomato market. It can be said that the compensatory quotas are responsible for curbing or discouraging an unfair practice in foreign trade by charging tariffs that apply to goods imported under conditions of price discrimination or a subsidy in the country of origin, which are considered unfair practices, all this as established in the Foreign Trade Law. This research had as objective to measure the demand price elasticity of tomato on two periods: with tariff and without tariff. The price elasticity was calculated using a logarithm regression, it was shown that the tomato price elasticity was elastic in both cases but higher with tariff. That means that if the tomato price changes the demand quantity falls down more than proportional. At the beginning the producers, exporters and both economies will lose, because the consumer price of the product increases. Concluding that Mexico does not have a strong international commerce structure, and as a recommendation should diversify its world market, facing out for instance Asian markets.

Keywords Tomato · Elasticity · Logarithmic regression · Tariff

M. E. Estrada-Chavira (✉) · S. V. Guerrero-García · M. R. Hernández-Velázquez · G. Arredondo-García
Universidad Politécnica de Tecámac, Tecámac, Mexico
e-mail: eugenia_estrada@uptecamac.edu.mx

S. V. Guerrero-García
e-mail: syljav_guerrero@uptecamac.edu.mx

M. R. Hernández-Velázquez
e-mail: maribel_hernandez@uptecamac.edu.mx

G. Arredondo-García
e-mail: guillermo_arredondo@uptecamac.edu.mx

E. León-Castro et al. (eds.), *Soft Computing and Fuzzy Methodologies in Innovation Management and Sustainability*, Lecture Notes in Networks and Systems 337,
https://doi.org/10.1007/978-3-030-96150-3_12

1 Introduction

The importance of tomato production worldwide can be exposed by its amount in production and consumption. It is considered that, worldwide it is one of the most demanded vegetables and is native to the lower Andes and in pre-Hispanic times was cultivated by the Aztecs in Mexico. Its name changes depending on the geographical area and to identify it is necessary to know the root of both names: red tomato and tomato (Agri-Food and Fisheries Information Service 2021).

The tomato was domesticated by Mesoamerican peoples 2,600 years ago, so in Mexico we enjoy a huge variety of both size and colors and of course flavors, some examples are: round (ball), saladette or guajillo, pear, cherry, or cherry, in addition for its quality positions us as world leaders in export [16].

Currently 100% of national requirements are supplied with domestic production. The main producers on 2019 were the states of Michoacán, Durango, Coahuila and Baja California [17]; Global imports have also increased and it is estimated that by 2030 it showed cumulative growth of 32.10% which has led to an increase in international market share of 25.11% in Mexican exports mainly to the United States because it absorbs 95% of Mexican exports and is the world's largest importer [14].

The tomato is one of the most important vegetables in Mexico and the world for its economic importance, as well as its nutritional value and gastronomic staple for families. According to that, it is important to measure the tomato price elasticity if a tariff is imposed.

The objective of this research is to give an academic explanation about the effects that brought the tax imposed by the government of the USA to the Mexican exports of tomato because of the dumping practice that Mexico was accused, during May 2019 to September, 19th, 2019. That is to measure the price elasticity of an historic period without tariff against a period whit the imposed tariff.

First, it is stated this brief introduction after that it is presented the literature review, where is found the academic basis for the study, focusing in the different theories that take place for the research. Second, is presented the method that it was used to measure the demand price elasticity and the effects that brought in the domestic and importing country. Last, it is presented the results or findings of the research, where is explained the values of the logarithm regression and the contribution in the study of tomato, giving some conclusions at the end.

2 Theoretical and Conceptual Framework

2.1 Tomato Suspension Agreement

On April 18, 1996, an anti-dumping investigation was promoted under section 732 of the Tariffs Act of 1930, to determine whether imports into the United States of fresh tomato from Mexico were sold or were likely to be sold less than of fair value.

On May 16, 1996, the United States International Trade Commission (ITC) announced that there was an anti-dumping, but there was a request from Mexican producers to postpone the preliminary determination, it was given a limit on October 28, 1996 for a final determination, the date on which the United States Department of Commerce and Mexican tomato producers signed the "Agreement to suspend the investigation" that entered into force on November 1, 1994.

The tariff fractions and subheadings according to the Harmonized System of Tariff Fractions that cover this suspension were: 0702.00.20, 0702.00.40, 0702.00.60 and 9906.07.01 to 9906.07.09, depending on the seasonal period in which they occur.

To avoid this 17.5% tariff on the sale price, the Agreement has been revised and renewed in 2002, 2008 and 2013 which was valid for 5 years.

However, at the request of the Florida producers, the President of the United States, Donald Trump and his work team, reviewed and published in the Federal Register of the United States Government's Journal dated May 13, 2019: "they reported that on February 06 of 2019 (ninety days before) that they intended to withdraw the 2013 Agreement, and that the U.S. Department of Commerce intended to issue its final determination 135 days after the effective date of withdrawal, this on September 19th, 2019, unless a new suspension agreement was effective before or on that day".

On May 7, 2019, the Tomato Suspension agreement was terminated. With what imposes a tariff of 17.5% on sale price to exports from Mexico to the United States of tomato. But The USA made a new agreement that started on September 19th, 2019, where Mexican tomato growers and exporters were exceeded to pay the 17.5% tariff, but included to review the tomato quality of the Mexican tomatoes sent, (Office of the FederalRegister (US), 2019).

3 From NAFTA to USMCA

International treats are important since they are those that promote the access of products from a foreign country to a national one, thus allowing the increase of commercialization, improving cooperation between countries, generating fair connectivity, promoting market security between the two countries leaving differences to join and this way always offer the best.

Tomato is one of the main export products to the United States of America, it is therefore important to maintain or improve the conditions of access of this product to the USA in the USMCA negotiations. According to [19], in May 2019, 148,436,059 kg of tomatoes were exported to the USA.

On August 16, 2017, negotiations began to modernize NAFTA in the city of Washington, with the governments of Mexico and the United States wishing to end the process in early 2018, before the elections that would be held in both nations. The negotiation process was terminated on September 30, 2018, where NAFTA was replaced by the Treaty between Mexico, the United States and Canada (T-MEC or USMCA). It was ratified and signed on December 10th, 2019. Put it into effect on July1st, 2020. Active.

3.1 Articles About Tomato

In foreign trade the models can be distinguished on the one hand free trade agreements (FTA) and other regional economic integration.

The purpose is to reach agreements that allow the movement of goods and services with a criterion of commercial freedom, avoiding tariff and non-tariff barriers to trade. Tariff rates tend to zero, although residual rates that are not merely protectionist can be established. Since in this case the FTA would be distorted. The content of the FTA can be extended to other areas of the economy, as a complement to the commercial objective, such as the homologation of professional titles or the establishment of financial and insurance entities [3].

Economic integration is a specific phenomenon of cooperation between autonomous units that must be inserted into a much broader and more meaningful process of growing relationship and independence between nations. This evolution, can be seen as a long process of gestation of a true political structure worldwide, has meant that exchanges between countries have increased and increased, therefore the dependence between one and the other [9].

The Agreement or economic integration among countries generates a commercial dependence that affects for instance Mexican vegetable exports in relation to the imposition of a 17.5% tariff percent applied on May 2019.

On the other hand, Sergio Esquer, Vice President of Fruits and Vegetables of the National Agricultural Council (CNA), suggested that this fact put at risk the export of other agricultural products to the United States such as eggplants, cucumbers, etc. (Mexico, 2019).

The tomato is the object of study by several researchers given the national and international problems in which it has been involved. The agricultural trade balance has resulted in a surplus in 2018, according to data from [15], Mexico's agri-food trade with the world has increased by 9% on an annual average since 1994. In 2018, 32,455 million were exported, and a historical surplus of 5,268 million dollars were obtained.

Some research articles referring to the problem of tomato in Mexico as: [8], who analyzes the competitiveness of exports of agricultural products from Mexico using the Revealed Comparative Advantage methodology, concluding that despite the fact that Production of agricultural products in Mexico has increased since the 1980s, it still does not have a comparative advantage worldwide in regards to the export of agricultural products, however when conducting the analysis by subgroups such as vegetables and fruits, there is a clear comparative advantage in these areas. Of which the USA absorbs 86% of the export of vegetables and fruits. However, despite the increase in the value of exports of agricultural products to Mexico, in the USA there is a decrease with respect to imports, despite having NAFTA in force.

In the other hand, with respect to demand price elasticity, [11], in his article about price flexibility and seasonal variations of major vegetables in Sindh Pakistan, found that demand was almost elastic for potato, tomato and onions while there was flexible price trend appeared in the production. Also, wrote that prices on agricultural products are mostly determined by supply and demand like (Vargas, 2004), the tomato demand elasticity in USA equals to the supply, is strongly affected by the before consumption and the consumer product price of the market.

Likewise, the process of exporting tomatoes from Mexico to the USA is of the most importance considering whether this process is profitable for Mexican exporters, so that [21], in his dissertation, mentions that the export of tomato brings foreign exchange to the country and determines that the demand trend is greater than the supply that exists in Mexico, so the tomato export process is profitable.

3.2 *Import Tariff, Demand, Demand Price Elasticity, the Log Linear Model*

3.2.1 Import tariff

According to, [13], import barriers can come in many forms, and countries have been increasingly creative in establishing new barriers over time. The most visible and trade barrier is the simple import tariff (tax). It can be a fixed amount per unit or a fixed percentage of imported good's price.

The effects of an import tariff under free competition, the economy face´s a perfect elastic supply curve from the rest of them (at P_w in Fig. 1). When this function is added to the domestic supply curve, the effective supply curve, S_e, is the darkened line that includes the domestic supply curve below pw and is horizontal at Pw. In equilibrium (autarkic price is above P_w), the country produces at Q_p, consumes at Q_c, and imports Q_c–Q_p. The importing price is P_d, which is identical to the world price. If the country imposes a specific of t per unit, the domestic price will move up to P_{w+} t

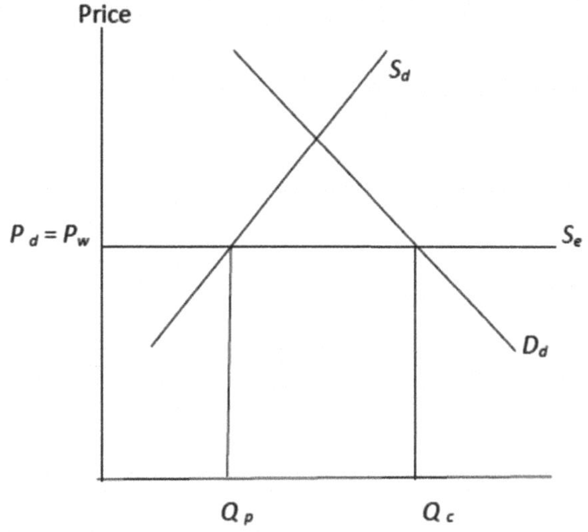

Fig. 1 International trade at autarky. *Source* [13]

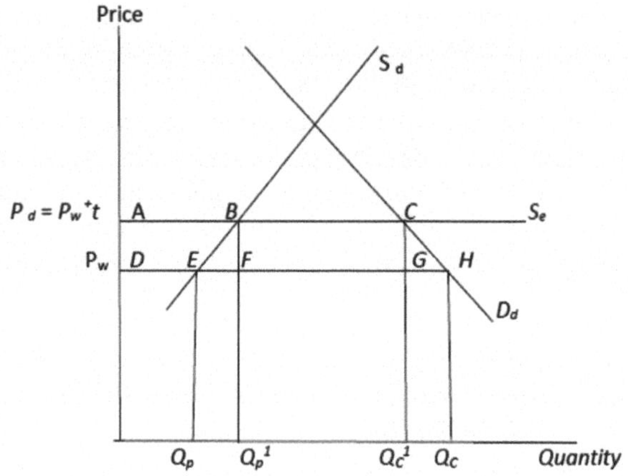

Fig. 2 International trade with an importing tariff. *Source* [13]

as long as the autarkic price is above P_{w+} t. Under perfect competition, importers are not making positive economic profits, so they must pass on the tariff fully to the importing country. Producers react to the new price and increase production to Q'_p, consumers react by decreasing consumption to Q'_c and imports fall to Q'_c–Q'_p.

In the next figure it can be seen the gains to producers, consumers, and the government. Domestic producers benefit from the tariff because they receive a higher price for their output, and this higher price also encourages them to produce more (area ABED. The government receives t for every unit imported, so it has revenue of BCGF. Consumers are the losers because they suffer an increase in price for the good, which also encourages them to consume less. The loss in consumer surplus is ACHD, an area that exceeds the other two areas combined.

The country's welfare falls by two triangles BFE and CHG. BFE called production loss because it measures inefficiencies caused by increased production by domestic producers. The tariff has encouraged production of the good by domestic producers, whose marginal cost is above that of the more efficient international producers. CHG, called the consumption loss because it measures inefficiencies caused by reduced consumption (Fig. 2).

3.2.2 Market Demand

The market demand function of a given good (Xi) is the sum of each individual's demand for that good:

$$X_i = \sum_{j=1}^{m} X_{ij} = D_i(P_1, \ldots\ldots.P_m, I_1 \ldots\ldots I_m) \tag{1}$$

The market demand function for the Xi good is derived from the Pi-varying demand function while keeping all of Xi's other determinants constant, if each individual's demand curve has a negative slope, this market demand curve will also have negative slopes.

3.2.3 Demand Price Elasticity

Changes in the price of a good (P) will cause changes in the quantity acquired (Q) and the elasticity of the demand price ($e_{Q,P}$) shall be defined as follows:

$$e_{Q,P} = \frac{Percentage\,change\,of\;Q}{Percentage\,change\,of\;P} = \frac{\partial Q}{\partial P} * \frac{P}{Q} \quad (2)$$

Shows how Q changes (in percentage terms) in response to a percentage change of P. Since ∂Q/∂P is usually negative (i.e. P and Q, move in the opposite direction, except in the case of Giffen's paradox). $e_{Q,P}$ is usually negative.

A distinction is often made between the values of $e_{Q,P,}$ depending on whether it is greater than, less than, or equal to 1. A curve is elastic >1, if in the event of an increase in price it causes a more than proportional reduction in the quantity. For a unit elasticity curve, the price increase and quantity reduction have the same proportional magnitude. For an inelastic curve, the price increases more than proportionally than the drop in the amount. If a curve is elastic the price will greatly affect the amount. If the curve is inelastic, the price will have no effect on the quantity demanded. Likewise, goods are classified by their price elasticity: elastic, unit elasticity and inelastic [10].

3.2.4 The Log Linear Model

Considering the following model, known as the exponential regression model:

$$Y_i = \beta_1 = X_i^{\beta_2} e^{u_i} \quad (3)$$

which may be expressed alternatively as:

$$lnY_i = ln\beta_1 + \beta_2 lnX_i + u_i \quad (4)$$

where:

$\propto = ln\beta_{1,}$ this model is linear in the parameters α and β2, linear in the logarithms of the variables Y and X and can be estimated by OLS (Ordinary Least Squares) regression. Because of the linearity such model is called log–log model. In the assumption of the classical linear regression model are fulfilled, the parameter of

$$lnY_i = ln\beta_1 + \beta_2 lnX_i + u_i. \quad (5)$$

can be estimated by OLS method letting $Y_i^* = \propto + \beta_2 X_i^* + u_i$ where, $Y_i^* = X_i^*$ and $X_i^* = lnX_i$. The OLS estimators α and β_2 obtained Will be the best linear unbiased estimatord of α and β_2, respectively.

One attractive feature of the log–log model, which has made it popular in applied work, is that the slope coefficient β2 measures the elasticity of Y with respect to X, that is, the percentage change in Y given (small) percentage change in X. Thus, if Y represents the quantity of a commodity demanded and X its unit Price, β2 measures the Price elasticity od demand, a parameter o economic interest [5].

4 Method

The method applied in this research was a mix of quantitative method and qualitative method. Qualitative method to investigate all the background around the topic. And quantitative method to calculate the price elasticity of the demand as follows:

Export data in the period of the tariff was taken and the demand price elasticity of tomato was calculated through a logarithm regression, with a model known as the exponential regression model:

$$Y_i = \beta_1 X_i^{\beta_2} e^{u_i} \quad (6)$$

where:
Y = Exports from Mexico to USA.
X = Price.
alternatively expressed as:

$$lnY_i = ln\beta_1 + \beta_2 \ln X_i + u_i. \quad (7)$$

We can write it as:

$$lnY_i = \propto + \beta_2 lnX_i + u_i \quad (8)$$

where $\alpha = \ln \beta_1$, this model is linear in the parameters α and β_2, linear in the logarithms of the variables Y and X and can be estimated by a regression.

The slope coefficient β_2 measures the elasticity of Y with respect to X, that is, the percentage change in Y for a given (small) percentage change in X.

Applied to this research first, it was obtained the tomato exports from Mexico to USA, quantity and price in dollars from 2014 to 2019th. Then it was obtained the price per unit dividing the quantity by money paid.

After that, the price in dollars was multiply by the exchange rate given on [2], historic series, 2014–2019, obtaining the tomato price by kg.

Consecutively, it was taken the period from May 2014th to August 2019th. It was worked the first logarithm regression in the Excel software with data from May–August 2014th to 2018th. Taken tomato exports to USA from [19] and price converted into thousands of tons, obtaining the first equation.

For the second log equation in the Excel software, the process was the same as before only that the period from May to August 2019th was taken, using the following data (Tables 1 and 2):

Table 1 Exports from Mexico to USA May 2014th–August 2018th

Month	X	Y
May-14	158.30	116.71
Jun-14	158.82	110.79
Jul-14	134.96	85.04
Aug-14	109.69	66.02
May-15	169.47	120.41
Jun-15	176.38	128.15
Jul-15	130.93	103.41
Aug-15	129.24	96.06
May-16	251.27	138.41
Jun-16	212.22	113.77
Jul-16	149.25	80.23
Aug-16	157.31	85.15
May-17	302.30	153.38
Jun-17	218.17	113.68
Jul-17	201.51	91.60
Aug-17	184.32	82.69
May-18	308.99	150.53
Jun-18	286.75	135.60
Jul-18	229.04	110.91
Aug-18	213.98	96.48

Source Own elaboration with data from SIAVI and Banco de Mexico, 2014–2018

Table 2 Exports from Mexico to USA May 2019th–August 2019th

Month	X	Y
May-19	148.436059	2680.43604
Jun-19	130.723235	2772.16835
Jul-19	121.523595	2625.21764
Aug-19	102.300689	1828.56000

Source of information Own elaboration with data from SIAVI and Banco de Mexico, 2019th

5 Results

Tomato is one of the main agri-food products for Mexican consumption, being the most cultivated vegetable in both greenhouse and open space, where Mexican exports, aided by national production, generate a direct and significant impact on the national economic fluctuation, giving as result the introduction a diverse amount of foreign exchange to the country.

The national production of tomatoes, whether red or green, has grown from 1993 to 2017 as a result of the increase in constant sales to the United States; resulting in the large participation by the state of Sinaloa where it has historically proven itself as the main producer, however other entities have been gaining participation thanks to the various production and financing plans currently available, which also determines the productivity index that will be reflected with the exports that were made.

Analyzing the tomato production in Mexico for 2019, the accumulated production was 1 million 902 thousand 276 tons, 23,390 thousand of tons less than the year before. And by Federal Entity, the states with the highest production were: Michoacan 23,797 tons (12.6%), Durango with 19,398 (60.8%), Coahuila 11,120 (10.2%) and Baja California 9,616 tons (10.4%). On the other hand, Jalisco, Nayarit, Sonora and San Luis Potosí, their production was down, reporting: 40,126 tons (29.2%), 24,909 (98.5), 22,194 (53.7%) and 14,501 tons (4.4%) respectively.

The production by region in 2019, was central west 42.7%, center 20.9%, northeast 20.6%, northwest 11.1% and south-southeast 4.7% of the national total, [17].

Mexican saladette tomato producers are facing difficult situations, as getting their product across the border is a difficult job because the U.S. government wants to inspect almost 100 percent of the vegetable before crossing its border.

Due to the collection of compensating quotas of 17.5% on Mexican tomato exports to the United States, the volume of shipments fell 20%. When conducting an average analysis of shipments of saladette tomato in the first two weeks of May was just over 15 thousand tons per week, while in the first of June the volume fell to 12 thousand tons a week, for a 20 percent decline, according to the data reported by the Mexican Association for Protected Horticulture, [1].

According to the data proportioned by the AMHPAC, 95% of the saladette tomato for exportation is destined to the United States, and 5% to Canada. Providing also, all the internal consumption, by which the saladette tomato sees itself affected due to its main market being the United States.

And being before of a 17.56% tariff, it makes the entire export process more expensive, which implies the payment of higher costs, on one hand affecting Mexican producers where the production and export of tomato is discouraged and on the other hand the product becomes more expensive for the American consumer. So, Mexico must diversify its market in the face of the threat of job losses and the negative effect it can have on the economy.

Maintaining the business relationship with the United States is of vital importance to Mexico as it is the main market, if this would die off, there would be other countries

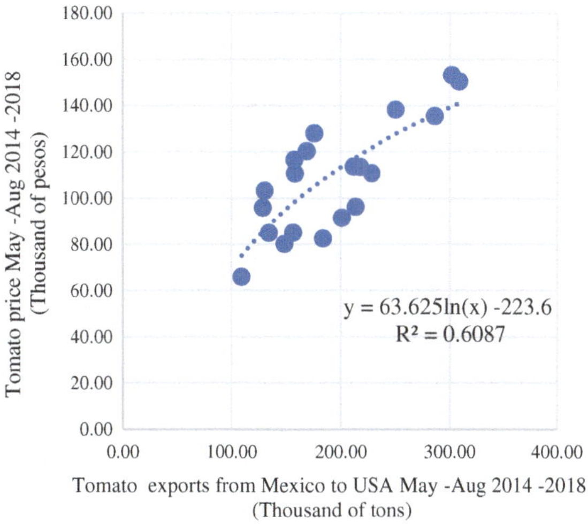

Fig. 3 Logarithmic regression tomato elasticity May-Aug 2014–2018. *Source* Proper elaboration with data from SIAVI and Banco de Mexico

that took advantage of the situation to start business with Mexico. However, options are analyzed as an area of opportunity in the industrialization of the tomato, thus opening commercial relations with other countries [22].

Analyzing two periods: without tariff May–Aug 2014–2018 and with tariff May–Aug 2019, using a logarithm regression it is stated that the tomato price elasticity is elastic, that means than if the price is raised the quantity is reduced more than the proportional quantity. The logarithm regression gave as a result an elastic price demand of tomato of 63.62 more than 1 with a reliability of 61% during the period of May to August 2014–2018. Comparing the same period in 2019 it was found that the elasticity price tomato demand was elastic (2400) with a reliability of 74%, as follows:

In both equations with tariff and without tariff it is stated that the tomato price elasticity is higher that one, however the tariff increased all the export process, having to pay high costs, affecting Mexican producer and exporters, but also, as consequence the tomato price product was more expensive to the USA consumers, due to during May to August 2019th, period where USA commerce committee imposed a 17.56% tariff to Mexican tomato exports, it were down 40%. For that reason, it is recommended that Mexico should face out other international markets for its products, even though they are far away (Figs. 3, 4).

It was found that there are two articles with related results, first [11], found that the tomato demand is almost elastic in some products as potato, tomato, and onions. Also [20], said that the tomato demand elasticity is USA equal to the supply, was strongly affected by the before consumption and the consumer product price of the market.

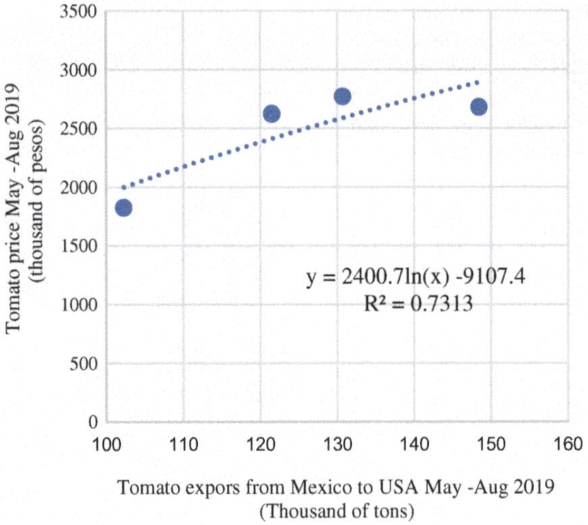

Fig. 4 Logarithmic regression tomato elasticity. *Source* Proper elaboration with data from SIAVI and Banco de Mexico

6 Conclusions

1. The tomato is an elastic product with an elasticity more than one, indicating that if there is an increase in the price the demanded amount will decrease more than proportional.
2. The results for the tomato elasticity were elastic in both logarithm regressions: without tariff for the period May–Aug 2014–2018 and with tariff May–Aug 2019. Also, with the imposition of the tariff the amount demanded was decreased and the US consumer price increased.
3. Given the uncertainty that Mexico was experiencing with the export of tomato to the USA and that the USA intends to verify every shipment of tomato that enters its territory results in economic losses since the product in question is very perishable and with it decreases the entry of foreign exchange into the country. Therefore, it is recommended the diversification of international markets such as newly expanding markets like the Asian market.

References

1. AMHPAC (28 de junio de 2019) Asociacion Mexicana de Horticultura Protegida. Recuperado el 8 de julio de 2019, de Agricultura Protegida en Mexico: http://www.amhpac.org/es/index.php/informacion/horticultura-en-mexico
2. Banco de México (Mayo de 2020) Tipo de cambio pesos/dólar. Obtenido de https://www.banxico.org.mx/tipcamb/llenarTiposCambioAction.do?idioma=sp
3. Coll P (2015) Manual de gestión aduanera. Marge, Barcelona
4. FIRA (2019) Panorama agroalimentario. México
5. Gujarati D (2004) Basic Econometrics. The McGraw Hill Companies

6. Krugman PR (2007) Introducción a la Economía, Microeconomía. Reverté S.A. Obtenido de Int, Barcelona
7. Krugman PR (2007) Introducción a la Esconomía, Microeconomía. Barcelona, España, Reverté SA
8. Málaga JE (2010) La competitividad de México en la exportación de productos agrícolas. Revista MExicana de Agronegocios, 295–309
9. Muns J (2005) Lecturas de integración económica. La Unión Europea. Universitat de Barcelona, Barcelona
10. Nicholson W (2008) Teoría Microeconómica. Principios básicos y ampliaciones. CENGAGE Learnings
11. Noonari e. a. (2015) Price flexibility and seasonal variations of major vegetables in Sindh Pakistan. J Food Process Technol 524
12. Office of the FederalRegister (US) (13 de Mayo de 2019). Obtenido de https://www.federalregister.gov/documents/2019/05/13/2019-09786/fresh-tomatoes-from-mexico-termination-of-suspension-agreement-rescission-of-administrative-review
13. Reed MR (2001) International Trade in Agricultural Products. Prentice Hall, New Jersey
14. SAGARPA. (2017). Planeación Agrícola Nacional 2017–2030. Obtenido de https://www.gob.mx/cms/uploads/attachment/file/257077/Potencial-Jitomate.pdf
15. SAGARPA (2018) Análisis de la Balanza comercial Agroalimentaria. Obtenido de https://www.gob.mx/cms/uploads/attachment/file/395303/Balanza_Comercial_Agropecuaria_y_Agroindustrial_enero_2018.pdf
16. SAGARPA (18 de Abril de 2021) Gobierno de México. Obtenido de Secretaría de Agricultura y Desarrollo Rural: https://www.gob.mx/agricultura/articulos/el-jitomate-hortaliza-mexicana-de-importancia-mundial?idiom=es
17. Secretaría de Agricultura y Desarrollo Rural (Febrero de 2020) Boletín mensual de producción tomate rojo (jitomate). Obtenido de https://www.gob.mx/cms/uploads/attachment/file/539442/Bolet_n_avance_producci_n_tomate_rojo_enero_2020.pdf
18. Servicio de información Agroalimentaria y Pesquera (18 de Abril de 2021) Servicio de información Agroalimentaria y Pesquera. Obtenido de https://www.gob.mx/siap/articulos/tomate-rojo-o-jitomate-como-lo-llaman-donde-radicas#:~:text=El%20tomate%20rojo%20cerr%C3%B3%202016%20con%203.3%20millones%20de%20toneladas&text=Es%20originario%20de%20los%20bajos,nombres%3A%20tomate%20rojo%20y%20jitomate.
19. SIAVI (2014–2019) Obtenido de http://www.economia-snci.gob.mx/
20. Vargas O a (2004) Un modelo econométrico del mercado del jitomate (Licopersicon esculentum Mill) en México, 1970–1994. Comunicaciones En socioeconomía, Estadística e informática, 8(2):115–133
21. Vázquez VA (08 de 08 de 2015) Repositorio Digital. Universidad Autónoma Agraria Antonio Narro (CID-UAAAN). Obtenido de http://repositorio.uaaan.mx:8080/xmlui/handle/123456789/6853
22. Verdín R (16 de abril de 2019) El debate. Recuperado el 8 de julio de 2019, de Tomateros, a la expectativa de los planes de Trump: https://www.debate.com.mx/sinaloa/Tomateros-a-la-expectativa-de-los-planes-de-Trump-20190416-0147.html

Tourism Competitiveness Theory Perspectives Through the Application of Counter-Expertise Methods

Miriam Edith Pérez-Romero, Martha Beatriz Flores-Romero, and Víctor G. Alfaro-García

Abstract The objective of this study is to compare the opinions of two groups of experts regarding the relationship between causes and effects of the touristic destination competitiveness, through the use of fuzzy techniques. The methodology consisted of the formation of two groups of experts, a group of experts who investigate the touristic activity and other group of experts who work in the tourism sector; both groups carried out the evaluation of cause-effect relationships using a forgotten effects theory; then the results were compared using the counter-expertise technique; finally, it was grouped in a frequency distribution to facilitate its interpretation and analysis. The application of counter-expertise showed the distance between the opinions of the two groups of experts; with values ranging from 0.00% to 18.75% in direct incidences and values from 0.00% to 16.25% in indirect incidences. The frequency distribution showed that the data is highly skewed to the right, which indicates that most of the results are in the lowest values, that is, the distance between the opinions of the two groups of experts is small in most cases, and therefore, the information provided by the experts appears to be valid. In both, direct incidents and forgotten effects, appear to be environmental commitment cause with the biggest difference of opinion between the experts.

Keywords Tourist destination competitiveness · Destination competitiveness · Fuzzy logic · Expertons · Expertise

1 Introduction

Tourism is an important contributor to employment and to Gross Domestic Product in many countries and regions [29], it is considered one of the most solid alternatives to achieve the economic development and social well-being of nations around the world, but especially for those that are developing or less developed [26], infrastructures

M. E. Pérez-Romero (✉) · M. B. Flores-Romero · V. G. Alfaro-García
Universidad Michoacana de San Nicolás de Hidalgo, Facultad de Contaduría y Ciencias Administrativas , Morelia, Mexico
e-mail: miromero@umich.mx

E. León-Castro et al. (eds.), *Soft Computing and Fuzzy Methodologies in Innovation Management and Sustainability*, Lecture Notes in Networks and Systems 337,
https://doi.org/10.1007/978-3-030-96150-3_13

created for tourism purposes contribute to local development, while jobs created or maintained can help counteract industrial or rural decline [23]. Since the 80s and 90s, tourism has become a generalized economic phenomenon worldwide, a situation that has led to the search for competitive models that explain what makes a destination more attractive than other [6], however, new market trends in tourist destinations and their own life cycle, lead to the need to plan and undertake renovation processes [28], in addition, the resources that today position a destination as the most competitive may not have any importance in the future [18].

Competitiveness is an ability of the country, region, or an enterprise to create wealth greater than the competitors and the competitive system on the global market [21]. The tourism sector has shown a growing interest in competitiveness, mainly due to the diversification that has occurred in this activity, there are destinations competing for more tourist arrivals or more tourist spending, indicators that are reflected in the economic prosperity of its residents [6]. Then, to turn tourism into an element of economic development, tourism competitiveness is shown as a necessary and key tool [17, 22, 26], it offers to countries the possibility of maintaining their leading position in tourism [8] or gain a favorable position [13].

Research on tourism competitiveness has been booming in recent years, one of its areas of interest is the identification of the factors that affect it [26], so the theory around this concept involves multiple factors and it has an effect in multiple situations. From the above, it is possible to establish causes and effects that competitiveness brings to a tourist destination, however, the question that arises is if experts who investigate the touristic activity and experts who work in the tourism sector have the same opinion about causes and effects of destination competitiveness and its relationship?

The objective of this study is to compare the opinions of two groups of experts regarding the relationship between causes and effects of the tourist destination competitiveness, through the use of fuzzy techniques. This paper is structured in five sections. The first section shows the importance of the tourism competitiveness and the objective of the research. The second section, named preliminaries, mentions the theory of expertons, forgotten effects theory and the expertise and counter-expertise methods. The third section presents the methodology that has been followed to achieve the research objective, it consisted of the formation of two groups of experts who evaluated cause-effect relationship and then the fuzzy tools described in the previous section were applied. The fourth section presents the results, where few discrepancies are observed among the experts, mainly on the subject of environmental commitment and sustainable development. And, finally, in the fifth section the conclusions are shown.

2 Preliminaries

In this section, we present the fuzzy logic tools used for the development of this research work. Working with models derived from fuzzy logic allows an inherent treatment of uncertainty in the analysis of the phenomenon under study [3] by including quantitative data and the human perspective [11]. Furthermore, studies with a fuzzy-oriented standpoint have proven efficacy while dealing with complex phenomena [5].

2.1 Theory of Expertons

An experton is an extension of the concept of a random set (probabilistic set) where an interval of probabilities is obtained for each level $\alpha \in (0,1)$ from valuation of several experts [10], it arises as a result of a procedure of aggregation of opinions of various experts regarding the degree of veracity of a statement and offers the percentage of experts who agree that the veracity of the statement is at least the given value [14]. Expertons ease group decision-making by providing quantitative data retrieved directly from the dialog maintained with several experts surrounding a certain phenomenon [4].

The theory of expertons is defined as follows [9]:

Definition 1 Let D be a finite set of certain alternatives. A group of q experts is asked to express their subjective opinion on each element of D in the form of a confidence interval:

$$\forall d \in D : \left[\alpha^{j*}(a)\right], \left[\alpha^{j^*}(a)\right] \subset [0, 1]$$

where $\subset$ is an inclusion set and j is the expert. The property is monotonous, not horizontally and vertically down, with a distribution network configuration Λ and $\vee$, as its shown in the following:

$$\forall \alpha \in [0, 1] a_1(\alpha) in [a_1(\alpha), a_2(\alpha)]$$

$$\forall \alpha, \alpha' \in [0, 1] : \left(\alpha' > \alpha\right) \Rightarrow (a_1(\alpha) \leq a_1\left(\alpha'\right), a_2(\alpha) \leq a_2\left(\alpha'\right))$$

$$(\alpha) = 0 \Rightarrow (a_1(\alpha) = 1, a_2(\alpha) = 1)$$

2.2 Theory of Forgotten Effects

The theory of forgotten effects is an extension of the applications of fuzzy logic [30], allows an approach to the objective of globalizing direct and indirect incidents between a set of causes and effects [9], suggests that all events, phenomena and facts that surround people start from some type of system or subsystem, therefore, they are subject to some type of cause—effect relationship, with the possibility that some relationships will not be voluntarily or involuntarily directly perceived [24], in this way, it is not uncommon to find hidden effects of the analyzed phenomena due to the incidence of other effects or outcomes [1]. This tool quantifies the opinion of experts in order to provide elements for better decision-making in the design of strategies, actions and public policies, to mention a few examples [2].

The theory of forgotten effects is defined as follows [9]:

Definition 2 The existence of two sets is assumed $A = \left\{\frac{a_i}{i} = 1, 2, \ldots, n\right\}$ and $B = \left\{\frac{b_j}{j} = 1, 2, \ldots, m\right\}$. It is conjectured that an incidence of a_i on b_j if the value of the membership feature $\left(a_i, b_j\right)$ is evaluated in a range [0,1], that is to say:

$$\forall(a_i, b_j) \Rightarrow \mu(a_i, b_j) \in [0, 1]$$

The set of pairs of elements evaluated is the direct incidence matrix (M), it shows the cause-effect relationship in different degrees caused by the corresponding set A (causes) and set B (effects), as presented below:

$$M = \begin{array}{c|c|c|c|c|c|} & b_1 & b_2 & b_3 & \cdots & b_m \\ \hline a_1 & \mu_{a_1b_1} & \mu_{a_1b_2} & \mu_{a_1b_3} & \cdots & \mu_{a_1b_m} \\ a_2 & \mu_{a_2b_1} & \mu_{a_2b_2} & \mu_{a_2b_3} & \cdots & \mu_{a_2b_m} \\ a_3 & \mu_{a_3b_1} & \mu_{a_3b_2} & \mu_{a_3b_3} & \cdots & \mu_{a_3b_m} \\ \vdots & \cdots & \cdots & \cdots & \cdots & \cdots \\ a_n & \mu_{a_nb_1} & \mu_{a_nb_2} & \mu_{a_nb_3} & \cdots & \mu_{a_nb_m} \end{array}$$

Now we proceed to the detection of the forgotten effects, for this it is assumed that there is a third set of elements, called C, expressed as follows, $C = \left\{\frac{C_k}{k} = 1, 2, \ldots, k\right\}$. This set consists of elements that are effects of set B and has an incidence matrix expressed as follows:

N =

	c_1	c_2	c_3	…	c_k
b_1	$\mu_{b_1c_1}$	$\mu_{b_1c_2}$	$\mu_{b_1c_3}$	…	$\mu_{b_1c_k}$
b_2	$\mu_{b_2c_1}$	$\mu_{b_2c_2}$	$\mu_{b_2c_3}$	…	$\mu_{b_2c_k}$
b_3	$\mu_{b_3c_1}$	$\mu_{b_3c_2}$	$\mu_{b_3c_3}$	…	$\mu_{b_2c_k}$
⋮	…	…	…	…	…
b_m	$\mu_{b_mc_1}$	$\mu_{b_mc_2}$	$\mu_{b_mc_3}$	…	$\mu_{b_mc_k}$

There are two incident matrices, M and N, both have element B in common, this relationship is expressed:

$$M \subset AxB \; and \; N \subset BxC$$

Next, the max–min operator (convolution) is used to detect the relationship between the sets A and C using B, the result is a new incidence matrix expressed by:

$$M \circ N = P$$

$$P \subset AxC$$

This new relationship is formulated like this:

$$\forall(a_i, c_z \in AxC)$$

$$\mu(a_i, c_z)M \circ N = \forall_{bj}\left(\mu_M\left(a_i, b_j\right) \wedge \mu_N\left(b_j, c_z\right)\right)$$

The matrix that results from the max–min operation is:

P =

	c_1	c_2	c_3	…	c_k
a_1	$\mu_{a_1c_1}$	$\mu_{a_1c_2}$	$\mu_{a_1c_3}$	…	$\mu_{a_1c_k}$
a_2	$\mu_{a_2c_1}$	$\mu_{a_2c_2}$	$\mu_{a_2c_3}$	…	$\mu_{a_2c_k}$
a_3	$\mu_{a_3c_1}$	$\mu_{a_3c_2}$	$\mu_{a_3c_3}$	…	$\mu_{a_2c_k}$
⋮	…	…	…	…	…
a_n	$\mu_{a_nc_1}$	$\mu_{a_nc_2}$	$\mu_{a_nc_3}$	…	$\mu_{a_nc_k}$

The matrix P defines the causal relationships between the elements of the sets A and C, this in the intensity or degree that B allows.

2.3 The Expertise and the Counter-Expertise

Expertise is the process of consulting a specific group of experts in relation to a defined topic, with the purpose of limiting uncertainty ([15]) and counter-expertise establishes a comparison or reaffirms the results between the estimates of two or more groups of experts, it can become an effective aid for decision-making, since in a certain way, it mitigates subjectivity, and like expertise, reduces uncertainty in most cases [7].

The counter-expertise is calculated by the next formulation:

$$d(E[A], E[B]) = \frac{|E_I(A) - E_I(B)| + |E_D(A) - E_D(B)|}{2}$$

where:

E (A) is the mathematical expectation of the experton A;

E (B) is the mathematical expectation of the experton B.

Please note, that the mathematical expectation is directly obtained by adding left (resp. right) bounds except the level 0 and dividing the result by 10 [10], in this case only one of the mathematical expectations (left or right) was used since the information was had in individual data.

3 Methodology

The methodology that has been followed consists of a review of the literature to identify causes and effects of tourism competitiveness; see Table 1.

Subsequently, two groups of experts were formed, a group of experts who investigate the tourist activity and a group of experts who work in the tourism sector; next, the experts were asked to evaluate the cause-effect, cause-cause and effect-effect relationships using an endecadary scale such as that proposed by Kaufmann [9].

Following, the expertons were obtained, through the next 3 steps: the calculation of the absolute frequencies, the calculation of the relative frequencies and the accumulation of relative frequencies [16, 15, 14, 27]. The theory of forgotten effects was applied to identify indirect impact between the relationships cause-effect, for each group of experts we constructed M matrix, A matrix and B matrix; immediately the matrices were convolved and then the forgotten effects matrix was calculated.

Next, the counter-expertise was obtained, by means of the average of the difference in the mathematical expectations of the experts obtained, as presented in the preliminaries section, and finally, the results were grouped in a frequency distribution to facilitate their interpretation and analysis.

Table 1 Causes and effects of tourism competitiveness

Causes	Effects
a. Environmental commitment	A. Sustainable development
b. Legacy resources	B. Tourist demand
c. Resources created	C. Customer satisfactions
d. General infrastructure	D. Customer loyalty
e. Quality of service	E. Tourist spending
f. Accesibility	F. Profitability
g. Hospitality	G. Prosperity
h. Destination management	H. Arrival of tourists
i. Ties to the market	I. Economic growth
j. Location	
k. Security	
l. Price	
m. The microenvironment	
n. The macroenvironment	
o. Demand conditions	

Source [19]

4 Results

The results show that there are 15 main causes that make a tourist destination competitiveness, in turn, there are 9 effects that competitiveness brings to the destination, thus generating a total of 135 cause-effect relationships. The results obtained from the counter-expertise between experts are shown in Tables 2, 3, 4, 5. Table 2 shows the counter-expertise between the two groups of experts in terms of the evaluation of the relationship between causes and effects of tourist competitiveness, these are direct incidents. The application of counter-expertise to the 135 cause-effect relationships, showed that the distance between the opinions of the two groups of experts has a range of 0.00% to 18.75%.

The frequency distribution of direct incidences showed that the data is highly skewed to the right, indicating that most of the results are at the lowest values, that is, the distance between the opinions of the experts is small, therefore, validates the information provided by the experts, see Fig. 1. Regarding the results found in the tail of the frequency distribution, the greatest difference between the experts arises in the relationship between environmental commitment (cause)—tourism demand (effect), the distance in this case being 18.75%, followed by environmental commitment (cause)—tourist spending (effect) with a value of 17.50% and hospitality (cause)—sustainable development (effect) with a value of 16.25%.

Table 2 Counter-expertise between experts in the relationship cause-effect

Cause/Effect	A (%)	B (%)	C (%)	D (%)	E (%)	F (%)	G (%)	H (%)	I (%)
a	3.75	18.75	15.00	0.00	17.50	1.25	2.50	2.50	6.25
b	10.00	5.00	6.25	2.50	11.25	5.00	11.25	5.00	5.00
c	3.75	0.00	2.50	0.00	3.75	7.50	5.00	5.00	2.50
d	5.00	0.00	2.50	5.00	2.50	1.25	1.25	1.25	3.75
e	13.75	1.25	1.25	1.25	7.50	1.25	5.00	1.25	7.50
f	15.00	2.50	1.25	0.00	1.25	1.25	5.00	7.50	2.50
g	16.25	1.25	0.00	1.25	5.00	6.25	3.75	1.25	3.75
h	12.50	1.25	10.00	3.75	1.25	2.50	3.75	1.25	1.25
i	5.00	0.00	0.00	3.75	8.75	1.25	5.00	1.25	5.00
j	6.25	0.00	1.25	2.50	7.50	0.00	1.25	3.75	0.00
k	13.75%	2.50	0.00	0.00	0.00	0.00	3.75	5.00	1.25
l	13.75	6.25	2.50	1.25	8.75	5.00	2.50	0.00	3.75
m	6.25	0.00	2.50	2.50	3.75	1.25	1.25	1.25	3.75
n	2.50	5.00	5.00	2.50	2.50	1.25	2.50	1.25	3.75
o	1.25	8.75	5.00	8.75	1.25	5.00	3.75	3.75	11.25

Source Own elaboration

In Table 3, the counter-expertise between the two groups of experts is presented regarding the evaluation of the relationship between causes of tourism competitiveness. The highest counter-expertise, with a value of 26.25%, occurs in the relationship between General Infrastructure and Location, again in this relationship it was the experts who have investigated the tourist activity who evaluated in a higher way, with respect to the evaluations made by the experts who work in the tourism sector. In this matrix there is a diagonal line of the relationships in which the distance is zero, however, this occurs because the evaluation of the cause-cause relationship when it is the same cause has a value of 1.

In Table 4, the counter-expertise between the two groups of experts is presented regarding the evaluation of the relationship between effects of tourism competitiveness. The greatest counter-expertise, with a value of 25.00%, occurs in the relationship that exists between Customer Loyalty and Sustainable Development, on this occasion it was also experts who have investigated the tourist activity who evaluated the relationship higher. Observe that in this matrix there is also a diagonal line of relations in which the distance is zero, for the same reasons as in the previous table.

Finally, in Table 5, the counter-expertise between the two groups of experts is presented regarding the evaluation of the forgotten effects of tourism competitiveness. The application of counter-expertise to the forgotten effects showed that the distance between the opinions of the two groups of experts has a range of 0.00% to 16.25%. The frequency distribution of indirect incidents showed that the data is highly skewed to the right, indicating that most of the results are at the lowest values, that is, the distance between the forgotten effects of the experts is small, therefore, validates the

Table 3 Counter-expertise between experts in the relationship cause-cause

Cause/Cause	a (%)	b (%)	c (%)	d (%)	e (%)	f (%)	g (%)	h (%)	i (%)	j (%)	k (%)	l (%)	m (%)	n (%)	o (%)
a	0.00	1.25	3.75	1.25	5.00	10.00	17.50	7.50	5.00	13.75	16.25	12.50	11.25	3.75	5.00
b	1.25	0.00	0.00	2.50	3.75	5.00	12.50	3.75	2.50	13.75	8.75	1.25	1.25	7.50	7.50
c	5.00	8.75	0.00	2.50	5.00	1.25	3.75	2.50	1.25	10.00	2.50	1.25	2.50	3.75	7.50
d	12.50	15.00	16.25	0.00	16.25	8.75	22.50	16.25	21.25	26.25	8.75	11.25	8.75	22.50	16.25
e	11.25	13.75	8.75	6.25	0.00	11.25	2.50	13.75	10.00	13.75	7.50	3.75	5.00	7.50	6.25
f	11.25	10.00	6.25	7.50	6.25	0.00	7.50	7.50	11.25	10.00	10.00	3.75	3.75	3.75	6.25
g	15.00	12.50	6.25	10.00	1.25	3.75	0.00	0.00	2.50	16.25	10.00	5.00	3.75	7.50	3.75
h	17.50	7.50	0.00	11.25	2.50	5.00	13.75	0.00	3.75	11.25	3.75	0.00	7.50	1.25	6.25
i	7.50	5.00	5.00	8.75	0.00	5.00	1.25	3.75	0.00	11.25	7.50	3.75	2.50	3.75	6.25
j	7.50	3.75	1.25	2.50	3.75	1.25	12.50	3.75	3.75	0.00	8.75	2.50	2.50	3.75	5.00
k	12.50	8.75	0.00	13.75	0.00	12.50	2.50	2.50	1.25	12.50	0.00	5.00	3.75	2.50	6.25
l	11.25	13.75	3.75	6.25	1.25	1.25	3.75	5.00	2.50	8.75	6.25	0.00	2.50	0.00	8.75
m	8.75	8.75	3.75	6.25	1.25	6.25	2.50	2.50	3.75	8.75	1.25	2.50	0.00	2.50	6.25
n	0.00	1.25	2.50	1.25	2.50	1.25	6.25	6.25	6.25	8.75	16.25	10.00	3.75	0.00	6.25
o	0.00	2.50	0.00	0.00	1.25	5.00	1.25	2.50	5.00	8.75	3.75	2.50	2.50	6.25	0.00

Source own elaboration

Table 4 Counter-expertise between experts in the relationship effect-effect

Effect/Effect	A (%)	B (%)	C (%)	D (%)	E (%)	F (%)	G (%)	H (%)	I (%)
A	0.00	1.25	2.50	5.00	1.25	3.75	3.75	2.50	1.25
B	2.50	0.00	2.50	6.25	16.25	15.00	11.25	23.75	22.50
C	7.50	6.25	0.00	3.75	1.25	1.25	3.75	1.25	3.75
D	25.00	15.00	2.50	0.00	12.50	16.25	20.00	2.50	17.50
E	10.00	6.25	1.25	8.75	0.00	0.00	3.75	7.50	1.25
F	1.25	5.00	3.75	7.50	6.25	0.00	3.75	2.50	6.25
G	3.75	3.75	0.00	5.00	6.25	10.00	0.00	7.50	1.25
H	11.25	3.75	6.25	7.50	3.75	2.50	0.00	0.00	1.25
I	10.00	5.00	6.25	10.00	16.25	8.75	20.00	10.00	0.00

Source own elaboration

Table 5 Counter-expertise in the forgotten effects

Cause-Effect	A (%)	B (%)	C (%)	D (%)	E (%)	F (%)	G (%)	H (%)	I (%)
a	0.00	11.25	16.25	3.75	13.75	6.25	3.75	2.50	1.25
b	7.50	0.00	1.25	3.75	7.50	2.50	10.00	0.00	6.25
c	2.50	2.50	0.00	2.50	0.00	5.00	5.00	2.50	2.50
d	10.00	2.50	0.00	2.50	2.50	3.75	2.50	3.75	1.25
e	8.75	2.50	0.00	0.00	5.00	0.00	0.00	0.00	6.25
f	7.50	2.50	2.50	6.25	5.00	6.25	13.75	7.50	7.50
g	8.75	2.50	0.00	2.50	5.00	5.00	0.00	0.00	5.00
h	7.50	1.25	6.25	0.00	1.25	0.00	0.00	2.50	1.25
i	1.25	0.00	1.25	2.50	10.00	3.75	3.75	1.25	3.75
j	2.50	0.00	3.75	2.50	7.50	0.00	1.25	2.50	0.00
k	8.75	3.75	0.00	0.00	0.00	1.25	0.00	3.75	0.00
l	10.00	5.00	1.25	0.00	2.50	1.25	1.25	1.25	2.50
m	7.50	2.50	2.50	1.25	3.75	0.00	1.25	0.00	2.50
n	2.50	1.25	11.25	7.50	8.75	2.50	3.75	2.50	2.50
o	3.75	2.50	2.50	5.00	1.25	1.25	0.00	2.50	12.50

Source own elaboration

information provided by the experts, see Fig. 2. Regarding the results found in the tail of the frequency distribution, the greatest difference between the experts arises in the relationship between environmental commitment (cause)—customer satisfaction (effect) with a value of 16.25%, followed by environmental commitment (cause)—tourist spending (effect) with a value of 13.75% and accessibility (cause)—prosperity (effect) with a value of 13.75%.

Hence, tourism leaves its mark on all dimensions of Sustainable Development; one of its challenges is the conservation of natural and cultural resources that constitute the

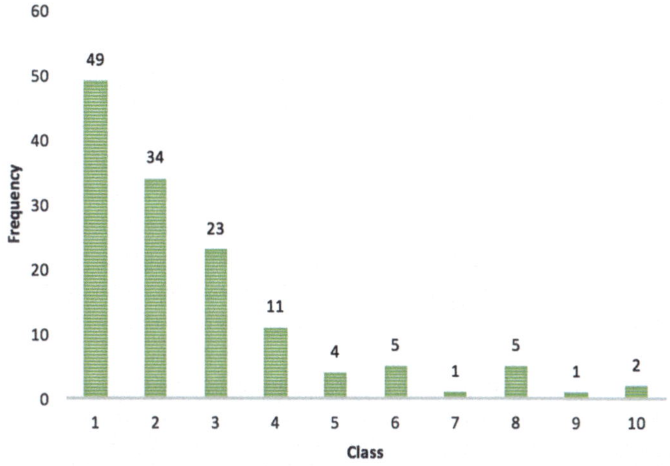

Fig. 1 Frequency distribution of counter-expertise of Cause-Effect. *Source* own elaboration

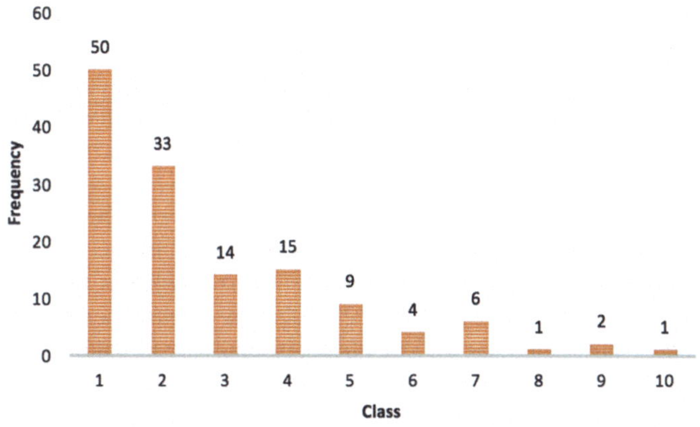

Fig. 2 Frequency distribution of counter-expertise of forgotten effects. *Source* own elaboration

basis of the activity, which will guarantee its sustainability avoiding the decline of the tourist destination and competitiveness as a complementary concept to environmental quality and sustainability [20]. Achieving a link between the tourist activity and the sustainability of the resources that support it, will ensure that quality tourism products and services are maintained, accompanied by adequate management instruments to contribute to the conservation of the environment and the socioeconomic well-being of the communities where it develops [25].

5 Conclusions

This research work began by exposing the importance of the tourism in the economy and the growing interest of the tourism sector in the competitiveness. The use of fuzzy techniques such as the forgotten effects theory allowed to evaluate cause-effect relationships and to find indirect incidents in this evaluation; counter-expertise reaffirmed the causes and effects that tourism competitiveness entails, as well as the relationships between them and detected the relationships in which there is a difference of opinion, making it possible to compare the opinions of two groups of experts regarding the relationship between causes and effects of the tourist destination competitiveness.

It was found that the opinion of experts who investigate the tourist activity is similar to the opinion of experts who work in the tourism sector in most cases with respect to cause-effect relationship of tourism competitiveness and forgotten effects. In the counter-expertise of direct incidents, stands out the relationship between environmental commitment-tourism demand, environmental commitment-tourist spending and hospitality-sustainable development with the biggest difference of opinion; where the experts who investigated the tourist activity gave valuations above experts who work in the tourism sector. In the counter-expertise of forgotten effects stands out the relationship between environmental commitment-customer satisfaction, environmental commitment-tourist spending and accessibility-prosperity with the biggest difference of opinion, in the first two relationships experts who investigated the tourist activity gave valuations above experts who work in the tourism sector, but in the relationship accessibility—prosperity was the opposite.

It is observed that the experts who investigate the tourist activity recognize directly the effect of environmental commitment in variables as tourism demand and tourist spending, although they forgot the effect that environmental commitment has on customer satisfaction. On the other hand, by applying forgotten effects theory emerged other difference of opinion between the experts, the relationship accessibility—prosperity, when analyzing this we can observe that experts who work in the tourism sector recognize the effect of accessibility on the prosperity of the destination.

The analysis carried out made it possible to know the elements that must be improved from the sector that provides the tourist service to boost the competitiveness of the destinations, being in this case, the main variable the environmental commitment, followed by accessibility. Moreover, it was also possible to detect some variables that require attention and to be studied by researchers, which are accessibility and prosperity. One of the limitations that we find in this work is the number of experts that make up each group, it would be beneficial to expand the number of experts. For future research, we consider extending the study to more groups of experts, for example, to the government that, among other things, directs actions in favor of the tourism sector. We also consider it appropriate to analyze the barriers that have prevented the tourism sector from implementing actions in favor of environmental commitment.

References

1. Alfaro-Calderón GG, Zaragoza A, Alfaro-García VG, Gil-Lafuente AM (2020) Innovation capabilities and innovation systems: a forgotten effects analysis of their components. In: Ferrer-Comalat J, Linares-Mustarós S, Merigó J, Kacprzyk J (eds) Modelling and simulation in management sciences. MS-18 2018. Advances in intelligent systems and computing, vol 894, Springer, Cham
2. Alfaro-Calderón GG, Godínez-Reyes NL, Gómez-Monge R, Alfaro-García V, Gil-Lafuente AM (2019) Forgotten effects in the valuation of the social well-being index in Mexico's sustainable development. Fuzzy Econ Rev 24(1):67–81
3. Alfaro-García VG (2019) Toma de decisiones en la incertidumbre: técnicas y herramientas ante escenarios altamente desafiantes. Inquietud Empresarial 19(2):1–3
4. Alfaro-García VG, Gil-Lafuente AM, Alfaro Calderón GG (2017) A fuzzy methodology for innovation management measurement. Kybernetes 46(1):50–66
5. Alfaro-García VG, Gil-Lafuente AM, Alfaro-Calderón GG (2015) A fuzzy logic approach towards innovation measurement. Global J Bus Res, 9 (3), 53–71
6. García A, Siles D (2015) Cómo mejorar la competitividad turística de un destino: análisis del mediterráneo español y recomendaciones a los gestores de los destinos. Revista de Análisis Turístico 19(1):1–11
7. Gil-Lafuente AM (2005) Fuzzy logic in financial analysis. Springer-Verlag, Berlin/Heidelberg
8. Ibáñez RM (2011) Diagnóstico de la calidad y competitividad del sector turístico en México. Cuadernos de turismo 28:121–143
9. Kaufmann A (1988) Theory of expertons and fuzzy logic. Fuzzy Sets Systems 28(3):295–304
10. Kaufmann A (1990) Expert appraisements and counter-appraisements with experton processes. IEEE 619–624
11. Kaufmann, A.J. and Gil-Aluja, J. (1993). *Técnicas especiales para la gestión de expertos*. Vigo: Milladoiro.
12. Kaufmann A, Gil Aluja J (1988) Models for the Research of Forgotten Effects. España, Milladoiro
13. Leung SY, Baloglu S (2013) Tourism competitiveness of Asia Pacific Destinations. Tour Anal 18:371–384
14. Linares-Mustarós S, Ferrer-Comalat JC, Corominas-Coll D, Merigó JM (2018) The ordered weighted average in the theory of expertons. Int J Intell Syst 1–21
15. Luna Altamirano K, Tinto Arandes J, Sarmiento Espinoza W, Cisneros Quintanilla D (2017) Tratamiento de impagos bajo el enfoque de la incertidumbre con la aplicación de redes neuronales. Rev CiencS Pedagógicas E Innovación, 5(1):61–70
16. Merigó JM, Casanovas M, Yang JB (2014) Group decision making with expertons and uncertain generalized probabilistic weighted aggregation operators. Eur J Oper Res 215–224
17. Miguel AE, Solís N, Torres JC (2014) El impacto territorial del turismo en el desarrollo sostenible: el caso de las regiones de México 2000–2010. Revista de Turismo y Patrimonio Cultural 12:357–368
18. Pascarella R, Fontes JR (2010) Competitividad de los destinos turísticos, modelo de evaluación basado en las capacidades dinámicas y sus implicancias en las políticas públicas. Estudios y Perspectivas en Turismo 19:1–17
19. Pérez-Romero ME, Flores-Romero MB, Alfaro-García VG (2021) Tourism and destination competitiveness analysis applying the forgotten effects theory. J Intell Fuzzy Syst 40:1795–1804
20. del Pérez S, V. (2017) La planificación y prevención de los impactos ambientales del turismo como herramienta para el desarrollo sostenible: caso de estudio Timotes, Venezuela. Revista interamericana de ambiente y turismo 13(2):164–183
21. Plummer P, Tonts M, Martinus K (2014) Endogenous growth, local competitiveness and regional development: Western Australia's Regional Cities, 2001–2011. Journal of Economic & Social Policy 16:146–176

22. Pulido JI, Sánchez M (2010) Competitividad versus crecimiento en destinos turísticos. Un análisis mediante técnicas multivariantes. Cuadernos de Economía 22:159–182
23. Rodríguez C, Florido C, Jacob M (2020) Circular economy contributions to the tourism sector: a critical literature review. Sustainability 12(11):4338
24. Saldaña CX, Guamán GA (2019) Análisis financiero basado en la técnica Fuzzy Logic, como instrumento para la toma de decisiones en la empresa Italimentos Cia. Ltda. Rev Econ Y Política XV(30)
25. F, Rebolloso, MP, Salcedo (2007) Turismo, sustentabilidad y certificación: un reto global. Revista del centro de Investigación 7(27):77–91
26. Sánchez M (2012) Análisis cuantitativo del impacto económico de la competitividad en destinos turísticos internacionales. Revista de Economía Mundial 32:103–125
27. Sarmiento W, Luna K, Cisneros D (2017) Estimación del ciclo de caja a partir de un sistema de subconjuntos borrosos (Caso artesanos de calzado Cantón Gualaceo Provincia del Azuay). Revista Ciencias Pedagógicas e Innovación 5(2):91–97
28. Vera JF, Baños CJ (2010) Renovación y re-estructuración de los destinos turísticos consolidados del litoral: las prácticas recreativas en la evolución del espacio turístico. Boletín de la Asociación de Geógrafos Españoles 53:329–353
29. World Tourism Organization UNWTO Tourism Highlights (2017) Edition. https://www.unwto.org/
30. Zadeh LA (1965) Fuzzy sets. Information and control 8:338–353

Sustainable Value: An Empirical Research on Large Firms

Norma Laura Godínez-Reyes, Rodrigo Gómez-Monge, Gerardo Gabriel Alfaro-Calderón, and Argelia Calderón-Gutiérrez

Abstract The objective of this research is to determine the extent to which environmental, social and governance indicators can explain corporate efficiency, given by three indicators: return on assets (ROA), return on equity (ROE) and return on sales (ROS). In order to determine the validity of corporate efficiency measurement a model of three linear regressions with panel data was designed. This model tests whether the environmental (E), social performance (S) and corporate governance (G) variables of the studied companies are able to explain their profitability for the period 2014–2017. Results show that sustainable value (S) explains profitability—or return on sales (ROS)—with a high degree of significance, whereas it explains corporate efficiency to a moderate extent. Likewise, corporate governance (G) is the most significant variable for the generation of sustainable value. We conclude that generating sustainable value allows firms to meet profitability goals and to mitigate environmental and social impacts. Thus, the model is useful for measuring corporate efficiency in sustainable companies.

Keywords Sustainable value · Corporate efficiency · Linear regression with panel data · ESG

1 Introduction

Health, economics and climate change are undoubtedly the main challenges firms are facing nowadays. The Covid-19 economic impact on global economies has urged firms to alter their priorities, understand new rules, set more ambitious goals, adapt ways of working and accelerate the transition into technological innovation [12].

N. L. Godínez-Reyes (✉) · G. G. Alfaro-Calderón · A. Calderón-Gutiérrez
Researcher at Facultad de Contaduría Y Ciencias Administrativas, Universidad Michoacana de San Nicolás de Hidalgo, Morelia, Mexico
e-mail: lgodinez@umich.mx

R. Gómez-Monge
Facultad de Economía "Vasco de Quiroga", Universidad Michoacana de San Nicolás de Hidalgo, Morelia, Mexico

E. León-Castro et al. (eds.), *Soft Computing and Fuzzy Methodologies in Innovation Management and Sustainability*, Lecture Notes in Networks and Systems 337,
https://doi.org/10.1007/978-3-030-96150-3_14

Certainly, the pandemic has also highlighted the importance of having solid risk management protocols, in order to strengthen a firm's capacity to respond to adversities [8]. More than ever, companies can choose to design strategies to strengthen the indicators that contribute to improving economic, environmental, social and corporate governance (ESG) practices. According to the Word Economic Forum [26] and the Mexican Stock Exchange [3] having ESG practices directly impacts on a firm's sustainable value creation, generating confidence and good reputation, within and outside the organization.

Sustainable value can be defined as a way for companies to manage, measure, evaluate and improve their corporate performance, according the [9] triple bottom line (TBL) approach the Corporate Social Responsibility (CSR) framework. This measurement aims to mitigate risks, save resources, optimize processes, generate innovation, increase productivity and promote the generation of economic value with social and environmental awareness, and corporate ethics [6].

In Mexico, firms that embrace those principles are awarded the Socially Responsible Company annual badge. However, only 1,012 large companies and 987 small and medium enterprises received the 2020 badge according to the Mexican Center for Philanthropy [5] and the Alliance for Corporate Social Responsibility (AliaRSE). The number is really low when compared to a total of 5,487,061 registered established businesses in Mexico [23]. Some businesspeople argue that adopting CSR principles would imply making capital investments that might not necessarily have sufficient returns on the investment, which in turn would result in less profitable and productive companies. Nonetheless, such an argument is no longer valid since, different analyses of empirical and scientific research confirm that the long term investment in social responsibility (social investment) is profitable.

For [21] social investment, under the principle of shared value, which involves creating economic value in a way that also creates value for society, has a positive effect on financial performance of companies in the medium and long term. In Mexico, Alonso-Armeida et al. [1] also found a positive relationship between CSR and financial variables in the companies they analyzed. Ojeda-Hidalgo et al. [19] studied 671 companies in the state of Guanajuato, Mexico, showing that there is a positive effect between actions aimed at respecting and taking care of the environment, and ethical behavior with clients, and the company's overall performance.

Evidence shows that having CSR practices is profitable and sustainable over time, hence the relevance of modeling corporate efficiency so that companies, regardless of their size, may identify the variables that will generate economic value in the short term if they commit to sustainability. Therefore, this study is aims at determining the impact of sustainable value indicators (ESG) on a firm's profitability and use them as a model of corporate efficiency. Sustainability and profitability ratings of a group of 18 large socially responsible companies were analyzed using a model of three linear regressions with panel data. All these enterprises are listed on the Mexican Stock Exchange (BMV), and the analyzed period was 2014–2017. Results show that creating sustainable value reports a positive impact on a firm's productivity (ROS) and that the ESG variable that explains profitability with a greater degree is corporate government (G).

This paper proceeds as follows: First, the theoretical framework is presented, followed by an explanation of the methodology used. In third section we describe and discuss the results of the model. Finally, we present our conclusions of this research.

2 Literature Review

2.1 Social Responsibility and Sustainable Value Creation

At the end the twentieth century, businesses faced the Climate Change crisis, which forced them to evolve from paradigm of creation of economic value to the creation of sustainable value, focusing on achieving sustainable development. With this new approach, Corporate Social Responsibility (CSR) strategies have been increasingly incorporated into business practice. CSR has three essential management pillars: economic, environmental and social (Fig. 1), in order to fulfill stakeholders' needs [10, 16, 20].

CSR is therefore understood as a firm's deliberate commitment to take accountability for its business impacts on society, the environment and the economy, considering its stakeholders' expectations, showing an ethical and transparent behavior that contributes to achieving the objectives of sustainable development [14].

Thus, business sustainability lies in developing a strategy based on CSR theories [13] so that a company assumes social and environmental commitments to minimize its negative impacts on the environment and the society, aligning its objectives with

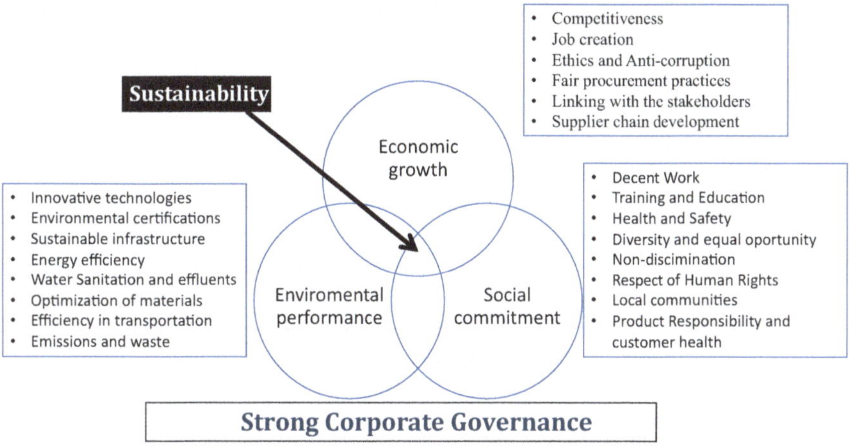

Fig. 1 Corporate social responsibility model. *Source* Authors

those of its stakeholders. By doing so, a firm's generation of economic value becomes a means for the creation of sustainable value.

The design of a company's sustainability strategy is related to its disclosure, both internally and externally, using Sustainability or Compliance Reports. These reports are based on Elkington's Triple Bottom Line (TBL) model (2006), according to which value generation in companies consists of three dimensions: economic, social and environmental. The TBL dimensions are valued to indicate a firm's score in the creation of sustainable value (ESG), and each dimension includes different indicators that allow for its measurement, implementation and improvement.

3 Methodology

The scope of this research is explanatory and evaluative, with a longitudinal design (2014–2017). A linear regression model with panel data technique was developed in order to support the degree of explanation of other models used for the measurement of empirical data built using mathematical economics and to obtain numerical results, which allow the analysis of economic, social and business information and have a greater certainty in the results that companies use to make the decisions of sustainable value generation [18].

The selected sample consisted of 18 sustainable large companies listed on the Mexican Stock Exchange. Sustainability ratings for each company were obtained from *Sustainalytics, Inc.* a database available on the Yahoo Finance website. Such sustainability ratings were issued by the Center of Excellence in Corporate Governance (CEGC) of Universidad Anáhuac México Sur, a sustainability rating agency affiliated to the Global Reporting Initiative (GRI). These ratings are determined based on public and open information released by companies, carrying out an exhaustive evaluation of long-term practices on environmental and social performance, as well as corporate governance [22]. Then, 600 observations of firms' Environment, Social and Governance performance ratings (ESG) were used to determine sustainable efficiency.

On the other hand, the series for economic value financial indicators—return on assets (ROA), return on equity (ROE), and return on sales (ROS)—were calculated from each company's Annual Report for the years 2014, 2015, 2016, and 2017. For this purpose, seventy-two Integrated Reports were reviewed.

In this study, corporate efficiency of sustainable companies is explained by means of a classic linear regression econometric model using the panel data technique. Therefore, this model tests, for the period 2014–2017, whether the environment, social performance and corporate governance variables explain the studied firms' sustainable value and profitability (Fig. 2).

For the estimation of the proposed model, three linear regressions with panel data were run, using three panels of 72 observations each (Table 1). The calculated financial ratios—ROA, ROE and ROS—were used as dependent variables (Y), and environment (E), social performance (S) and corporate governance (G) ratings of

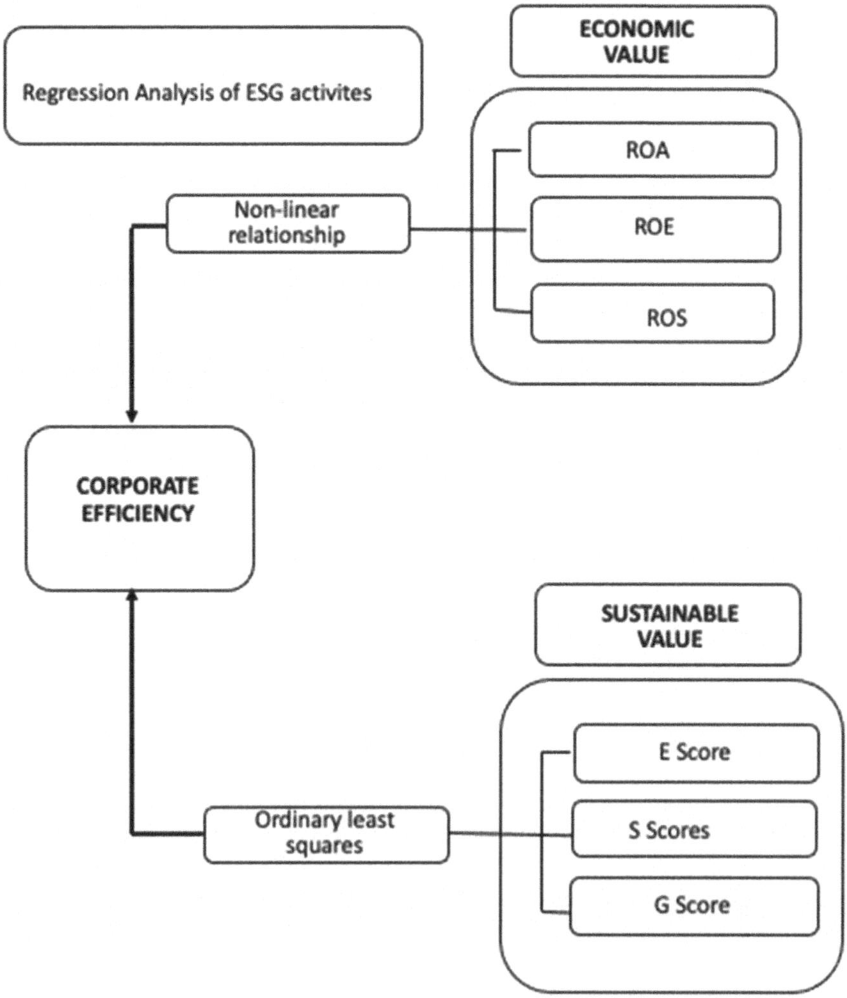

Fig. 2 The linear regression model used to explain corporate efficiency. *Source* Authors

Table 1 Series of linear regressions using panel data

Regression	Dependent variable	Independent variable
First	ROA (Return on Assets)	Sustainable value (ESG ratings)
Second	ROE (Return on Equity)	
Third	ROS (Return on Sales)	

Source Authors

companies were used as independent (X) in each of the three panels. Additionally, the statistical software *Eviews 7* was used with the objective of validating that social performance (X_1), environmental performance (X_2) and corporate governance (X_3) can actually explain economic value (Y) of the 18 analyzed companies in the years 2014, 2015, 2016 and 2017.

A random effects (RE) regression model for panel data was used in each of the three regressions. RE models consider that individual effects are not independent from each other but are randomly distributed around a given value. The random effects panel model, unlike the fixed effects panel, assumes that the explanatory variables (profitability) do not affect the cross-sectional units (firms) equally. Thus, each of the components of the total error follows a normal distribution with zero-mean, it is not correlated with itself, and it is not correlated with any of the explanatory variables (X). The general panel data model is presented below [18].

$$Y_{it} = \alpha_{it} + \beta X_{it} + u_{it} \tag{1}$$

where:

- i study unit of the cross section (companies).
- t time dimension (studied periods).
- β vector of k parameters (y).
- y_{it} dependent or explained variables (returnees).
- x_{it} independent or explanatory variables (regressor).
- $_{it}$ error term.

Finally, in order to determine the quality of the information used, the degree of confidence and its significance, the following validation tests were conducted: Hausman Test, coefficient of determination (R-squared), adjusted coefficient of determination (adjusted R-squared), p-value and Durbin-Watson Test.

4 Results

Results of the three regression models for panel data are presented in this section. Such models were run to determine the relationship between the dependent variable *economic value*—ROA, ROE and ROS-, and the independent variable *sustainable value*—environmental, social and corporate governance performance—of the 18 studied companies.

1st Regression: Return on Assets versus Sustainable Value.

Dependent variable: Return on Assets (ROA).

Method: Pooled EGLS (cross-section random effects).

Period: 2014–2017.

Cross-sections included: 18.

Included observations: 4

Total pool observations: 72.

According to this model, sustainable value variables explain to a very low degree the return on assets, since the obtained R-squared and adjusted R-squared values are 5% and 1% respectively. Therefore, it can be stated that the return on investments of sustainable companies is not explained by the weighted statistics ratings obtained from their ESG ratings (sustainable value generation).

In terms of the observed weighted probability, the model shows a significance level of 29%. However, when observed independently, corporate governance has significance level in the ROA of 9%. The model also shows that there is no autocorrelation between dependent and independent variables, since the result of the Durbin-Watson Test is between 1,703 (du) and 2,297 (4-dl), at a 5% confidence level.

2nd Regression: Return on Equity versus Sustainable Value.

Dependent variable: Return on Equity (ROE).

Method: Pooled EGLS (cross-section random effects).

Period: 2014–2017.

Cross-sections included: 18.

Included observations: 4

Total pool observations: 72.

This model, similarly to the ROA model, has a very low explanatory level of the return on capital (ROE) variable, since variables of sustainable value explain the profitability in a degree of 2.2% (R-squared) and −2% (adjusted R-squared). Also, the weighted probability has a significance level of 52%, therefore the regressor variables do not significantly explain the variable profitability. However, individually, corporate governance is again the only variable with some degree of significance 24.6%, a high value, but much lower than the value of social and environmental variables, 91%. As for the autocorrelation of the variables, according to the results, there is no autocorrelation between the variables of the sustainable value with the ROE.

3rd Regression: Return on Sales versus Sustainable Value.

Dependent variable: Return on Sales (ROS).

Method: Pooled Least Squares (cross-section fixed effects).

Period: 2014–2017.

Cross-sections included: 18.

Included observations: 4

Total pool observations: 72.

The return on sales model is a much more significant model, the level of distribution of sustainable value variables explains the return on sales by 84% (R-squared) and 78% (adjusted R-squared). This model shows that the return on sales for the 18 companies is explained in those percentages by weighted results of social, environmental and corporate governance performance. Similarly, the weighted significance level of the model is zero, which implies that it is a fully reliable model. As for the autocorrelation of its variables, the Durbin-Watson Test confirms that there is no autocorrelation, since the value 1.739447 is among the parameters for 72 observations of 1.703 du and 2.297 4-dl. This implies that this model does have significance and quality to explain the dependent variable from the independent ones.

With these model results, we can say that the social, environmental and corporate governance performance variables in the 18 studied companies during the period 2014–2017 have an adequate level of significance to explain the sustainable value of the 18 companies. Likewise, it is shown that out of the three explanatory variables, the one that has the highest explanatory level in two of the models (1st and 2nd regression) is corporate governance (G), according to its individual *p-value*, which is a critical test to evaluate the significance of the model (Table 2).

These results show that the variables of sustainable value (ESG), explain with a high degree of significance, the return on sales (ROS). These explanatory variables (ESG), on the other hand, do not have a direct linear relationship with return on assets (ROA) or return on equity (ROE). These latter results are consistent with the lack of conclusiveness of several studies that claim that companies that are socially responsible generate better economic results than those that are not [4, 7, 17, 24]. Although most companies generate economic value, in the analyzed period, economic benefit in the generation of sustainable value of companies was not conclusive (Tables 3, 4, 5, 6 and 7).

In spite of the low significance of the explanatory variables of the linear regression model with panel data, results of the second regression suggest with a certain degree of validity, that there is a positive relationship between the development of initiatives of Social Responsibility of companies and their economic results. This is because the costs involved in such initiatives are offset by the benefits they may bring. For this reason, it can also be stated that it is worth investing in the generation of sustainable value from the generation of economic value. This vision is related to the theories so [13] called "instrumental theories", which consider CSR as a strategic tool to obtain

Table 2 Results of the first panel data regression

Variable	Coefficient	Standard error	t-Statistic	Probability
C	0.095586	0.032406	2.949651	0.0044
Social	−0.000317	0.000442	−0.718257	0.4751
Environment	0.000506	0.000438	1.15475	0.2522
Corporate Governance	−0.000865	0.000502	−1.722862	0.0895

Source Authors, using Eviews 7 software, data processed on Nov. 7, 2019

Table 3 Weighted statistics of the first panel data regression

Weighted statistics	
R-squared	0.052935
Adjusted R-squared	0.011153
F-statistic	1.266932
Prob (F-statistic)	0.292698
Durbin-Watson stat	1.856938

Source Authors, using Eviews 7 software, data processed on Nov. 7, 2019

Table 4 Results of the second panel data regression

Variable	Coefficient	Standard error	t-Statistic	Probability
C	0.237034	0.094637	2.504673	0.0147
Social	0.00013	0.001201	0.107905	0.9144
Environment	0.000124	0.001218	0.101798	0.9192
Corporate Governance	−0.001714	0.001465	−1.170119	0.246

Source Authors, using Eviews 7 Software, processed on November 7, 2019

Table 5 Weighted statistics of the second panel data regression

Weighted statistics	
R-squared	0.022682
Adjusted R-squared	−0.020435
F-statistic	0.52606
Prob (F-statistic)	0.665863
Durbin-Watson stat	1.726478

Source Authors, using Eviews 7 Software, processed on November 7, 2019

Table 6 Results of the third panel data regression

Variable	Coefficient	Standard error	t-Statistic	Probability
C	0.05562	0.095286	0.583719	0.562
Social	−0.001213	0.001168	−1.037969	0.3042
Environment	0.000862	0.001227	0.702271	0.4857
Corporate Governance	0.000977	0.001603	0.60937	0.545

Source Authors, using Eviews 7 Software, processed on November 7, 2019

economic benefits, based on the premise that companies must have a social purpose which is consistent with their economic interests in the long term [20, 25]

Table 7 Weighted statistics of the third panel data regression

Weighted statistics	
R-squared	0.845096
Adjusted R-squared	0.784349
F-statistic	13.91181
Prob. (F-statistic)	0
Durbin-Watson stat	1.739447

Source Authors, using Eviews 7 Software, processed on November 7, 2019

5 Conclusions

The ESG components of CSR that generate sustainable value in companies were analyzed using panel data regression. After running three different models, it can be concluded that generating sustainable value in the studied firms produces a positive effect on the firms' corporate efficiency. Moreover, our findings show that Corporate Governance (G) is the dimension with the greatest weight to generate value. Furthermore, there is no correlation between the variables according to the obtained probability result (*p-value*); hence, the most significant variables of sustainable value are environment (E) in the first year, and corporate governance (G) in the first, second and third years.

Another outstanding finding is that the least valued dimension for the generation of sustainable value is, both theoretically and empirically, social performance (S). Despite being the dimension that has the greatest number of indicators in most of the models used internationally to assess firms' sustainability, social performance continues to have the least valued dimension when it comes to deciding business sustainability strategies. As companies do not find a direct benefit in investing on social practices, they rather invest in actions to improve their environmental (E) or corporate governance (G) indicators. Indeed, results from implementing improvement practices on corporate or environmental governance can easily be reflected in lower costs, or a more efficient use of resources. Social dimension (S) is then a variable that should be studied on its own, since it affects general welfare as well as the common good of the communities, they are located in.

Our findings suggest that sustainable value creation of socially responsible companies, measured in terms of their profitability, is a sound business strategy towards which large companies are directing their efforts. There are two well-founded reasons for this: firms expect this strategy to bear economic fruit in the long term, and/or they hope to avoid market or institutional risks, before the possibility that more restrictive measures might be implemented to deal with climate change and the social risks firms currently face. Our results should be taken cautiously, though, since the study focuses on a relatively small sample, Likewise, it was demonstrated that the first mechanism that companies have to reach sustainability is corporate governance.

Corporate government allows companies to monitor, control, optimize and systematize their processes, thus generating value towards the objectives pursued by the firm. The three basic instruments of corporate governance that guide companies towards corporate efficiency are: codes of ethics and social responsibility, sustainability reports, and audits. The first one refers to the elaboration of codes that regulate ethical behavior of a firm's staff and its stakeholders on the whole. The second instrument is the preparation of sustainability reports that stick to business models that support the design of sustainable competitiveness strategies. The third instrument, the audit, allows for monitoring sustainable value creation activities, to verify that firms keep track of the set goals and are continuously improving. A final consideration is that one of the major challenges companies face is to maintain a far-reaching strategy that allows them to generate value while committing to contribute to social and environmental solutions.

References

1. Alonso-Armeida M, Rodríguez M, Aimer K, Abreu J (2012) La responsabilidad social corporativa y el desempeño financiero: un análisis en empresas mexicanas que cotizan en la bolsa. Revista Contaduría y Administración, FCA-UNAM. vol 57, núm. 1, ener-marzo, 53–77 México. https://doi.org/10.22201/fca.24488410e.2018.1214
2. Becerril A (16 de mayo de 2020) Thomas Piketty: la desigualdad en tiempos del Covid-19. El Economista, México
3. BMV (19 de octubre de 2020) El camino correcto. La importancia de los factores ESG en las orgaizaciones. Recuperado de: https://blog.bmv.com.mx/2020/04/el-camino-correcto-la-importancia-de-los-factores-esg-en-las-organizaciones/
4. Cabeza-García L, Martínez-Campillo A, Marbella-Sánchez F (2010) Análisis de la relación entre los resultados sociales y económicos de las cajas de ahorro españolas: un vaso de responsabilidad social corporativa. Rev Innovar J 20(37) mayo-agosto 2010 33–46
5. Cemefi (2020, November 23) Empresas ganadoras del distintivo de RSE en el 2020. https://www.cemefi.org/esr/
6. Cruz M (2018) Generación de valor sustentable. Publicaciones Empresariales UNAM, FCA Publishing, Repensar la valuación empresarial
7. De la Torre OV, Galeana E, Aguilasocho D (2015) The use of the sustainable investment against the broad market one. A first test in the Mexican Stock Market. Investigaciones Europeas de Dirección y Economía de la Empresa, Elsevier.es https://doi.org/10.1016/j.iedee.2015.08.002
8. Delloite (2011) Sustentabilidad como herramienta para reducir los riesgos. Boletín Gobierno Corporativo. Retrived February 2018
9. Elkington J (2006) Governance for Sustainability. Corporate Governance: An International Review. 14:522–529. https://doi.org/10.1111/j.1467-8683.2006.00527
10. Freeman RE (1994) The Politics of Stakeholder Theory: Some Future Directions. Bus Ethics Q 4(4):409–421. https://doi.org/10.2307/3857340
11. Friedman M (1970) La responsabilidad social de la empresa es incrementar ganancias. The New York Times Magazine. September, 122–126
12. García-Sánchez IM, García-Sánchez A (2020) Corporate Responsibility during COVID-19 Pandemic. J Open Innov Technol Mark Complex 6:125 https://doi.org/10.3390/joitmc6040126
13. Garriga E, Melé D (2013) Corporate social responsibility theories: mapping the territory. J Bus Ethics 53(1). https://doi.org/10.1023/b:busi.0000039399.90587.34

14. Godínez-Reyes NL (2020) Empresas sustentables en México: Una medición de su eficiencia a través del análisis envolvente de datos, 2014–2017 [tesis de doctorado no publicada, Universidad Michoacana de San Nicolás de Hidalgo]
15. Gómez-Monge R (2012) El Sector Bancario en México, los depósitos a plazo y las cuentas de ahorro: un análisis de eficiencia durante el periodo de internacionalizaciones a través de la envolvente de datos (DEA). Revista Ciencias Estratégicas 20(27), enero-Julio. 12–38. Medellín, Colombia
16. Keinert C (2008) Corporate Social Responsability as an International Strategy. Springer. https://doi.org/10.1007/978-3-7908-2024-9
17. Kent W, And ZZ, Bring Y (2016) The angel-halo effect: How increases in corporate social responsibility and irresponsibility relate to firm performance. Eur Bus Rev 28(6):709–722. https://doi.org/10.1108/EBR-11-2015-0139
18. Mayorga M, Muñoz E (2000) La Técnica de datos panel. Una guía para su uso e interpretación. Banco Central de Costa Rica, División Económica, Departamento de investigaciones económicos. Septiembre, 2000, Costa Rica
19. Ojeda-Hidalgo J, López-Salazar A, Álvarez-Orozco D (2019) ¿Influye la Responsabilidad Social en el desempeño Organizacional?. *Estudios de Administración, 26*(1). doi:https://doi.org/10.5354/0719-0816.2019.55404
20. Porter ME, Kramer MR (2006) Estrategia y sociedad: el vínculo entre la ventaja competitiva y la responsabilidad social corporativa. Harv Bus Rev América Lat
21. Porter M, Kramer M (2011) La Creación de Valor Compartido. Harvard Business Review Latinoamérica, Septiembre-Octubre
22. S&P Dow Jones Indices (2019, January 19). S&P/BMV Índices Dow Jones Merodología. https://espanol.spindices.com/indices/equity/sp-bmv-ipc-sustainable
23. INEGI (2020, April 17). Directorio Estadístico de Unidades Económicas (DENUE). México: Instituto nacional de Estadística y Geografía. https://www.inegi.org.mx/app/mapa/denue/
24. Wang H, Choi J, Li J (2008) Too little or too much? Untangling the relationship between Corporate Philanthropy and firm financial performance. Organization Science. Vol 19 N° 1. January-February 2008:1443–2159
25. Wood DJ (2010) Measuring corporate social performance: a review. Blackwell Publishing Ltd and British Academy of Management. United Kingdom. Int J Manag Rev, 50–84
26. World Economic Forum (2020) Measuring Stakeholder Capitalism. Towards Common Metrics and Consistent Reporting of Sustainable Value Creation. WEF, Reports. Davos Retrieved September 29, 2020, from: https://www.weforum.org/reports/measuring-stakeholder-capitalism-towards-common-metrics-and-consistent-reporting-of-sustainable-value-creation

Bibliometric analyses of main research directions

A Bibliometric Analysis of Corporate Social Responsibility and Competitiveness

Marlenne Gisela Velázquez-Cázares, Sergio Alvarado-Altamirano, Ernesto León-Castro, and Fabio Blanco-Mesa

Abstract Corporate Social Responsibility and its relationship with Competitiveness has become a relevant field of study, mainly for the potential strategic and beneficial impact on firms and society. The objective of this paper is to present an overview of the main contributions in these subjects by a bibliometric review approach. The work covers a wide range of bibliometric indicators base on the Web of Science (WoS) Core Collection and it is complemented with the Scopus database. The leading journals, authors, institutions, countries, and articles are considered. The results highlight the increasing relevance of the research field and exhibit Porter and Kramer as the most influential authors and Seuring and Gallardo-Vazquez as the most productive authors on the field. The United States of America is leading the field with the largest number of publications and citations. The relationship between Corporate Social Responsibility and Competitiveness is a rising field of study that contributes to the development of new theories, approaches, and applications.

Keywords Corporate Social Responsibility · Competitiveness · Bibliometric

1 Introduction

The increasing awareness of the role of companies in society, the globalization, and the constant social and environmental change have pressured companies towards Corporate Social Responsibility (CSR). CSR is a complex and dynamic concept,

M. G. Velázquez-Cázares (✉) · S. Alvarado-Altamirano
Universidad of Occidente, Blvd. Lola Beltrán s/n esq. Circuito Vial, Culiacán 80200, México
e-mail: marlenne.velazquez@udo.mx

E. León-Castro
Faculty of Economics and Administrative Science, Universidad Católica de La Santísima Concepción, 4070129 Concepción, Chile

F. Blanco-Mesa
Facultad de Ciencias Económicas Y Administrativas, Escuela de Administración de Empresas, Universidad Pedagógica Y Tecnológica de Colombia, Av. Central del Norte, 39-115, 150001 Tunja, Colombia

E. León-Castro et al. (eds.), *Soft Computing and Fuzzy Methodologies in Innovation Management and Sustainability*, Lecture Notes in Networks and Systems 337,
https://doi.org/10.1007/978-3-030-96150-3_15

and there is not consensus on its definition [73]. The real academic debate about CSR arose in the early 1950s with Bowen's work "Social Responsibilities of the Businessman". However, this topic still attracts the attention of companies, society, practitioners, and academics, with potential areas left for future research [2].

The European [36] has defined CSR as "the responsibility of the enterprises for their impact on the society", and highlights its influence in the sustainability and competitiveness of the enterprises and the economy. Weber [102] mentions the competitive advantages that CSR activities contribute to the economic success of the firms, and global competitiveness [13, 44]. The importance of CSR lies in its relationship with a substantial list of benefits and positive performance outcomes [99], such as financial performance [57, 63] competitiveness [71, 107], firm valued [23] innovation [72], reputation [88] sustainability [8], benefits on customer loyalty and satisfaction [68], stakeholder relationship [31], productivity [92], attraction of investors [16, 56], motivation and engagement of employees [40], cost reduction [52, 102], etc.

In the long run, CSR is crucial to obtain market success, customer satisfaction, reputation and positive financial performance [88]. Firms that commit with CSR, not only reinforce strategy and reputation but also it turns into a long-term competitive advantage [74]. For this reason, firms spend massive amounts of resources on CSR activities [66, 68], supporting most CEO's reliance on CSR as vital for firm success [19]. Moreover, global consumers have officially embraced CSR, not only as a business's responsibility but also as a personal responsibility [24].

CSR studies have been conducted in several research fields [106], and an important number of this studies focus on the link between CSR and financial performance [82], with heterogeneous [42] and inconclusive results [65]. Nevertheless, the majority of the studies point to a positive relationship with financial performance [63, 80, 93]. However, in the past few years the debate of the strategic potential of CSR and its possible link with Competitiveness (CSR-C) has become an increasingly relevant issue [7], essentially because CSR is a source of competitive advantage [85].

The competitiveness concept took strength with Porter's work in [83]. The focus on business competitiveness has increased, becoming a multidimensional, relative and broad concept that refers to the ability of companies to create sustainable competitive advantages [71, 100]. Garelli [47] mention that competitiveness is an effective tool for nations achieving prosperity, economic and social well-being and even happiness, analyzing, integrating and evaluating the totality of a nations' performance. And the business world does not escape from it, as it is highly complex, dynamic and extremely competitive, companies must make a great effort to obtain competitive advantages and achieve prosperity. Firms competitiveness go beyond economic strength, rests on its adaptability and ability to compete through a large number of tangible and intangible assets, allowing the company to increase its market share, differentiation, cost reduction, and its sustainability. In this sense, the firm´s CSR strategy is a highlighted element that enables enterprises to achieve a large number of benefits as the competitiveness, becoming a direct or indirect motor in the creation of a large number of tangible and intangible assets that increase long-term competitive advantages.

Therefore, CSR influence in competitiveness has become increasingly important for organizations as well as understanding the impact and their relationship [100]. CSR has the challenge of influence in the core business activities and processes, as well as being part of the firm´s strategy in order to become more competitive, create value and improve society and environment in which operates. Researchers study a direct and/or indirect relationship between CSR and competitiveness (CSR-C), and some consider intangible assets as mediating variables [7, 60, 105]. For instance, [100] propose reputation as a key driver of CSR that generates innovation and impacts on the competitiveness of the firm. There are potential variables and elements that need to be understood to complete the CSR-C link, in order to create value for firms and society. Some of the key determinants of the company's competitiveness focus on issues such as reputation, innovation, productivity, culture, human resource, stakeholders, to name a few, and these issues are strongly influenced by CSR.

For all the above, the number of publications and research on CSR-C is significant and has increased in an accelerated way the past few years, with crucial outcomes. These results show a special relationship between CSR and competitiveness (CSR-C), with interesting findings, effects, theories, strategies, methods and a large number of considerations that allow us to know more precisely how this relationship occurs and its impact.

With the propose to analyze increasing amount of the literature about CSR and Competitiveness (CSR-C), this research utilizes bibliometric analysis to evaluate the publications related to CSR-C. Bibliometric research methods can help to analyze the literature effectively since bibliometric is a powerful tool to handle massive published literature enabling the full scientific structure of the research topic [108], providing quantitative and a more structured overview of a field extracted from bibliographic data. In this case, there are few bibliometric analyses about CSR, most of these bibliometric studies focuses on epistemology and concept evolution [27], and the relation with other fields or scientific disciplines such as supply chain management [38], international business journals [106], among others. One of the first and most known bibliometric studies on CSR was done by [27]. The authors evaluated 30 years of CSR research, the epistemological orientation, the evolution of the concept of CSR and CSP (Corporate Social Performance), authors, journals and citations, on the basis of data from Web of Science (WoS).

The main aim of this work is to present an overview of these two topics together CSR-C, from 1975 to 2018, their evolution and identification of the most relevant journals, authors, institutions, countries, papers and different documents (e.g. books, proceedings, etc.). The results disclose the increasing relevance of CSR-C as a research field. The most productive journals are the Journal of Business Ethics, Journal of Cleaner Production and Sustainability. Porter and Kramer [84, 85] papers are two of the most cited papers along with Brown and Dacin [17], and Garriga and Melé [48]. The main authors for their productivity are Seuring, Gallardo-Vazquez, and Husted. The University of Manchester, University of Western Ontario, Hong Kong Polytechnic University and Bucharest University of Economic Studies are the institutions with more publications in the topics, and The United States of America, Spain and England are the main countries that studies CSR and Competitiveness.

For this study, we considered articles, reviews, conference proceedings and books, this last two may show important contributions [10], since authors used them to publicize their investigation, supporting their theories, views, ideas, arguments, citations, and references, etc. The bibliometric analysis considered the Web of Science (WoS) and Scopus databases, with the limitation that some important contributions may not appear if they are not incorporated in these databases, because of the evaluations and indexation of the database and journals, incomplete coverage, errors in names of authors, titles, institutions and the level of complementarity and coincidence of the databases.

This article is structured as follows: Sect. 2 describes the methodology, and presents the evolution of CSR-C, the most important journals, papers, conferences proceedings, books, authors, institutions and countries. Section 3 develops a graphical visualization of the results using the VOS viewer software. Section 4 summarizes the main conclusions of the paper.

2 Methodology

As mentioned above, this study is focused on CSR and its relationship on Competitiveness (CSR-C). The bibliometric study provides a complete analysis of the topic and most important indicators related to authors, papers, institutions, journals, countries, conference proceedings, books, plus a graphical visualization of the topic. The work considers two databases Web of Science (WoS) Core Collections and Scopus Complete Collections. It is important to indicate that WoS core collections is the main database from which the most important data was obtained (e.g. journals, papers, authors), essentially because this database has an exceptional recognition in the scientific community [11, 77]. Meanwhile, Scopus is used to broaden the analysis and give complementary information.

The keywords selected to obtain the sample of this study are "corporate social responsibility*" AND "competitive*" (topic), from 1975 and 2018. CSR and competitiveness are used in a wide variety of associated terms, for example, corporate social performance, sustainability, competitive advantage, competitive success, competitive performance, for that reason and for the propose of this work those keywords are selected. This may lead to some research limitations since important studies may not be included or highlighted by the established specifications.

The sample is obtained in May 2019, with 1,047 publications from WoS and 796 documents from Scopus. Using WoS core collection Table 1 shows the number of document types found in CSR-C and each total citations: the documents published sum 30,943 citations, the articles represent the 84.67% of the citations, follows by reviews with 10.34% and proceedings papers with 3.26%.

There are some papers that stand out because of their high citation index. [85] paper have been cited 2,426 times, [17] have total citations of 1419, and a review by [48] has been cited 1,150 times. This is important to mention, Table 2 presents the citation structure of the documents and articles in CSR-C; 0.32% of the papers

Table 1 Document types in CSR-C in WoS core collection

Document types	Number	Total citations	Citations (%)
Article	882	26,199	84.67
Review	51	3201	10.34
Proceedings paper	16	1008	3.26
Book	6	143	0.46
Book chapter	76	129	0.42
Editorial material	8	257	0.83
Letter	7	3	0.01
Reprint	1	3	0.01
Total	1047	30,943	100.00

Source Own elaboration based on WoS 2019

Table 2 Citation structure in CSR-C in WoS core collection

Number of citations	# Documents	Documents (%)	# Articles and reviews	Articles (%)
≥ 2000	1	0.10	1	0.11
≥ 1000	2	0.19	2	0.21
≥ 500	3	0.29	3	0.32
≥ 250	12	1.15	10	1.07
≥ 100	44	4.20	42	4.50
≥ 50	55	5.25	53	5.68
≥ 25	120	11.46	111	11.90
< 25	810	77.36	711	76.21
Total	1047	100	933	100

Source Own elaboration based on WoS 2019

published exceed 1000 citations and 5.89% have at least 100 citations. The articles between 25 and 99 citations represent 17.58% of the published papers, and 76.21% have less than 25 citations. It is important to mention that Table 2 shows that articles and documents have a similar distribution of their percentage of citations, this is because articles and reviews are the two documents that contribute the most to this indicator.

The evolution of the number of publications in CSR-C is an important aspect to be highlighted. During 1996 and 2018, publications in CSR-C has considerably increased, as mentioned this the research in CSR and its influence or relationship with Competitiveness is sort of new, and only in the last ten years, publications have exponentially grown. Figure 1 shows the rise of publications in a five year periods, standing out the last period (1Q) with 706 publications, 67.43% of the total contributions on CSR-C. Period 1Q almost triplicates period 2Q, furthermore these two periods gather 92.17% of all publications. Evidently, this development manifests the positive interest of the researchers in CSR and competitiveness.

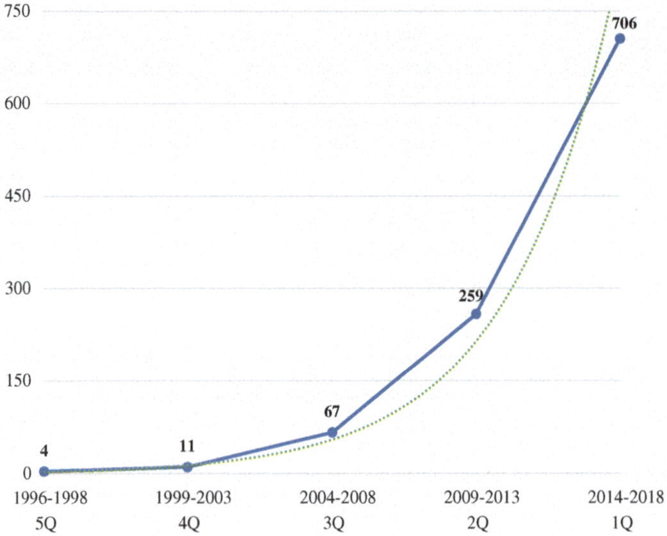

Fig. 1 Evolution of the number of publications on CSR-C in WoS since 1996

In order to present the analysis of the journals, papers, authors, countries, and institutions, it is important to mention the different indicators that have been calculated, such as productivity, total citations, and h-index. According to [10] h-index is related to the number of citations and the published documents, and may represent the importance of journals, institutions, author, and countries. Productivity shows the number of publications of journals, authors, institutions or countries, highlighting the leading ones in the field. Additionally, the total number of citations is the number of citations obtained by the publication. Finally, it has used the VOS viewer to generate graphics maps to show the bibliometric information, co-citations, co-occurrence of keywords and bibliographic coupling.

2.1 The 50 Most Prominent Journals in CSR-Competitiveness (CSR-C)

Journals have the important work of disseminating the contributions and research of CSR, hence it is essential to locate the most influential journals in the field. Table 3 contains the list of the most productive and prominent journals in CSR-C. The leading journal is JBE it has 98 publications about CSR-C, it is the most cited journal with 5313 total citations and it has the highest h-index of 36. In addition, this journal has 8 papers ranked in the 50 most cited papers in CSR-C (this list is presented in the next section). In second place is JCP, a very productive journal with 56 publications, most of them published in period 1Q, this means that this journal is very interested

Table 3 Top journals in CSR-C

R	J	PCSRC	PCSRC (%)	H-CSRC	TC	IF	CP	N50	≥ 1000	≥ 500	≥ 250	≥ 100	≥ 50	≥ 25	1Q	2Q	3Q	4Q	5Q
1	JBE	98	9.36	36	5313	2.917	54.2	8	1	–	2	7	14	25	40	41	11	5	1
2	JCP	56	5.35	20	1602	5.651	28.6	2	–	–	1	1	6	8	51	4	1	–	–
3	SUST	40	3.82	7	183	2.075	4.6	–	–	–	–	–	–	1	40	–	–	–	–
4	CSREM	26	2.48	11	460	4.918	17.7	1	–	–	–	1	1	4	18	8	–	–	–
5	SMJ	18	1.72	12	1424	5.482	79.1	3	–	–	2	1	3	4	12	4	1	–	1
6	BSE	18	1.72	9	283	5.355	15.7	–	–	–	–	–	1	2	14	4	–	–	–
7	SRJ	15	1.43	4	33	–	2.2	–	–	–	–	–	–	–	15	–	–	–	–
8	IJPE	14	1.34	10	1221	4.407	87.2	3	–	1	–	2	1	3	9	5	–	–	–
9	MD	14	1.34	7	201	1.525	14.4	–	–	–	–	–	–	4	7	4	3	–	–
10	JBR	11	1.05	10	624	2.509	56.7	2	–	–	–	2	1	2	6	4	1	–	–
11	BS	11	1.05	6	335	3.214	30.5	1	–	–	–	1	–	2	8	2	1	–	–
12	HBR	9	0.86	2	3279	4.374	364.3	2	1	1	–	–	–	–	–	–	8	1	–
13	JAMS	9	0.86	9	858	8.488	95.3	2	–	–	–	5	2	–	1	8	–	–	–
14	BIJ	9	0.86	3	32	–	3.6	–	–	–	–	–	–	–	9	–	–	–	–
15	BEER	8	0.76	8	339	3.029	42.4	1	–	–	–	1	–	4	2	5	1	–	–
16	AMP	7	0.67	6	754	4.686	107.7	2	–	–	1	1	2	–	3	3	1	–	–
17	EMJ	7	0.67	5	92	2.369	13.1	–	–	–	–	–	–	2	6	1	–	–	–
18	AE	7	0.67	4	58	0.664	8.3	–	–	–	–	–	–	1	1	5	1	–	–
19	JGR	7	0.67	2	11	–	1.6	–	–	–	–	–	–	7	–	–	–	–	–
20	IJTM	6	0.57	6	109	0.869	18.2	–	–	–	–	–	–	1	–	6	–	–	–
21	SD	6	0.57	5	68	2.75	11.3	–	–	–	–	–	–	1	4	2	–	–	–

(continued)

Table 3 (continued)

R	J	PCSRC	PCSRC (%)	H-CSRC	TC	IF	CP	N50	≥ 1000	≥ 500	≥ 250	≥ 100	≥ 50	≥ 25	1Q	2Q	3Q	4Q	5Q
22	AJBM	6	0.57	3	28	1.105	4.7	–	–	–	–	–	–	–	–	6	–	–	–
23	JM	5	0.48	5	605	8.08	121.0	2	–	–	–	3	–	1	2	2	1	–	–
24	IJPDLM	5	0.48	5	592	4.215	118.4	1	–	–	1	–	1	2	2	3	–	–	–
25	JMS	5	0.48	4	547	5.329	109.4	2	–	–	–	3	1	–	1	1	2	–	1
26	LRP	5	0.48	5	389	3.221	77.8	1	–	–	–	1	2	–	2	–	2	1	–
27	JSBM	5	0.48	5	94	3.248	18.8	–	–	–	–	–	1	–	3	2	–	–	–
28	IEEE	5	0.48	5	90	0.709	18.0	–	–	–	–	–	–	1	1	2	2	–	–
29	IJCHM	5	0.48	2	61	2.874	12.2	–	–	–	–	–	–	–	4	1	–	–	–
30	CMS	5	0.48	3	20	0.875	4.0	–	–	–	–	–	–	–	4	1	–	–	–
31	JMK	4	0.38	4	1532	7.338	383.0	1	1	–	–	–	1	1	2	–	1	–	1
32	IJHM	4	0.38	4	248	3.445	62.0	1	–	–	–	1	1	1	2	2	–	–	–
33	JPSM	4	0.38	4	246	3.667	61.5	1	–	–	–	1	1	1	1	3	–	–	–
34	JSM	4	0.38	4	223	2.169	55.8	–	–	–	–	1	1	–	–	4	–	–	–
35	JIBS	4	0.38	4	213	6.198	53.3	1	–	–	–	1	–	2	3	1	–	–	–
36	RCR	4	0.38	3	164	5.12	41.0	1	–	–	–	1	–	1	4	–	–	–	–
37	SCMIJ	4	0.38	3	101	3.833	25.3	–	–	–	–	–	1	1	3	1	–	–	–
38	IJPR	4	0.38	3	90	2.623	22.5	–	–	–	–	–	1	–	3	1	–	–	–
39	ABM	4	0.38	3	26	1.179	6.5	–	–	–	–	–	–	–	–	3	1	–	–
40	QQ	4	0.38	3	18	1.072	4.5	–	–	–	–	–	–	–	3	1	–	–	–
41	EBR	4	0.38	2	11	–	2.8	–	–	–	–	–	–	–	4	–	–	–	–
42	IC	4	0.38	1	9	–	2.3	–	–	–	–	–	–	–	4	–	–	–	–

(continued)

Table 3 (continued)

R	J	PCSRC	PCSRC (%)	H-CSRC	TC	IF	CP	N50	≥ 1000	≥ 500	≥ 250	≥ 100	≥ 50	≥ 25	1Q	2Q	3Q	4Q	5Q
43	MRR	4	0.38	3	9	–	2.3	–	–	–	–	–	–	–	4	–	–	–	–
44	CGIJBS	4	0.38	2	8	–	2.0	–	–	–	–	–	–	–	4	–	–	–	–
45	AMJ	3	0.29	3	709	6.7	236.3	2	–	1	–	1	–	–	1	1	–	1	–
46	JEMS	3	0.29	2	529	1.163	176.3	1	–	–	1	–	1	–	–	2	–	1	–
47	JSCM	3	0.29	3	460	6.105	153.3	2	–	–	–	2	–	1	1	2	–	–	–
48	AMR	3	0.29	3	434	8.855	144.7	–	–	–	1	–	1	2	1	–	–	–	–
49	IJRM	3	0.29	2	419	2.593	139.7	1	–	–	1	–	–	1	1	1	1	–	–
50	EJOR	3	0.29	3	142	3.428	47.3	1	–	–	–	1	–	–	3	–	–	–	–

Source Own elaboration based on WoS 2019. R: Ranking; J: Journal; PCSRC: Total number of documents published in CSR-C; %PCSRC: Percentage of articles published in CSR-C; H-CSRC: H-index only with CSR-C; TC: Total citations; IF: Impact factor; CP: Citations by paper; N50: Number of publications within the 50 most cited papers; ≥ 1000: number of articles with more than 1000 citations; ≥ 500: number of articles with more than 500 citations; ≥ 250: number of articles with more than 250 citations; ≥ 100: number of articles with more than 100 citations; ≥ 50: number of articles with more than 50 citations; ≥ 25: number of articles with more than 25 citations; 1Q: 2014–2018; 2Q: 2009–2013; 3Q: 2004–2008; 4Q: 1999–2003; 5Q: 1996–1998. Journal abbreviations are: JBE: Journal of Business Ethics; JCP: Journal of Cleaner Production; SUST: Sustainability; CSREM: Corporate Social Responsibility and Environmental Management; SMJ: Strategic Management Journal; BSE: Business Strategy and the Environment; SRJ: Social Responsibility Journal; IJPE: International Journal of Production Economics; MD: Management Decision; JBR: Journal of Business Research; BS: Business & Society; HBR: Harvard Business Review; JAMS: Journal of the Academy of Marketing Science; BIJ: Benchmarking an International Journal; BEER: Business Ethics a European Review; AMP: Academy of Management Perspectives; EMJ: European Management Journal; AE: Amfiteatru Economic; JGR: Journal of Global Responsibility; IJTM: International Journal of Technology Management; SD: Sustainable Development; AJBM: African Journal of Business Management; JM: Journal of Management; IJPDLM: International Journal of Physical Distribution & Logistics Management; JMS: Journal of Management Studies; LRP: Long Range Planning; JSBM: Journal of Small Business Management; IEEE: Inzinerine Ekonomika - Engineering Economics; IJCHM: International Journal of Contemporary Hospitality Management; CMS: Chinese Management Studies; JMK: Journal of Marketing; IJHM: International Journal of Hospitality Management; JPSM: Journal of Purchasing and Supply Management; JSM: Journal of Sport Management; JIBS: Journal of International Business Studies; RCR: Resources Conservation and Recycling; SCMIJ: Supply Chain Management an International Journal; IJPR: International Journal of Production Research; ABM: Asian Business Management; QQ: Quality & Quantity; EBR: European Business Review; IC: Intangible Capital; MRR: Management Research Review; CGIJBS: Corporate Governance the International Journal of Business in Society; AMJ: Academy of Management Journal; JEMS: Journal of Economics & Management Strategy; JSCM: Journal of Supply Chain Management; AMR: Academy of Management Review; IJRM: International Journal of Research in Marketing; EJOR: European Journal of Operational Research;

in the field at the moment. Ranked in third place we find SUST, this journal has 40 publications in CSR-C, however, it has a low h-index of 7 and 183 total citations, in fact, any of its articles exceed 40 citations. Moreover, HBR and JMK journals have low h-index and few publications, but their number of total citations by paper are highlighted, each journal has an article with more than 1000 citations. It is important to mention that there are 20 more journals tied with 3 publications, but the list is limited to the top 50 journals selecting the ones with 3 publications based on their total citations. Scopus database does not show major variations with the top journals listed.

2.2 The 50 Most Significant Papers in CSR-C

There are some articles that have a remarkable number of citations obtained, their influence impacts significantly to this prolific field. Table 4 shows the seminal papers related to CSR and Competitiveness. The principal and maybe the most essential work was presented by [85] titled "Strategy and Society: The link between Competitive Advantage and Corporate Social Responsibility" with 2426 of total citations in the WoS core collection. This study highlights the importance and priority of CSR for the enterprises, but they mention that unfortunately, CSR efforts are disconnected from strategy. They have proposed that CSR is more than a cost or obligation, actually, it is a potential source of innovation and competitive advantage, a new way to look the relationship between business and society, a win–win relationship, a mutual benefit, and value creation. CSR rather than as damage control or PR campaign is an opportunity and an important key for competitive success. Porter and Kramer [84] is another article titled "The competitive advantage of corporate philanthropy" with 850 total citations (TC) that mention the importance of strategic corporate philanthropy to improve the competitive context of the enterprises.

Brown and Dacin [17] submitted their paper titled "The company and the product: Corporate associations and consumer product responses". This study analyses the effects of CSR associations or perceptions and the corporate´s capabilities to produce products, the results show that what consumers know about a company affects their attitudes toward new products of that company. Besides, there are different responses of consumers to products from companies with negative associations. In the other hand, [48] presented a seminal work "Corporate Social Responsibility Theories: Mapping the territory". Due to the vast number of approaches, concepts, and theories that surround CSR, the authors classified all the theories into four large groups, ranging from the responsibility of maximizing benefits for shareholders up to the common good: instrumental theories, political theories, integrative theories, ethical theories. The listed studies established the principals and foundations of CSR and Competitiveness as a field of research, and the development of the idea that CSR impacts the competitive context of the firms, or that it is a source of competitive advantages.

Table 4 Top 50 papers in CSR-C

R	Article	Author	J	TC	PY
1	Strategy and society: the link between competitive advantage and corporate social responsibility	Porter, ME; Kramer, MR	HBR	2426	2006
2	The company and the product: corporate associations and consumer product responses	Brown, TJ; Dacin, PA	JM	1419	1997
3	Corporate social responsibility theories: mapping the territory	Garriga, E; Mele, DN	JBE	1150	2004
4	The competitive advantage of corporate philanthropy	Porter, ME; Kramer, MR	HBR	850	2002
5	An organizational theoretic review of green supply chain management literature	Sarkis, J; Zhu, QH; Lai, KH	IJPE	585	2011
6	The impact of environmental technologies on manufacturing performance	Klassen, RD; Whybark, DC	AMJ	545	1999
7	Corporate responsibility and financial performance: the role of intangible resources	Surroca, J; Tribo, JA; Waddock, S	SMJ	491	2010
8	A literature and practice review to develop sustainable business model archetypes	Bocken, NMP; Short, SW; Rana, P; Evans, S	JCP	486	2014
9	Private politics, corporate social responsibility, and integrated strategy	Baron, DP	JEMS	448	2001
10	Sustainable supply chain management: evolution and future directions	Carter, CR; Easton, PL	IJPDLM	444	2011
11	Blackett memorial lecture—humanitarian aid logistics: supply chain management in high gear	Van Wassenhove, LN	JORS	441	2006
12	Corporate social responsibility and resource-based perspectives	Branco, MC; Rodrigues, LL	JBE	419	2006
13	Does it pay to be green? a systematic overview	Ambec, S; Lanoie, P	AMP	397	2008
14	Beyond dichotomy: the curvilinear relationship between social responsibility and financial performance	Barnett, ML; Salomon, RM	SMJ	382	2006

(continued)

Table 4 (continued)

R	Article	Author	J	TC	PY
15	Reaping relational rewards from corporate social responsibility: the role of competitive positioning	Du, SL; Bhattacharya, CB; Sen, S	IJRM	377	2007
16	The contribution of corporate social responsibility to organizational commitment	Brammer, S; Millington, A; Rayton, B	IJHRM	344	2007
17	Research on organizations and the natural environment, 1992-present: a review	Etzion, D	JM	242	2007
18	Corporate social responsibility: three key approaches	Windsor, D	JMS	242	2006
19	Corporate social responsibility reports: a thematic analysis related to supply chain management	Tate, WL; Ellram, LM; Kirchoff, JF	JSCM	226	2010
20	Sustainable operations: their impact on the triple bottom line	Gimenez, C; Sierra, V; Rodon, J	IJPE	214	2012
21	Strategic explanations for the early adoption of ISO 14001	Bansal, P; Hunter, T	JBE	214	2003
22	Creating and capturing value: strategic corporate social responsibility, resource-based theory, and sustainable competitive advantage	McWilliams, A; Siegel, DS	JM	201	2011
23	Sustainable global supplier management: the role of dynamic capabilities in achieving competitive advantage	Reuter, C; Foerstl, K; Hartmann, E; Blome, C	JSCM	197	2010
24	Strategic corporate social responsibility and environmental sustainability	Orlitzky, M; Siegel, DS; Waldman, DA	BS	192	2011
25	Challenges of the "green imperative": a natural resource-based approach to the environmental orientation-business performance relationship	Menguc, B; Ozanne, LK	JBR	188	2005
26	How does corporate social responsibility contribute to firm financial performance? The mediating role of competitive advantage, reputation, and customer satisfaction	Saeidi, SP; Sofian, S; Saeidi, P; Saeidi, SP; Saaeidi, SA	JBR	180	2015

(continued)

Table 4 (continued)

R	Article	Author	J	TC	PY
27	Sustainable development and corporate performance: A study based on the Dow Jones sustainability index	Lopez, MV; Garcia, A; Rodriguez, L	JBE	180	2007
28	Green marketing strategies: an examination of stakeholders and the opportunities they present	Cronin, JJ; Smith, JS; Gleim, MR; Ramirez, E; Martinez, JD	JAMS	172	2011
29	Supply chain management and retailer performance: emerging trends, issues, and implications for research and practice	Ganesan, S; George, M; Jap, S; Palmatier, RW; Weitz, B	JR	168	2009
30	Sustainable supply chain management practices and dynamic capabilities in the food industry: A critical analysis of the literature	Beske, P; Land, A; Seuring, S	IJPE	164	2014
31	Researching corporate social responsibility: An agenda for the twenty-first century	Godfrey, PC; Hatch, NW	JBE	160	2007
32	Green management matters only if it yields more green: an economic/strategic perspective	Siegel, DS	AMP	158	2009
33	Strategic corporate social responsibility and value creation among large firms—lessons from the Spanish experience	Husted, BW; Allen, DB	LRP	152	2007
34	Corporate social responsibility and shareholder reaction: the environmental awareness of investors	Flammer, C	AMJ	151	2013
35	Managing supplier sustainability risks in a dynamically changing environment-Sustainable supplier management in the chemical industry	Foerstl, K; Reuter, C; Hartmann, E; Blome, C	JPSM	151	2010
36	A 'business opportunity' model of corporate social responsibility for small- and medium-sized enterprises	Jenkins, H	BEER	148	2009
37	From green to sustainability: information technology and an integrated sustainability framework	Dao, V; Langella, I; Carbo, J	JSIS	143	2011

(continued)

Table 4 (continued)

R	Article	Author	J	TC	PY
38	Exploring the nature of the relationship between CSR and competitiveness	Vilanova, M; Lozano, JM; Arenas, D	JBE	139	2009
39	Doing good to do well? Corporate social responsibility reasons, practices and impacts in small and medium accommodation enterprises	Garay, L; Font, X	IJHM	133	2012
40	The structure of sustainability research in marketing, 1958–2008: a basis for future research opportunities	Chabowski, BR; Mena, JA; Gonzalez-Padron, TL	JAMS	132	2011
41	An agenda for 'Green' information technology and systems research	Jenkin, TA; Webster, J; McShane, L	IO	131	2011
42	A study of management perceptions of the impact of corporate social responsibility on organisational performance in emerging economies: the case of Dubai	Rettab, B; Ben Brik, A; Mellahi, K	JBE	130	2009
43	With greater power comes greater responsibility? Takeover protection and corporate attention to stakeholders	Kacperczyk, A	SMJ	125	2009
44	Linking supply chain strength to sustainable development: a country-level analysis	Vachon, S; Mao, ZM	JCP	125	2008
45	Multinationals and corporate social responsibility in host countries: does distance matter?	Campbell, JT; Eden, L; Miller, SR	JIBS	124	2012
46	A literature review and perspectives in reverse logistics	Agrawal, S; Singh, RK; Murtaza, Q	RCR	120	2015
47	Corporate Reputation: a Combination of Social Responsibility and Industry	Melo, T; Garrido-Morgado, A	CSREM	120	2012
48	Designing a sustainable closed-loop supply chain network based on triple bottom line approach: a comparison of metaheuristics hybridization techniques	Devika, K; Jafarian, A; Nourbakhsh, V	EJOR	119	2014
49	How corporate social responsibility engagement strategy moderates the CSR-financial performance relationship	Tang, Z; Hull, CE; Rothenberg, S	JMS	116	2012

(continued)

Table 4 (continued)

R	Article	Author	J	TC	PY
50	On the determinants of corporate social responsibility: international evidence on the financial industry	Chih, HL; Chih, HH; Chen, TY	JBE	116	2010

Source Own elaboration based on WoS 2019. R: Ranking; TC: Total citations; PY: Publication year; J: Journal Abbreviations that are not available in the previous table: JORS: Journal of the Operational Research Society; IJHRM: International Journal of Human Resource Management; JR: Journal of Retailing; JSIS: Journal of Strategic Information Systems; IO: Information and Organization

It is interesting to mention that WoS and Scopus have very little variation with the top papers. Actually, WoS has more articles with high total citations, but in Scopus you can find the studies presented by Manaktola and Jauhari [70] with 341 citations, and Dusuki and Abdullah [32] with 183 citations, Werther and Chandler [103] with 141 citations and Thompson and Doherty [96] with 139 total citations.

2.3 *Most Important Proceedings Papers and Books in CSR-C*

There are some valuable and interesting contributions in the field of CSR-C such as conference proceedings and books. Tables 5 and 6 present the Top 3 books and the Top 10 conference proceedings in WoS and Scopus. Books are scientific publications that expose significantly and compiled information of a field. In this case, and even when there are numerous books of CSR and Competitiveness as separated topics, WoS core collection only gathers 6 books of CSR-C and Scopus lists 22 books. Table 5 lists the top 3 books of CSR-C with more than 30 total citations. The book presented by Ehnert, I., and Ehnert, I. [33] titled "Sustainable human resource management: A conceptual and exploratory analysis from a paradox perspective" stands up with 117 citations. Conference proceedings are another type of publication, they are short versions that may be broadened to deeper research and published later in a journal, for example, a study presented by Dorobantu et al. [29], Scopus shows it as a conference proceeding but in WoS is a published article. As it is possible to see in Table 6, there are some conference proceedings with high total citations such as Mackey and Mackey [69] with 328, and [78] with 316 citations.

2.4 *The 50 Most Influential Authors in CSR-C*

In the development of this field, authors are vital since their ideas and research contribute to the proliferation of CSR-C theories, approaches, and future investigations. Table 7 reveals the most influential and productive researchers in the field of

Table 5 Top 3 books in CSR-C

R	Book	Authors	PU	TC	PG	PY	DB
1	Sustainable human resource management: a conceptual and exploratory analysis from a paradox perspective	Ehnert, I	SPRINGER	117	289	2009	WoS
2	The structure of regulatory competition: corporations and public policies in a global economy	Murphy D.D	Oxford University Press	54	336	2012	Scopus
3	Corporate social strategy: stakeholder engagement and competitive advantage	Husted B.W., Allen D.B	Cambridge University Press	37	348	2010	Scopus

Source Own elaboration based on WoS 2019 and Scopus 2019. R: Ranking; PU: Publishing company; TC: Total citations; PY: Publication year; DB: Database

CSR-C. Even when there are almost 70 authors with 3 publications in CSR-C, the list is limited to 50 authors by the most total citation number. The most productive authors with 6 publications are Seuring and Gallardo-Vazquez. Ranked number one, Seuring has 372 TC and one study within the 50 most cited papers in CSR-C, while Gallardo-Vazquez works are recent and only has 72 total citations. They are followed by Husted with 5 publications, total citations of 335 and he has one study within the 50 most cited papers in CSR-C.

However, maybe the most noteworthy author is Zhu with only 3 publications has the highest number of total citations 659, citations by paper of 219.67, and one study within the 50 most cited papers in CSR-C. Actually, many authors with 4 and 3 publications have a significant number of total citations, Siegel production is highlighted since he only has 3 publications, 551 total citations and 3 papers within the 50 most cited papers in CSR-C.

In addition, even when Porter and Kramer are not ranked in this list since their level of productivity is low, these two authors have 3276 total citations received and 655.20 citations by paper, i.e., their two publications are very prominent. Furthermore, these fifty authors compile half of the 50 most cited papers, besides 57% of their studies were published in period 1Q, and 31% in 2Q, this demonstrates that most productive authors are still interested in this topic.

Table 6 Top 10 Conference Proceedings in CSR-C

R	Title	Authors	J	Conference	CY	TC	PY	DB
1	Corporate social responsibility and firm performance: investor preferences and corporate strategies	Mackey, A; Mackey, TB; Barney, JB	AMR	Conference on business as an agent of world benefit	2006	328	2007	WOS
2	Environmental marketing: A source of reputational, competitive, and financial advantage	Miles, MP; Covin, JG	JBE	Marketing exchange colloquium	1998	316	2000	WOS
3	Corporate social responsibility for developing country multinational corporations: lost war in pertaining global competitiveness?	Gugler, P; Shi, JYJ	JBE	2nd International conference on globalization and the good corporation	2007	112/ 135	2009	WOS/Scopus
4	The reasons behind non-ethical behaviour in business and entrepreneurship	Fassin, Y	JBE	Annual confernece of the european-business-ethics-network	2004	49	2005	WOS
5	Corporate social strategy in multinational enterprises: antecedents and value creation	Husted, BW; Allen, DB	JBE	19th Annual conference of the European-business-ethics-network (EBEN)	2006	45	2007	WOS

(continued)

Table 6 (continued)

R	Title	Authors	J	Conference	CY	TC	PY	DB
6	Globalization and the good corporation: a need for proactive co-existence	Sethi, SP	JBE	International conference on business ethics in the knowledge economy	2002	42/ 56	2003	WOS/Scopus
7	Nonmarket strategy research through the lens of new institutional economics: an integrative review and future directions	Dorobantu S., Kaul A., Zelner B	SMJ		2016	39	2017	Scopus
8	Stakeholder perceptions of green marketing: the effect of demand and supply integration	Kirchoff, JF; Koch, C; Nichols, BS	IJPDLM	5th Annual symposium on supply chain management and industrial distribution (SCMID)/conference of the society-of-marketing-advances (SMA)	2010	34	2011	WOS
9	The link of environmental and economic performance: drivers and limitations of sustainability integration	Wagner, M	JBR	Conference on competing in China—local firms, multinationals, and alliances	2013	30	2015	WOS
10	Evaluating community investments in the mining sector using multi-criteria decision analysis to integrate SIA with business planning	Esteves, AM	EIAR	3rd International conference on sustainable development indicators in the minerals industry	2007	30	2008	WOS

Source Own elaboration based on WoS 2019 and Scopus 2019. R: Ranking; CY: Conference year; TC: Total citations; PY: Publication year; DB: Database. J: Journal Abbreviations that are not available in the previous table: EIAR: Environmental Impact Assessment Review

Table 7 Top authors in CSR-C

R	Authors	C	PCSRC	CSRC (%)	H-CSRC	TC	CP	N50	Q1	Q2	Q3	Q4	Q5
1	Seuring S	DEU	6	0.57	6	372	62.00	1	4	2	–	–	–
2	Gallardo–Vazquez D	ESP	6	0.57	2	72	12.00	–	5	1	–	–	–
3	Husted BW	MEX	5	0.48	5	335	67.00	1	1	1	2	1	–
4	Doh JP	USA	5	0.48	4	144	28.80	–	2	3	–	–	–
5	Sanchez–Hernandez MI	ESP	5	0.48	1	55	11.00	–	5	–	–	–	–
6	Toppinen A	FIN	5	0.48	4	46	9.20	–	4	1	–	–	–
7	Du SL	USA	4	0.38	4	508	127.00	1	2	1	1	–	–
8	Blome C	GBR	4	0.38	4	440	110.00	2	2	2	–	–	–
9	Flammer C	USA	4	0.38	4	246	61.50	1	3	1	–	–	–
10	Govindan K	DNK	4	0.38	4	171	42.75	–	4	–	–	–	–
11	Gunasekaran A	USA	4	0.38	3	129	32.25	–	3	1	–	–	–
12	Perry P	GBR	4	0.38	3	68	17.00	–	2	2	–	–	–
13	Martinez–Martinez D	ESP	4	0.38	3	51	12.75	–	4	–	–	–	–
14	Jorge ML	ESP	4	0.38	3	51	12.75	–	4	–	–	–	–
15	Madueno JH	ESP	4	0.38	3	51	12.75	–	4	–	–	–	–
16	Pekovic S	MNE	4	0.38	3	35	8.75	–	4	–	–	–	–
17	Sun WB	USA	4	0.38	2	34	8.50	–	4	–	–	–	–
18	Bernal–Conesa JA	ESP	4	0.38	2	18	4.50	–	4	–	–	–	–
19	Briones–Penalver AJ	ESP	4	0.38	2	18	4.50	–	4	–	–	–	–
20	Nieves–Nieto C	ESP	4	0.38	2	18	4.50	–	4	–	–	–	–
21	Romero–Fernandez PM	ESP	4	0.38	2	14	3.50	–	4	–	–	–	–

(continued)

Table 7 (continued)

R	Authors	C	PCSRC	CSRC (%)	H-CSRC	TC	CP	N50	Q1	Q2	Q3	Q4	Q5
22	Lopez–Perez ME	ESP	4	0.38	2	12	3.00	–	2	2	–	–	–
23	Zhu QH	CHN	3	0.29	3	659	219.67	1	2	1	–	–	–
24	Siegel DS	USA	3	0.29	3	551	183.67	3	–	3	–	–	–
25	Carter CR	USA	3	0.29	3	527	175.67	1	–	3	–	–	–
26	Bhattacharya CB	USA	3	0.29	3	491	163.67	1	1	1	1	–	–
27	Sen S	USA	3	0.29	3	491	163.67	1	1	1	1	–	–
28	Brammer S	AUS	3	0.29	3	486	162.00	1	–	2	1	–	–
29	Hartmann E	DEU	3	0.29	3	353	117.67	2	1	2	–	–	–
30	Bansal P	CAN	3	0.29	3	336	112.00	1	1	1	–	1	–
31	Etzion D	CAN	3	0.29	2	316	105.33	1	1	1	1	–	–
32	Allen DB	USA	3	0.29	3	214	71.33	1	1	–	2	–	–
33	Walker M	USA	3	0.29	3	158	52.67	–	1	2	–	–	–
34	O'Donohue W	AUS	3	0.29	2	148	49.33	–	–	3	–	–	–
35	Torugsa NA	THA	3	0.29	2	148	49.33	–	–	3	–	–	–
36	Wagner M	DEU	3	0.29	3	122	40.67	–	1	2	–	–	–
37	Hull CE	USA	3	0.29	2	121	40.33	1	2	1	–	–	–
38	Tang Z	USA	3	0.29	2	121	40.33	1	2	1	–	–	–
39	Kaufmann L	DEU	3	0.29	3	120	40.00	–	1	2	–	–	–
40	Boehe DM	NZL	3	0.29	3	111	37.00	–	–	2	1	–	–
41	Cruz LB	CAN	3	0.29	3	111	37.00	–	–	2	1	–	–
42	Delmas MA	USA	3	0.29	3	98	32.67	–	2	1	–	–	–

(continued)

Table 7 (continued)

R	Authors	C	PCSRC	CSRC (%)	H-CSRC	TC	CP	N50	Q1	Q2	Q3	Q4	Q5
43	Patari S	FIN	3	0.29	3	72	24.00	–	2	1	–	–	–
44	Jamali D	LBN	3	0.29	2	58	19.33	–	2	1	–	–	–
45	Jantunen A	FIN	3	0.29	3	52	17.33	–	1	2	–	–	–
46	Tuppura A	FIN	3	0.29	3	48	16.00	–	2	1	–	–	–
47	Ayuso S	ESP	3	0.29	3	38	12.67	–	2	1	–	–	–
48	Aravind D	USA	3	0.29	2	35	11.67	–	2	1	–	–	–
49	Arevalo JA	USA	3	0.29	2	35	11.67	–	2	1	–	–	–
50	Peng MW	USA	3	0.29	2	30	10.00	–	3	–	–	–	–

Source Own elaboration based on WoS 2019; R: Ranking; C: Country; PCSRC: Total number of documents published in CSR-C; %PCSRC: Percentage of articles published in CSR-C; H-CSRC: H-index only with CSR-C; TC: Total citations; CP: Citations by paper; N50: Number of publications within the 50 most cited papers; 1Q: 2014–2018; 2Q: 2009–2013; 3Q: 2004–2008; 4Q: 1999–2003; 5Q: 1996–1998

2.5 The 50 Most Influential Institutions in CSR-C

Institutions such as Universities support their researchers and enable the research in certain fields. Table 8 displays the 54 most notable universities that have promoted the development of CSR-C. The University of Manchester leads this rank with 13 publications, although it only has 112 total citations and most of its works are from 1Q period. The University of Western Ontario has 11 publications, 1169 total citations, 3 publications in the 50 most cited articles and an h-index of 9. It is followed by Hong Kong Polytechnic University and Bucharest University of Economic Studies with 10 publications.

Another interesting indicator is the total cited (TC) and the citations by paper (CP), in this criteria Harvard University is the most prominent institutions due to the work of [84, 85], these two papers give to this university most of its TC of 3364 and 560.67 CP. Texas A&M University College Station has 7 publications, 1598 TC and 228.29 CP. Moreover, the University of Cambridge with only 5 publications, has 879 TC and 175.8 citations by paper. Besides more than 91% of the works were published in 1Q and 2Q periods.

There are some University Systems that contribute to the research of CSR-C such as the State University System of Florida, University of North Carolina, Pennsylvania Commonwealth System of Higher Education, University of Texas System, University of London, all of them with 10 publications or more.

2.6 The 50 Most Influential Countries in CSR-C

Knowledge allows countries to position, develop, grow, and improve their economy and social welfare. This knowledge can be achieved through scientific research, which is why many countries strive to promote and support scientific development. Particularly in this field, there are several countries that have shown interest, as it is possible to see in Fig. 2, these countries are highlighted with green. Countries like the United States, Spain, China, and England show a stronger green color, being the countries with the highest number of publications in CSR-C, more than 100 works. As the intensity of color decreases, countries with fewer publications are found.

Table 9 presents the 50 most influential countries in the CSR-C field. In first place is The United States of America (USA) as the most productive country with 262 publications, which represents 25.02% of the total productivity on CSR-C. The USA has the highest total citations 15,373, a citation index of 50, and 58.68 citations by paper. Besides this country possess 26 articles within the 50 most cited papers. Next, we find Spain (ESP) with 121 publications, 3972 total citations and 9 articles within the N50. The United Kingdom is ranked third with 106 publications and 3420 total citations. Countries such as China, Canada, Germany, and France have a high number of total citations. USA and China are the countries with more number of publications

Table 8 Top institutions in CSR-C

R	Institutions	C	PCSRC	PCSRC (%)	H-CSRC	TC	CP	N50	Q1	Q2	Q3	Q4	Q5
1	University of Manchester	GBR	13	1.24	6	112	8.62	–	12	1	–	–	–
2	University of Western Ontario	CAN	11	1.05	9	1168	106.18	3	5	4	–	2	–
3	Hong Kong Polytechnic University	CHN	10	0.96	6	735	73.50	1	8	2	–	–	–
4	Bucharest University of Economic Studies	ROU	10	0.96	5	117	11.70	–	2	7	1	–	–
5	Copenhagen Business School	DNK	9	0.86	7	136	15.11	–	5	4	–	–	–
6	Universidad de Extremadura	ESP	9	0.86	2	73	8.11	–	9	–	–	–	–
7	Universidad de Cadiz	ESP	8	0.76	5	65	8.13	–	8	–	–	–	–
8	Texas A&M University College Station	USA	7	0.67	5	1598	228.29	2	5	1	–	–	1
9	Arizona State University	USA	7	0.67	6	748	106.86	2	3	4	–	–	–
10	Hec Montreal (University of Montreal)	CAN	7	0.67	5	623	89.00	2	3	2	2	–	–
11	Universitat Kassel	DEU	7	0.67	7	419	59.86	1	5	2	–	–	–
12	IE University	ESP	7	0.67	5	384	54.86	1	2	2	3	–	–
13	Tecnologico de Monterrey	MEX	7	0.67	5	336	48.00	1	3	1	2	1	–
14	University of Granada	ESP	7	0.67	6	279	39.86	1	4	2	1	–	–
15	Penn State University	USA	7	0.67	5	248	35.43	1	4	3	–	–	–
16	University of Kent	GBR	7	0.67	6	221	31.57	–	6	1	–	–	–
17	University of Minnesota System (Twin Cities)	USA	7	0.67	5	215	30.71	–	4	1	2	–	–
18	University of Nottingham	GBR	7	0.67	4	144	20.57	–	5	1	1	–	–

(continued)

Table 8 (continued)

R	Institutions	C	PCSRC	PCSRC (%)	H-CSRC	TC	CP	N50	Q1	Q2	Q3	Q4	Q5
19	Lappeenranta University of Technology	FIN	7	0.67	5	135	19.29	–	5	2	–	–	–
20	City University of Hong Kong	CHN	7	0.67	5	120	17.14	–	7	–	–	–	–
21	University of New South Wales Sydney	AUS	7	0.67	3	76	10.86	–	6	1	–	–	–
22	Universidad Politecnica de Cartagena	ESP	7	0.67	4	58	8.29	–	6	1	–	–	–
23	University of Zaragoza	ESP	7	0.67	3	17	2.43	–	6	1	–	–	–
24	Harvard University	USA	6	0.57	6	3364	560.67	2	2	2	1	1	–
25	European Business School EBS University	DEU	6	0.57	5	563	93.83	2	2	4	–	–	–
26	Baruch College Cuny	USA	6	0.57	5	541	90.17	1	2	2	1	1	–
27	Villanova University	USA	6	0.57	5	241	40.17	–	2	3	1	–	–
28	Nanyang Technological University	SGP	6	0.57	4	214	35.67	–	5	1	–	–	–
29	Shanghai Jiao Tong University	CHN	6	0.57	5	126	21.00	–	6	–	–	–	–
30	University of Helsinki	FIN	6	0.57	4	48	8.00	–	5	1	–	–	–
31	Universidad De Jaen	ESP	6	0.57	3	36	6.00	–	4	1	1	–	–
32	Universiti Sains Malaysia	MYS	6	0.57	3	30	5.00	–	5	1	–	–	–
33	University of Cambridge	GBR	5	0.48	4	879	175.80	1	3	–	1	1	–
34	University of Bath	GBR	5	0.48	4	520	104.00	1	2	2	1	–	–
35	Ohio State University	USA	5	0.48	4	418	83.60	–	4	–	1	–	–
36	Universitat Ramon Llull (ESADE)	ESP	5	0.48	5	403	80.60	2	2	3	–	–	–
37	University of Florida	USA	5	0.48	4	349	69.80	1	2	3	–	–	–

(continued)

Table 8 (continued)

R	Institutions	C	PCSRC	PCSRC (%)	H-CSRC	TC	CP	N50	Q1	Q2	Q3	Q4	Q5
38	University of Tennessee Knoxville	USA	5	0.48	4	335	67.00	1	2	3	–	–	–
39	University of Melbourne	AUS	5	0.48	5	301	60.20	1	2	1	2	–	–
40	University of Michigan	USA	5	0.48	3	236	47.20	1	2	3	–	–	–
41	University of Southern Denmark	DNK	5	0.48	4	175	35.00	–	5	–	–	–	–
42	Bocconi University	ITA	5	0.48	5	143	28.60	–	2	1	2	–	–
43	Northeastern University	USA	5	0.48	4	143	28.60	–	4	1	–	–	–
44	University of Murcia	ESP	5	0.48	4	95	19.00	–	4	1	–	–	–
45	Aalto University	FIN	5	0.48	4	79	15.80	–	3	2	–	–	–
46	Southwestern University of Finance Economics	CHN	5	0.48	4	77	15.40	–	5	–	–	–	–
47	Fudan University	CHN	5	0.48	4	73	14.60	–	5	–	–	–	–
48	Pompeu Fabra University	ESP	5	0.48	5	57	11.40	–	4	1	–	–	–
49	York University Canada	CAN	5	0.48	3	52	10.40	–	2	3	–	–	–
50	University of Nottingham Ningbo China	CHN	5	0.48	3	40	8.00	–	5	–	–	–	–
51	Wageningen University Research	NDL	5	0.48	3	39	7.80	–	4	1	–	–	–
52	University of Sevilla	ESP	5	0.48	3	33	6.60	–	4	1	–	–	–
53	Curtin University	AUS	5	0.48	3	32	6.40	–	4	1	–	–	–
54	Tsinghua University	CHN	5	0.48	3	11	2.20	–	3	2	–	–	–

Source Own elaboration based on WoS 2019; R: Ranking; C: Country; PCSRC: Total number of documents published in CSR-C; %PCSRC: Percentage of articles published in CSR-C; H-CSRC: H-index only with CSR-C; TC: Total citations; CP: Citations by paper; N50: Number of publications within the 50 most cited papers; 1Q: 2014–2018; 2Q: 2009–2013; 3Q: 2004–2008; 4Q: 1999–2003; 5Q: 1996–1998

Fig. 2 Countries that support CSR-C research

in period 1Q. Furthermore, the 70% of the publications made by these countries are from the last 5 years (1Q) and 23% from 2Q, for a total of 93% in the last 10 years.

3 VOS Viewer

VOS viewer software is used in this section to analyze the bibliographic information in a graphical way, considering the co-occurrence of author keywords that measures the most common keywords that appear more frequently in the same papers, co-citation that occurs when two documents receive a citation from the same third document and the bibliographic coupling appears when two documents cite the same third document [10]. The co-occurrence of author keywords in CSR-C documents reveals the most common keywords that appear in the papers and the most frequent words in the same documents. Figure 3 shows the keywords considering a one hundred strongest co-occurrence links and a minimum threshold of ten occurrences. The words with the highest number of occurrence are Corporate Social Responsibility (CSR), sustainability, competitive advantage, competitiveness. Furthermore, there are some other words that are common to finds within these documents, such as sustainable development, performance, reputation, innovation, supply chain, stakeholder, ethics, governance, resource-based view. This result shows the different approaches and topics of interest involving this field. Some studies of CSR-C have considered reputation, innovation, and some others topics, presented here as keywords, as a mediating variables or outcomes. It is highlighted the colors of the links, this colors helps to understand the orientations for some potential areas for

Table 9 Top countries in CSR-C

R	C	PCSRC	PCSRC (%)	H-CSRC	TC	CP	N50	Q1	Q2	Q3	Q4	Q5
1	USA	262	25.02	50	15,373	58.68	26	161	74	20	6	1
2	ESP	121	11.56	22	3972	32.83	9	83	31	7	–	–
3	GB–ENG	106	10.12	28	3420	32.26	4	67	27	8	2	2
4	CHN	103	9.84	20	1611	15.64	1	93	9	1	–	–
5	CAN	65	6.21	25	2768	42.58	6	35	24	4	2	–
6	DEU	54	5.16	22	1663	30.80	3	31	22	1	–	–
7	AUS	52	4.97	14	871	16.75	1	35	11	6	–	–
8	ITA	48	4.58	14	626	13.04	–	36	9	3	–	–
9	TWN	36	3.44	10	495	13.75	1	28	8	–	–	–
10	FRA	33	3.15	15	1379	41.79	2	25	6	2	–	–
11	IND	31	2.96	8	385	12.42	1	27	3	1	–	–
12	KOR	30	2.87	7	228	7.60	–	29	1	–	–	–
13	POL	25	2.39	5	61	2.44	–	19	6	–	–	–
14	FIN	23	2.20	8	234	10.17	–	17	6	–	–	–
15	DNK	21	2.01	10	542	25.81	1	17	4	–	–	–
16	NLD	21	2.01	8	195	9.29	–	16	4	1	–	–
17	PRT	19	1.81	8	702	36.95	1	15	3	1	–	–
18	MYS	19	1.81	7	311	16.37	1	15	3	1	–	–
19	BRA	19	1.81	8	231	12.16	–	12	5	2	–	–
20	ROU	19	1.81	6	141	7.42	–	8	9	2	–	–
21	TUR	16	1.53	8	212	13.25	–	11	5	–	–	–
22	SWE	14	1.34	6	147	10.50	–	11	3	–	–	–
23	MEX	12	1.15	6	363	30.25	1	8	1	2	1	–
24	BEL	11	1.05	8	386	35.09	–	7	2	2	–	–
25	CHE	11	1.05	7	316	28.73	–	2	9	–	–	–
26	ARE	11	1.05	3	192	17.45	1	9	2	–	–	–
27	GB–SCT	11	1.05	5	158	14.36	–	6	4	1	–	–
28	LTU	10	0.96	7	156	15.60	–	6	2	2	–	–
29	AUT	10	0.96	4	110	11.00	–	6	3	1	–	–
30	GRC	10	0.96	5	63	6.30	–	7	3	–	–	–
31	ZAF	9	0.86	4	127	14.11	–	8	1	–	–	–
32	IRN	8	0.76	6	443	55.38	2	8	–	–	–	–
33	NZL	8	0.76	4	360	45.00	1	5	–	3	–	–
34	SGP	8	0.76	5	223	27.88	–	7	1	–	–	–
35	JPN	8	0.76	3	22	2.75	–	4	3	1	–	–

(continued)

Table 9 (continued)

R	C	PCSRC	PCSRC (%)	H-CSRC	TC	CP	N50	Q1	Q2	Q3	Q4	Q5
36	SVN	6	0.57	4	103	17.17	–	3	3	–	–	–
37	RUS	6	0.57	2	31	5.17	–	6	–	–	–	–
38	THA	6	0.57	2	11	1.83	–	6	–	–	–	–
39	NOR	5	0.48	3	80	16.00	–	4	–	1	–	–
40	GB–NIR	5	0.48	3	30	6.00	–	5	–	–	–	–
41	PAK	5	0.48	2	23	4.60	–	5	–	–	–	–
42	GHA	5	0.48	2	16	3.20	–	5	–	–	–	–
43	SRB	5	0.48	1	5	1.00	–	3	2	–	–	–
44	COL	4	0.38	2	35	8.75	–	4	–	–	–	–
45	CZE	4	0.38	1	10	2.50	–	4	–	–	–	–
46	IDN	4	0.38	1	2	0.50	–	4	–	–	–	–
47	GB–WLS	3	0.29	2	182	60.67	1	2	1	–	–	–
48	CYP	3	0.29	3	64	21.33	–	2	1	–	–	–
49	LBN	3	0.29	2	58	19.33	–	2	1	–	–	–
50	SVK	3	0.29	2	30	10.00	–	3	–	–	–	–

Source Own elaboration based on WoS 2019; R: Ranking; C: Country; PCSRC: Total number of documents published in CSR-C; %PCSRC: Percentage of articles published in CSR-C; H-CSRC: H-index only with CSR-C; TC: Total citations; CP: Citations by paper; N50: Number of publications within the 50 most cited papers; 1Q: 2014–2018; 2Q: 2009–2013; 3Q: 2004–2008; 4Q: 1999–2003; 5Q: 1996–1998

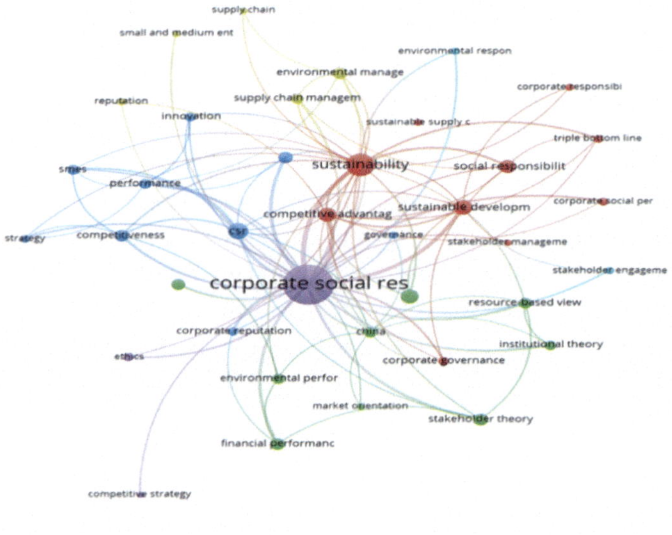

Fig. 3 Co-occurrence of author keywords in CSR-C

research. For example, sustainability has a large connections with many variables related with CSR-C.

Co-citation of journals and authors in CSR-C illustrate the most cited journals or authors, and those that repeatedly receive citations from the same sources. Figure 4 presents the co-citation of journals and Fig. 5 the co-citation of authors results considering 100 strongest co-citations links and a minimum threshold of 100 citations. As it is possible to see in Fig. 4 the Journal of Business Ethics is highlighted. Besides this figure shows four different approaches, first the journals colored in red related to business ethics, society and governance, led by the Journal of Business Ethics. Then the journals related to management and strategy are colored in green, and the principal journals are Strategic Management Journal and Academic Management Review. The journals colored in yellow are related to marketing and reputation and finally, the journals colored in blue are related to operation, production, and environment.

Figure 5 shows many of the most relevant and cited authors, Porter ME, Carroll AB, and McWilliams A. Furthermore, there are three principal slopes, the biggest one is colored in red and presents authors such as Porter, McWilliams, Husted, Freeman, and Orlitzky focused mainly in the strategic part of CSR-C and its performance and impact in the firms. The blue slope introduces a well-known author in the CSR field for its major contributions to the topic, Carroll, AB. The last slope colored in green, display authors as Barney, that focuses on Resource-based view, and other authors related to sustainable and environmental management.

Next, it analyzes bibliographic coupling in CSR-C publications. Figure 6 visualizes the results from the bibliographic coupling of journals, considering the journals with the highest number of publications and those that are frequently cited in the

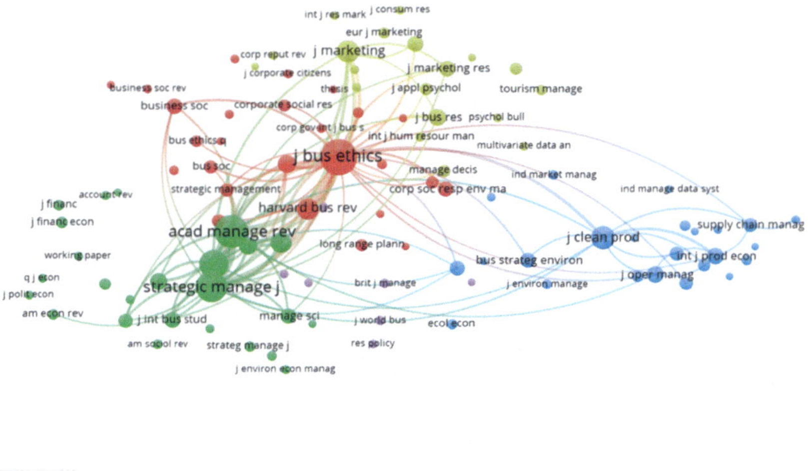

Fig. 4 Co-citation of journals among CSR-C publications

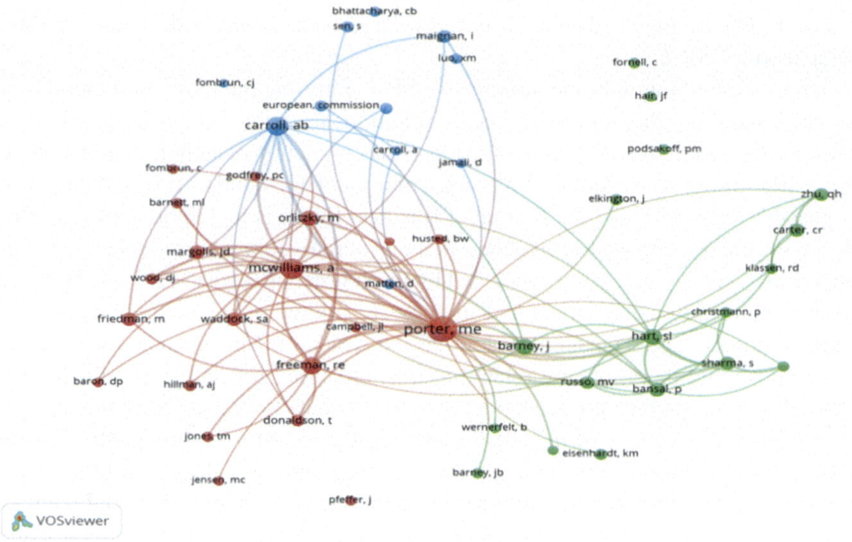

Fig. 5 Co-citation of authors among CSR-C publications

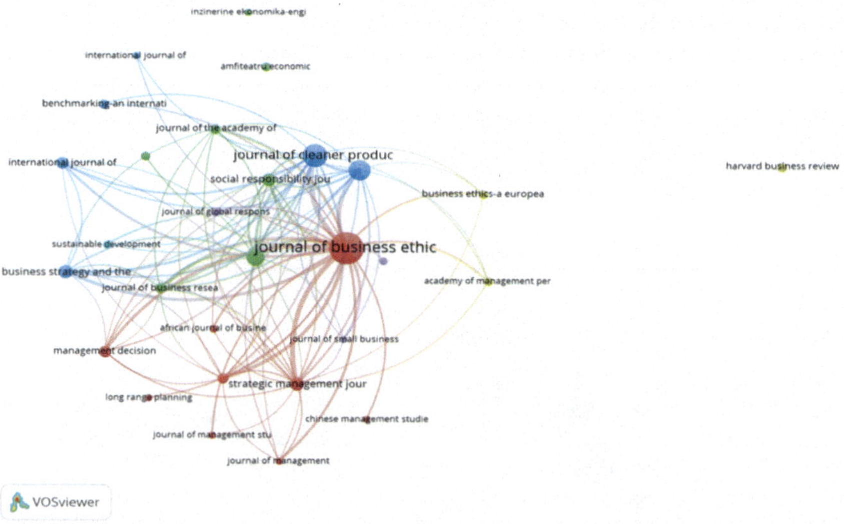

Fig. 6 Bibliographic coupling of journals in CSR-C publications

same sources, based on the one hundred strongest bibliographic coupling connections and a minimum threshold of 5 publications. The Journal of Business Ethics and Journal of Cleaner production are the most productive journals in CSR-C, this is consistent with the information presented in Table 3.

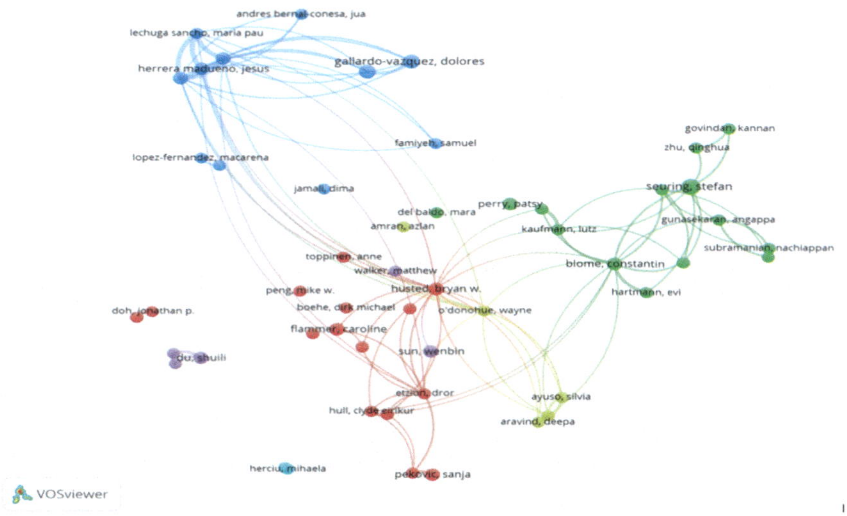

Fig. 7 Bibliographic coupling of authors in CSR-C publications

In this sense, a bibliographic coupling of authors maps the most productive authors in CSR-C and those that more frequently cite the same documents. Figure 7 presents the analysis of bibliographic coupling of authors based on 100 strongest bibliographic coupling connections and a threshold of 3 papers. The productivity level of publications by author in CSR-C is not very high, Seuring S and Gallardo-Vazquez D are the most productive authors with only 6 publications.

Bibliographic coupling of institutions is presented in Fig. 8, this map indicates the institutions where their authors more frequently publish documents in CSR-C, considering a threshold of 5 documents at least and one hundred strongest bibliographic coupling connections. The University of Manchester, Hong Kong Polytechnic University and the University of Western Ontario are the most productive institutions in CSR-C. Another interesting aspect is that American universities lead and highlight in Fig. 8, followed by Spanish universities.

Finally, the bibliographic coupling of countries, considering a 100 strongest bibliographic coupling links and a minimum threshold of 5 publications, the results are presented in Fig. 9. The USA, Spain, and England are the most productive countries in CSR-C field, nevertheless, the USA level of publications doubles the production of the other countries mentioned.

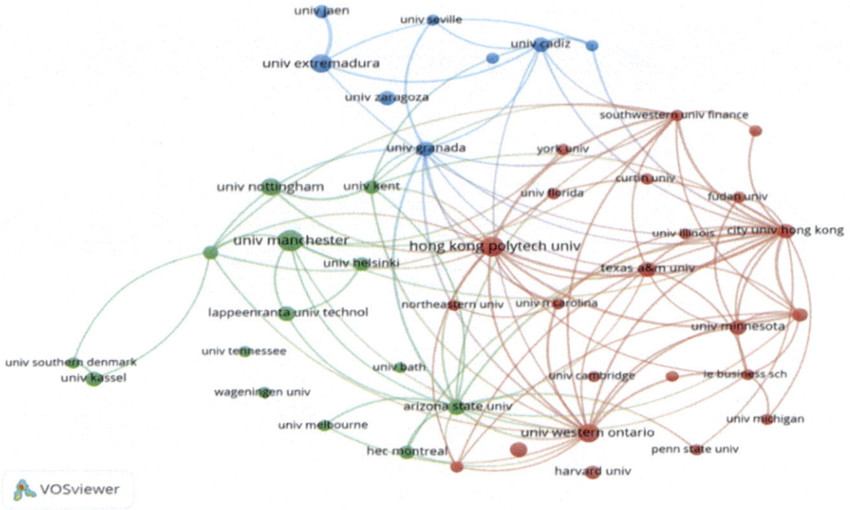

Fig. 8 Bibliographic coupling of institutions in CSR-C publications

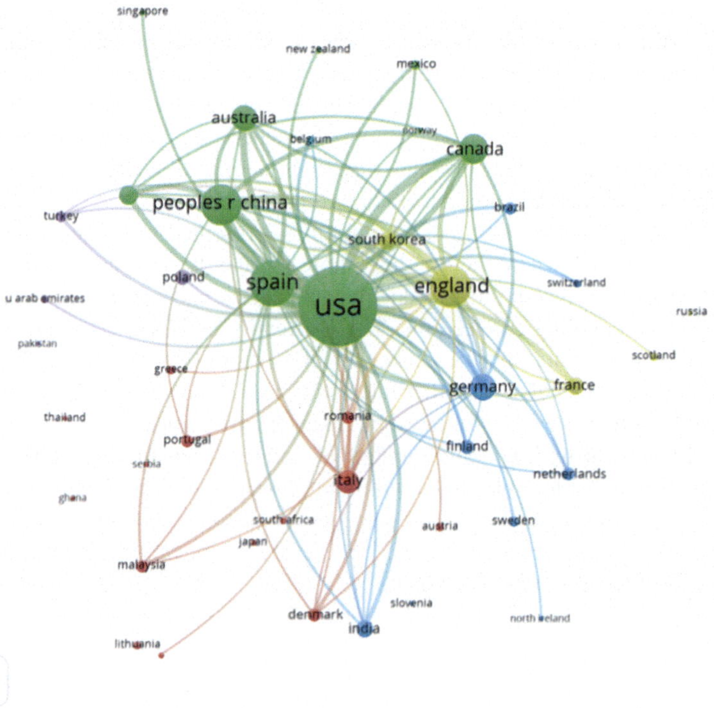

Fig. 9 Bibliographic coupling of countries in CSR-C publications

4 Conclusions

This work presents a bibliometric analysis, an overview of the CSR-C field based mainly on the Web of Science (Wos) database, with a sample of 1047 publications between 1975 and 2018. One of the most interesting findings is the accelerated growth of papers published in the last 5 years, which constitute more than 65% of the total publications, this indicates the importance and relevance of CSR-C at this moment. Another outstanding finding is the impact of the most cited article "Strategy and society: the link between Competitive Advantage and Corporate Social Responsibility". This article was published in 2006 by Porter and Kramer and has 2426 citations. It is interesting to mention that these two authors have only one more paper related to CSR-C named "The competitive advantage of corporate philanthropy", thus according to this sample, these two articles represent 10% of the total citations, granting to Porter and Kramer the recognition of the most influential authors in the field.

The analysis indicates that the journals with more publications in CSR-C are Journal of Business Ethics, Journal of Cleaner Production and Sustainability. The most cited journals are Journal of Business Ethics with 5313 citations and 8 papers rank in the 50 most significant papers, the second most cited journal is Harvard Business Review with 3279 citations. Citations by paper is another important indicator, the journal leading this factor is the Journal of Marketing with 1 paper with more than 1000 citations.

The analysis of the most cited papers shows that 48 of the top 50 papers rank were published from 2000 to 2015, besides more than the 30% of the total citations are from published since 2010. Four of the top 10 cited papers were published from 2010 to the present, indicating the current relevance of CSR-C.

The most important and productive authors are Seuring from Germany, Gallardo-Vazquez from Spain and Husted from Mexico. However, it is highlighted that the most cited author from this analysis is Zhu from China. In this regard, most of these top authors are from the United States and Spain, reinforcing their positions as the most influential countries in CSR-C. The analysis from the top institutions shows the same pattern with the important presence of institutions from the United States and Spain, followed by institutions from England and China. These results help academics and researchers to create connections and networks in the topic.

Another important contribution of this work is presented through the VOS viewer, which presents more graphically and visually the links, connections, and patterns between the different indicators. This analysis reinforces most of the results presented above, showing connections, and additional slopes, authors, and important keywords.

Finally, there are several limitations due to the methodology that was employed to obtain the results. The Web of Science (WoS) was the main database for the analysis and Scopus database only complement a small part of the results, especially with some important cited papers, books, and conference proceedings. Not all the journals or important contributions are incorporated in these databases, furthermore incomplete or incorrect coverage of publications in CSR-C field are another important limitation.

Future research will focus on some other issues related to CSR such as performance, innovation, reputation, among others.

References

1. Agrawal S, Singh RK, Murtaza Q (2015) A literature review and perspectives in reverse logistics. Resour Conserv Recycl 97:76–92
2. Aguinis H, Glavas A (2012) What we know and don't know about corporate social responsibility: a review and research agenda. J Manag 38(4):932–968
3. Ambec S, Lanoie P (2008) Does it pay to be green? A systematic overview. The Academy of Management Perspectives 45–62
4. Bansal P, Hunter T (2003) Strategic explanations for the early adoption of ISO 14001. J Bus Ethics 46(3):289–299
5. Barnett ML, Salomon RM (2006) Beyond dichotomy: the curvilinear relationship between social responsibility and financial performance. Strateg Manag J 27(11):1101–1122
6. Baron DP (2001) Private politics, corporate social responsibility, and integrated strategy. Journal of Economics & Management Strategy 10(1):7–45
7. Battaglia M, Testa F, Bianchi L, Iraldo F, y Frey M (2014) Corporate social responsibility and competitiveness within SMEs of the fashion industry: evidence from Italy and France. Sustainability 6:872–893
8. Bernal-Conesa JA, de Nieves Nieto C, Briones-Peñalver AJ (2017) CSR strategy in technology companies: its influence on performance, competitiveness and sustainability. Corp Soc Responsib Environ Manag 24(2):96–107
9. Beske P, Land A, Seuring S (2014) Sustainable supply chain management practices and dynamic capabilities in the food industry: A critical analysis of the literature. Int J Prod Econ 152:131–143
10. Blanco-Mesa, F., León-Castro, E., & Merigó, J. M. (2019). A bibliometric analysis of aggregation operators. *Applied Soft Computing*, 105488.
11. Blanco-Mesa F, Merigó JM, Gil-Lafuente AM (2017) Fuzzy decision making: A bibliometric-based review. Journal of Intelligent & Fuzzy Systems 32(3):2033–2050
12. Bocken NM, Short SW, Rana P, Evans S (2014) A literature and practice review to develop sustainable business model archetypes. J Clean Prod 65:42–56
13. Boulouta I, Pitelis CN (2014) Who needs CSR? The impact of corporate social responsibility on national competitiveness. J Bus Ethics 119(3):349–364
14. Bowen HR (1953) Social responsibility ofthe businessman. Harpers and Brothers, New York
15. Brammer S, Millington A, Rayton B (2007) The contribution of corporate social responsibility to organizational commitment. The International Journal of Human Resource Management 18(10):1701–1719
16. Branco MC, Rodrigues LL (2006) Corporate social responsibility and resource-based perspectives. J Bus Ethics 69(2):111–132
17. Brown TJ, Dacin PA (1997) The company and the product: Corporate associations and consumer product responses. J Mark 61(1):68–84
18. Campbell JT, Eden L, Miller SR (2012) Multinationals and corporate social responsibility in host countries: Does distance matter? J Int Bus Stud 43(1):84–106
19. Carroll AB, Shabana KM (2010) The business case for corporate social responsibility: A review of concepts, research and practice. Int J Manag Rev 12(1):85–105
20. Carter CR, Liane Easton P (2011) Sustainable supply chain management: evolution and future directions. Int J Phys Distrib Logist Manag 41(1):46–62
21. Chabowski BR, Mena JA, Gonzalez-Padron TL (2011) The structure of sustainability research in marketing, 1958–2008: a basis for future research opportunities. J Acad Mark Sci 39(1):55–70

22. Chih HL, Chih HH, Chen TY (2010) On the determinants of corporate social responsibility: International evidence on the financial industry. J Bus Ethics 93(1):115–135
23. Chung C, Jung S, Young J (2018) Do CSR activities increase firm value? Evidence from the Korean market. Sustainability 10(9):3164
24. Cone Communications/Ebiquity (2015). 2015 Cone Communications/Ebiquity–Global CSR study. http://www.conecomm.com/2015-cone-communications-ebiquity-global-csr-study-pdf
25. Cronin JJ, Smith JS, Gleim MR, Ramirez E, Martinez JD (2011) Green marketing strategies: an examination of stakeholders and the opportunities they present. J Acad Mark Sci 39(1):158–174
26. Dao V, Langella I, Carbo J (2011) From green to sustainability: Information Technology and an integrated sustainability framework. J Strateg Inf Syst 20(1):63–79
27. De Bakker FG, Groenewegen P, Den Hond F (2005) A bibliometric analysis of 30 years of research and theory on corporate social responsibility and corporate social performance. Bus Soc 44(3):283–317
28. Devika K, Jafarian A, Nourbakhsh V (2014) Designing a sustainable closed-loop supply chain network based on triple bottom line approach: A comparison of metaheuristics hybridization techniques. Eur J Oper Res 235(3):594–615
29. Dorobantu S, Kaul A, Zelner B (2017) Nonmarket strategy research through the lens of new institutional economics: An integrative review and future directions. Strateg Manag J 38(1):114–140
30. Du S, Bhattacharya CB, Sen S (2007) Reaping relational rewards from corporate social responsibility: The role of competitive positioning. Int J Res Mark 24(3):224–241
31. Du S, Bhattacharya CB, Sen S (2011) Corporate social responsibility and competitive advantage: Overcoming the trust barrier. Manage Sci 57(9):1528–1545
32. Dusuki AW, Abdullah NI (2007) Why do Malaysian customers patronise Islamic banks? International Journal of Bank Marketing 25(3):142–160
33. Ehnert I, Ehnert I (2009) Sustainable human resource management. Physica-Verlag
34. Esteves AM (2008) Evaluating community investments in the mining sector using multi-criteria decision analysis to integrate SIA with business planning. Environ Impact Assess Rev 28(4–5):338–348
35. Etzion D (2007) Research on organizations and the natural environment, 1992-present: A review. J Manag 33(4):637–664
36. European Commission (2011) Communication from the commission to the European parliament, the council, the European economic and social committee and the committee of the regions. European Commission, Brussels
37. Fassin Y (2005) The reasons behind non-ethical behaviour in business and entrepreneurship. J Bus Ethics 60(3):265–279
38. Feng Y, Zhu Q, Lai KH (2017) Corporate social responsibility for supply chain management: A literature review and bibliometric analysis. J Clean Prod 158:296–307
39. Flammer C (2013) Corporate social responsibility and shareholder reaction: The environmental awareness of investors. Acad Manag J 56(3):758–781
40. Flammer C, Luo J (2017) Corporate social responsibility as an employee governance tool: Evidence from a quasi-experiment. Strateg Manag J 38(2):163–183
41. Foerstl K, Reuter C, Hartmann E, Blome C (2010) Managing supplier sustainability risks in a dynamically changing environment—Sustainable supplier management in the chemical industry. J Purch Supply Manag 16(2):118–130
42. Galbreath J, Shum P (2012) Do customer satisfaction and reputation mediate the CSR–FP link? Evidence from Australia. Aust J Manag 37(2):211–229
43. Gallardo-Vazquez D, Sanchez-Hernandez I (2012) Information on corporate social responsibility and SME's environmental responsiveness: A regional study. Economics & Sociology 5(2):103
44. Galvão A, Mendes L, Marques C, Mascarenhas C (2019) Factors influencing students' corporate social responsibility orientation in higher education. J Clean Prod 215:290–304

45. Ganesan S, George M, Jap S, Palmatier RW, Weitz B (2009) Supply chain management and retailer performance: emerging trends, issues, and implications for research and practice. J Retail 85(1):84–94
46. Garay L, Font X (2012) Doing good to do well? Corporate social responsibility reasons, practices and impacts in small and medium accommodation enterprises. Int J Hosp Manag 31(2):329–337
47. Garelli S (2018) 30 years of competitiveness research. IMD international institute for management development. https://www.imd.org/contentassets/3c6bc732e8864fb0a1d7ee79d664fefb/tc036-18.pdf
48. Garriga E, Melé D (2004) Corporate social responsibility theories: Mapping the territory. J Bus Ethics 53(1–2):51–71
49. Gimenez C, Sierra V, Rodon J (2012) Sustainable operations: Their impact on the triple bottom line. Int J Prod Econ 140(1):149–159
50. Godfrey PC, Hatch NW (2007) Researching corporate social responsibility: An agenda for the 21st century. J Bus Ethics 70(1):87–98
51. Gugler P, Shi JY (2009) Corporate social responsibility for developing country multinational corporations: lost war in pertaining global competitiveness? J Bus Ethics 87(1):3–24
52. Husted BW, de Jesus Salazar J (2006) Taking Friedman seriously: Maximizing profits and social performance. J Manage Stud 43(1):75–91
53. Husted BW, Allen DB (2007) Corporate social strategy in multinational enterprises: Antecedents and value creation. J Bus Ethics 74(4):345–361
54. Husted BW, Allen DB (2007) Strategic corporate social responsibility and value creation among large firms: lessons from the Spanish experience. Long Range Plan 40(6):594–610
55. Husted BW, Allen, DB (2010). Corporate social strategy: stakeholder engagement and competitive advantage. Cambridge University Press
56. Husted BW, Jamali D, Saffar W (2016) Near and dear? The role of location in CSR engagement. Strateg Manag J 37(10):2050–2070
57. Jain P, Vyas V, Roy A (2017) Exploring the mediating role of intellectual capital and competitive advantage on the relation between CSR and financial performance in SMEs. Social Responsibility Journal 13(1):1–23
58. Jenkin TA, Webster J, McShane L (2011) An agenda for 'Green' information technology and systems research. Inf Organ 21(1):17–40
59. Jenkins H (2009) A 'business opportunity' model of corporate social responsibility for small- and medium-sized enterprises. Business ethics: A European review 18(1):21–36
60. Joo J, Eom M, Shin M (2017) Finding the missing link between corporate social responsibility and firm competitiveness through social capital: A business ecosystem perspective. Sustainability 9(5):707
61. Kacperczyk A (2009) With greater power comes greater responsibility? Takeover protection and corporate attention to stakeholders. Strateg Manag J 30(3):261–285
62. Kirchoff JF, Koch C, Satinover Nichols B (2011) Stakeholder perceptions of green marketing: the effect of demand and supply integration. Int J Phys Distrib Logist Manag 41(7):684–696
63. Kim KH, Kim M, Qian C (2018) Effects of corporate social responsibility on corporate financial performance: A competitive-action perspective. J Manag 44(3):1097–1118
64. Klassen RD, Whybark DC (1999) The impact of environmental technologies on manufacturing performance. Acad Manag J 42(6):599–615
65. Lee MDP (2008) A review of the theories of corporate social responsibility: Its evolutionary path and the road ahead. Int J Manag Rev 10(1):53–73
66. Lev B, Petrovits C, Radhakrishnan S (2010) Is doing good good for you? How corporate charitable contributions enhance revenue growth. Strateg Manag J 31(2):182–200
67. López MV, Garcia A, Rodriguez L (2007) Sustainable development and corporate performance: A study based on the Dow Jones sustainability index. J Bus Ethics 75(3):285–300
68. Luo X, Bhattacharya CB (2006) Corporate social responsibility, customer satisfaction, and market value. J Mark 70(4):1–18

69. Mackey A, Mackey TB, Barney JB (2007) Corporate social responsibility and firm performance: Investor preferences and corporate strategies. Acad Manag Rev 32(3):817–835
70. Manaktola K, Jauhari V (2007) Exploring consumer attitude and behaviour towards green practices in the lodging industry in India. Int J Contemp Hosp Manag 19(5):364–377
71. Marin L, Martín PJ, Rubio A (2017) Doing good and different! The mediation effect of innovation and investment on the influence of CSR on competitiveness. Corp Soc Responsib Environ Manag 24(2):159–171
72. Martinez-Conesa I, Soto-Acosta P, Palacios-Manzano M (2017) Corporate social responsibility and its effect on innovation and firm performance: An empirical research in SMEs. J Clean Prod 142:2374–2383
73. McWilliams A, Siegel DS, Wright PM (2006) Corporate social responsibility: Strategic implications. J Manage Stud 43(1):1–18
74. McWilliams A, Siegel DS (2011) Creating and capturing value: Strategic corporate social responsibility, resource-based theory, and sustainable competitive advantage. J Manag 37(5):1480–1495
75. Melo T, Garrido-Morgado A (2012) Corporate reputation: A combination of social responsibility and industry. Corp Soc Responsib Environ Manag 19(1):11–31
76. Menguc B, Ozanne LK (2005) Challenges of the "green imperative": A natural resource-based approach to the environmental orientation–business performance relationship. J Bus Res 58(4):430–438
77. Merigó JM, Gil-Lafuente AM, Yager RR (2015) An overview of fuzzy research with bibliometric indicators. Appl Soft Comput 27:420–433
78. Miles MP, Covin JG (2000) Environmental marketing: A source of reputational, competitive, and financial advantage. J Bus Ethics 23(3):299–311
79. Murphy DD (2006) The structure of regulatory competition: corporations and public policies in a global economy. OUP Oxford University Press
80. Orlitzky M, Schmidt FL, Rynes SL (2003) Corporate social and financial performance: A meta-analysis. Organ Stud 24(3):403–441
81. Orlitzky M, Siegel DS, Waldman DA (2011) Strategic corporate social responsibility and environmental sustainability. Bus Soc 50(1):6–27
82. Perrini F, Russo A, Tencati A, Vurro C (2011) Deconstructing the relationship between corporate social and financial performance. J Bus Ethics 102(1):59–76
83. Porter ME (1990) The competitive advantage of nations. Competitive Intelligence Review 1(1):14–14
84. Porter ME, Kramer MR (2002) The Competitive Advantage of Corporate Philanthropy. Harv Bus Rev 80:56
85. Porter ME, Kramer MR (2006) The link between competitive advantage and corporate social responsibility. Harv Bus Rev 84(12):78–92
86. Rettab B, Brik AB, Mellahi K (2009) A study of management perceptions of the impact of corporate social responsibility on organisational performance in emerging economies: the case of Dubai. J Bus Ethics 89(3):371–390
87. Reuter C, Foerstl KAI, Hartmann EVI, Blome C (2010) Sustainable global supplier management: the role of dynamic capabilities in achieving competitive advantage. J Supply Chain Manag 46(2):45–63
88. Saeidi SP, Sofian S, Saeidi P, Saeidi SP, Saaeidi SA (2015) How does corporate social responsibility contribute to firm financial performance? The mediating role of competitive advantage, reputation, and customer satisfaction. J Bus Res 68(2):341–350
89. Sarkis J, Zhu Q, Lai KH (2011) An organizational theoretic review of green supply chain management literature. Int J Prod Econ 130(1):1–15
90. Sethi SP (2003) Globalization and the good corporation: A need for proactive co-existence. J Bus Ethics 43(1–2):21–31
91. Siegel DS (2009) Green management matters only if it yields more green: an economic/strategic perspective. The academy of management perspectives, 5–16

92. Sun L, Stuebs M (2013) Corporate social responsibility and firm productivity: Evidence from the chemical industry in the United States. J Bus Ethics 118(2):251–263
93. Surroca J, Tribó JA, Waddock S (2010) Corporate responsibility and financial performance: The role of intangible resources. Strateg Manag J 31(5):463–490
94. Tang Z, Hull CE, Rothenberg S (2012) How corporate social responsibility engagement strategy moderates the CSR–financial performance relationship. J Manage Stud 49(7):1274–1303
95. Tate WL, Ellram LM, Kirchoff JF (2010) Corporate social responsibility reports: a thematic analysis related to supply chain management. J Supply Chain Manag 46(1):19–44
96. Thompson J, Doherty B (2006) The diverse world of social enterprise: A collection of social enterprise stories. Int J Soc Econ 33(5/6):361–375
97. Vachon S, Mao Z (2008) Linking supply chain strength to sustainable development: a country-level analysis. J Clean Prod 16(15):1552–1560
98. Van Wassenhove LN (2006) Blackett memorial lecturet humanitarian aid logistics: Supply chain. Journal of the Operational Research Society 57(5):475–489
99. Velazquez-Cazares MG, Leon-Castro E, Blanco-Mesa F, Alvarado-Altamirano S (2019) The ordered weighted average corporate social responsibility. Kybernetes
100. Vilanova M, Lozano JM, Arenas D (2009) Exploring the nature of the relationship between CSR and competitiveness. J Bus Ethics 87(1):57–69
101. Wagner M (2015) The link of environmental and economic performance: Drivers and limitations of sustainability integration. J Bus Res 68(6):1306–1317
102. Weber M (2008) The business case for corporate social responsibility: A company-level measurement approach for CSR. Eur Manag J 26(4):247–261
103. Werther WB Jr, Chandler D (2005) Strategic corporate social responsibility as global brand insurance. Bus Horiz 48(4):317–324
104. Windsor D (2006) Corporate social responsibility: Three key approaches. J Manage Stud 43(1):93–114
105. Yu HC, Kuo L, Kao MF (2017) The relationship between CSR disclosure and competitive advantage. Sustainability Accounting, Management and Policy Journal 8(5):547–570
106. Zhao H, Zhang F, Kwon J (2018) Corporate social responsibility research in international business journals: An author co-citation analysis. Int Bus Rev 27(2):389–400
107. Zhao Z, Meng F, He Y, Gu Z (2019) The Influence of Corporate Social Responsibility on Competitive Advantage with Multiple Mediations from Social Capital and Dynamic Capabilities. Sustainability 11(1):218
108. Zupic I, Čater T (2015) Bibliometric methods in management and organization. Organ Res Methods 18(3):429–472

A Bibliometric Analysis of Robustness and MCDA

Pavel López-Parra, Pavel A. Alvarez, Ernesto León-Castro, Marlenne Velázquez-Cázares, and Manuel Muñoz-Palma

Abstract Multicriteria Decision Aiding (MCDA) is an area of Operations Research that provides a methodological structure for the treatment of complex decision-making problems with multiple conflicting assessment criteria. Currently, robust analysis have been used to find a more precise solution within the methodologies of MCDA. For this reason, the objective of this work is to present an overview of this topics with a bibliometric analysis approach. Scopus and Web of Science databases are the two main sources of this study. A large number of indicators is presented to provide relevant information about journals, authors, countries, institutions, among others. The results highlight Kahraman as the most productive author in the field and The United States stands out has the leading country in this area of research. The association of this topics, in order to find a more desirable solutions have contributed to the development of theories and applications, and this field has had an important growth in the last years.

Keywords Robustness · MCDA · Bibliometric · Scopus · VOS viewer

---

P. López-Parra (✉) · M. Velázquez-Cázares
Doctoral Program of Management Science, Universidad Autónoma de Occidente, 80120 Culiacán, Sinaloa, México
e-mail: eleon@ucsc.cl

P. A. Alvarez
Management and Economic Science, Universidad Autónoma de Occidente, 80120 Culiacán, Sinaloa, México

E. León-Castro
Universidad Católica de La Santísima Concepción, 4070129 Concepción, Chile

M. Muñoz-Palma
Management department, Universidad de Sonora, Hermosillo, México

E. León-Castro et al. (eds.), *Soft Computing and Fuzzy Methodologies in Innovation Management and Sustainability*, Lecture Notes in Networks and Systems 337,
https://doi.org/10.1007/978-3-030-96150-3_16

1 Introduction

At present, the MCDA has had significant progress, from the theoretical and real-world application point of view. The MCDA has been dealing with multidimensional problems because this approach is more similar to the real problems found in society. If all the relevant factors for decision making are considered in a realistic approach, the decision-maker may face the difficulty of identifying feasible solutions for these types of problems.

Therefore, the perception of MCDA is related to how human beings have been making their decisions. Consequently, MCDA methods explicitly recognize subjectivity and uncertainty in decision making, providing a framework for sensitivity analysis and also offer support to generate consensus in group decision making [1]. In general, MCDA find methods that seek to find the optimal solution among an infinite set of alternatives and techniques that aim to classify a finite set of discrete alternatives under the subjectivity of the preferences of decision-makers [2, 3].

Despite the diversity of approaches, methods, and techniques of MCDA, the main basic components of MCDA are very simple in any real problem that a decision-maker faced. These components are: a finite or infinite set of actions (alternatives, courses of action, solutions, etc.); at least two criteria and one decision-maker. In other words, MCDA is an activity that helps the decision-maker to choose, classify or order the alternatives for a problem in question [4].

Within the field of study of the MCDA, the subject of robustness has been increasingly present in scientific journals. This issue has become of interest to companies where less formal jobs apply operational research tools to problems in decision-making aid [5].

A quick scan of the literature in different fields, it is possible to find that the definition of robustness has a deep meaning depending on the discipline in which it is implemented, reflecting their concerns and objectives to which they want to resolve. Therefore, the meaning of the different definitions that can be found of robustness seek to convey in terms of "ability to resist ...", "resistance to ...", "instability to ...", [6]. According to [7], defines robustness as the ability to resist "vague approximations" and/or "zones of ignorance" to avoid undesirable impacts, in particular, the degradation of the properties to be maintained. This results in a process that responds to a concern, where the result of it resists or self-protection against a change in its variables.

Therefore, two types of robustness can be found in the articles found, one that is a concern of robustness where a concern is often taken into account a priori when formulating the problem, or the robustness analysis that can be given to understand that the work done is a posteriori as it is the case of sensitivity analysis, for example[5].

Within the MCDA field of study, the subject of robustness has been increasingly presented in scientific research. This issue has gained interest in business world where less formal jobs apply operational research tools to problems in decision-making aid [5, 7]. There are important studies that focus on the application of robustness in the MCDA methods in areas such as computer sciences [8–10] engineering [11–13] administration [13, 14], decision science [15, 16], and environment [17, 18.

In order to evaluate the wide number of publications in this two topics, Robustness and MCDA (R-MCDA), a bibliometric analysis has been conducted. Some bibliometric studies have been carried out in the field of the MCDA with different approaches such as health care [19], finance [20], and fuzzy decision making [21]. Nevertheless, a bibliometric analysis in MCDA and robustness has not been made.

The main objective of the paper is to provide an overview of the contributions of robustness for MCDA methods (R-MCDA) from 1977 to 2018, the evolution of the research topics, and important indicators of journals, papers, books, authors, institutions, and countries. The results reveal an essential interest in R-MCDA as a research field. The most crucial magazine in this area is Expert Systems with Applications; the principal authors are Bernard Roy, Salvatore Greco, Thomas L. Saaty, and Roman Słowiński; the leading institutions are The University of Warwick, Istanbul Technical University and Galatasaray University. The central countries that publish R-MCDA are The United States and China. The conference papers and books were also considered to show their importance in this field.

The structure of this work is as follows: in Sect. 2 shows the methodology followed during the analysis, in Sect. 3 displays a graphical visualization of the principal results using VOS viewer software, and finally, in Sect. 4 presents the conclusions with the most relevant points of the investigation.

2 Methodology

This study is an overview of the contributions on MCDA and robustness through bibliometric approach. The bibliometric study has become an important element for the analysis of information in the different areas of research. This analysis presents the leading indicators related to authors, articles, magazines, journals, institutions, countries, books, etc., [22]. For this work, Scopus and Web of Science (WoS) databases are considered. It is essential to clarify that Scopus is the central database to obtain the information (authors, types of documents, institutions, to name a few). Meanwhile, WoS was used to expand the analysis and complement the information in the database.

The keywords for the sample of this research are "robust *" and "Decision analysis Multiple criteria *" OR "MULTICRITERIA DECISION-ANALYSIS *" OR "mcda *" OR "mcdm *" OR "MULTI$CRITERIA DECISION ANALYSIS * "OR" Multi$criteria decision making * "OR" Multiple criteria decision making * "(subject) from 1977 to 2018, these keywords used because there is a wide variety of associated terms, and those terms are those that taken consider. However, other keywords that did not represent a significant result used, so it excluded in the analysis; this may cause some limitations in the investigation because the established specifications may not include essential studies.

Table 1 Document types in R-MCDA in Scopus core collection

Documents types	Number	Total citations	Percentage citations
Articule	574	11,682	89.45
Review	24	602	4.61
Conference paper	156	594	4.55
Book chapter	20	143	1.09
Note	3	17	0.13
Articule in press	9	14	0.11
Editorial	1	7	0.05
Conference review	18	1	0.01
Book	1	0	0.00
Erratum	1	0	0.00
Total	807	13,060	100.00

Source Own elaboration, based on Scopus 2019

The sample was obtained in July 2019, with 807 documents found in Scopus and 497 WoS publications. The primary Scopus collection used, in Table 1 it shows the quantity of the different documents found of Robustes in MCDA (R-MCDA) and the total citations of the published documents total 13,060 citations, the articles represent 89.45% of the appointments, followed by the reviews with 4.61% and the conferences paper with 4.55%.

Within the analysis of the data, it can find certain documents with a high rate of citations. The Stewart document [14] has cited 395 times, [13] present 380 total citations and [8] has been cited 296 times. It is essential to mention this, because Table 2 presents a structure of citations of documents and articles in R-MCDA; Where 0.83% of published works exceed 200 citations, while 1.50% have at least 100 citations and between 50 and 99 citations represent 8.53% of these papers, and 89.13% have at least 25 citations. It is important to mention that Table 2 shows that

Table 2 Citation structure in R-MCDA in Scopus core collection

Number of citations	Number of documents	Percentage documents	# Articles and review	% Articles and review
≥300	2	0.25	2	0.33
≥200	3	0.37	3	0.50
≥150	3	0.37	3	0.50
≥100	7	0.87	6	1.00
≥50	52	6.44	51	8.53
<50	740	91.70	533	89.13
Total	807	100	598	100

Source own elaboration based on Scopus 2019

the articles and reviews represent more than half of the total number of documents published; this is because the most important documents for the study are articles and reviews.

An important aspect to highlight is the evolution of the number of publications in R-MCDA. In Fig. 1 it can be seen that from 1994 to 2018 the publications in R-MCDA have increased considerably, as mentioned earlier the investigations in MCDA have been concerned about presenting a robust result or process, so they have conducted research with the intention of including robustness in their articles making them more relevant than ever to provide reliable decision support through more powerful analytical tools [23]. Figure 1 shows the increase in publications for four-year periods, highlighting the last period (1Q) with 502 publications; this represents 62.21% of the total contributions in R-MCDA. If we take into account the 2Q period added to the 1Q period, this will represent 84.76% of all publications. These results show a real interest in research in R-MCDA.

Different indicators such as productivity, total citations, and the h index have been calculated to present a better analysis of the journals, articles, authors, countries, and institutions. Blanco-Mesa (2019) mention that the index "h" is the relationship between the number of citations and the number of publications made, this may represent the importance of magazines, articles, authors, countries and institutions, highlighting the main ones in the field of study. It should be noted that the total number of citations is the number of citations obtained per publication.

Finally, the software VOS viewer has been used to generate graphics, to display bibliometric information, bibliographic coupling, Co-citations, and Co-occurrence of keywords.

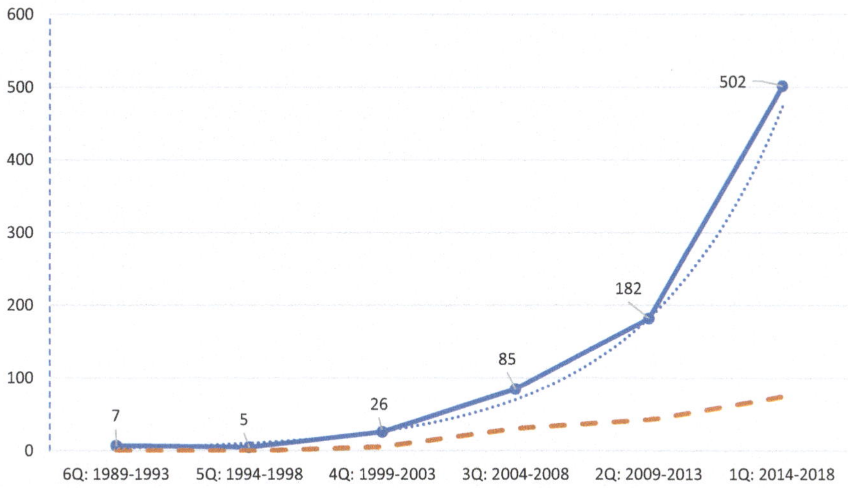

Fig. 1 Evolution of the number of publications on HR-MCDA in Scopus since 1989

2.1 The 30 Most Prominent Journals in R-MCDA

The journals have the important task of disseminating the research and contributions made in a specific subject, which in this case is R-MCDA; therefore, it is essential to locate the most important and influential journals in this area. Table 3 shows a list of the most productive and cited journals in R-MCDA. The leading magazine in this field is ESA, with a production of 22 publications with a total of 1396 citations, having the highest h index of 16. Besides, it presents 8 articles classified in the 30 most cited articles in R-MCDA (this list It will be presented in the next section), most of the publications presented are between the periods 1Q and 2Q, making it one of the journals with the highest scientific interest in this field. Secondly, this EJOR, with a production of 17 publications and with the h index of 11, although this magazine ranks as the third-best for the total citations with 536 citations, it can be said that it is a Good magazine of interest in the field. In third place is JCP, this magazine has 15 publications, most of them between the 1Q period, with an h index of 11 and a total of 250 citations, with this it can be said that this magazine has a presence in the community scientist interested in this field. In addition, the OME magazine has few publications and has a shallow h index, but stands out in the second place with respect to its number of total citations with 574 and within this magazine there are two articles within the top 30 of the most cited articles in R-MCDA and one of them is the most cited of the entire top 30. It is worth mentioning that there are 17 journals with 3 publications, but the list was limited to 30 journals. Therefore, the selection of the magazine with 3 publications was based on their total citations.

2.2 The 30 Most Significant Papers in R-MCDA

In this bibliometric, some articles present a notable number of citations; their influence impacts significantly in this field. Table 4 shows the most cited crucial documents in R-MCDA. The main and perhaps the most important work is by author [14] entitled "A critical survey on the status of multiple criteria decision-making theory and practice," with 395 total citations in the main Scopus collection. This study reviews and contrasts the main theoretical and practical trends of the Multiple Criteria Decision Making (MCDM). The objective of the article is to identify the main difficulties of the methods and to suggest which approaches are more robust and effective, especially for those people who are not experts in MCDM, even the author presents a discussion on those problematic areas of the MCDM that are required to investigate. [24] present other article entitled "Robustness of additive value function methods in MCDM" with 108 total citations where they mention how it can affect the outcome of a problem by making changes in the decision maker's preference points as a result of the decision maker's preference information types.

Wang et al. [13] presented a document entitled "Product-driven supply chain selection using integrated multi-criteria decision-making methodology". This study

Table 3 Top journals in R-MCDA

R	J	PR-MCDA	Percentage	HR-MCDA	TC	CS	CP	T30	≥200	≥100	≥50	1Q	2Q	3Q	4Q	5Q	6Q
1	ESA	22	3.48	16	1396	6.36	22.4	8	1	4	3	12	9	1	–	–	–
2	EJOR	17	2.69	11	536	4.98	25.1	2	–	–	3	8	3	5	1	–	–
3	JCP	15	2.37	11	250	7.32	21.6	–	–	–	–	13	2	–	–	–	–
4	JMCDA	12	1.90	7	378	1.32	32.9	3	–	1	2	2	6	1	2	1	–
5	ASCJ	10	1.58	8	159	6.27	16.3	–	–	–	–	8	2	–	–	–	–
6	IJPR	10	1.58	9	391	4.34	14.5	2	–	1	3	2	2	4	2	–	–
7	SUST (CHE)	9	1.42	5	56	3.01	27.6	–	–	–	–	9	–	–	–	–	–
8	IJITDM	8	1.26	3	57	2.49	23.3	–	–	–	–	6	1	1	–	–	–
9	JEM	8	1.26	7	284	5.32	39	–	–	–	2	4	3	1	–	–	–
10	CIE	7	1.11	6	87	4.68	6	–	–	–	–	6	–	1	–	–	–
11	IJAMT	7	1.11	5	198	3.04	21.8	–	–	–	2	7	–	–	–	–	–
12	STE	5	0.79	4	66	5.92	6.2	–	–	–	–	4	1	–	–	–	–
13	ENER	5	0.79	4	94	6.2	58.4	–	–	–	1	5	–	–	–	–	–
14	ISORMS	5	0.79	3	125	0.59	32.6	–	–	1	–	3	1	1	–	–	–
15	COR	5	0.79	5	253	4.56	13.4	1	1	–	–	3	–	1	1	–	–
16	RCR	5	0.79	5	121	6.82	58.4	–	–	–	–	5	–	–	–	–	–
17	DSS	5	0.79	5	135	5.97	27	–	–	–	–	1	3	1	–	–	–
18	AOR	5	0.79	2	22	2.58	56.6	–	–	–	–	4	1	–	–	–	–
19	IJPE	5	0.79	3	442	7.13	10.8	1	1	–	–	2	2	1	–	–	–
20	WRM	5	0.79	4	85	3.3	29.8	–	–	–	1	4	1	–	–	–	–
21	PLOS	4	0.63	2	16	3.02	10.8	–	–	–	–	4	–	–	–	–	–

(continued)

Table 3 (continued)

R	J	PR-MCDA	Percentage	HR-MCDA	TC	CS	CP	T30	≥200	≥100	≥50	1Q	2Q	3Q	4Q	5Q	6Q
22	OME	4	0.63	4	574	7.54	71.5	2	1	1	–	–	–	1	–	1	2
23	KBS	4	0.63	4	271	7.01	50.3	1	–	–	4	2	2	–	–	–	–
24	IJFS	4	0.63	3	43	2.97	40.8	–	–	–	–	4	–	–	–	–	–
25	OME (GBR)	4	0.63	4	129	7.54	31.3	1	–	–	1	4	–	–	–	–	–
26	IJMDM	4	0.63	2	8	0.68	32.3	–	–	–	–	2	2	–	–	–	–
27	RSER	4	0.63	4	138	12.21	32	–	–	–	2	4	–	–	–	–	–
28	JIFS	4	0.63	3	22	1.96	24.8	–	–	–	–	4	–	–	–	–	–
29	EMS	4	0.63	4	142	5.08	24.8	–	–	–	2	2	–	1	1	–	–
30	EI	3	0.47	3	277	5.06	22	1	1	–	–	3	–	–	–	–	–

Source Own elaboration based on Scopus 2019. R: Ranking; J: Journal; PR-MCDA: Total number of documents published in R-MCDA; %: Percentage of articles published in R-MCDA; HR-MCDA: H-index only with R-MCDA; TC: Total citations; CS: Cite score; CP: Citations by paper; T30: Number of publications within 30 most cited papers; ≥300: Number of articles with more than 300 citations; ≥200: Number of articles with more than 200 citations; ≥150: Number of articles with more than 150 citations; ≥100: Number of articles with more than 100 citations; ≥50: Number of articles with more than 100 citations; 1Q: 2014–2018; 2Q: 2009–2013; 3Q: 2004–2008; 4Q: 1999–2003; 5Q: 1994–1998; 6Q: 1989–1993. Journal abbreviation are: ESA, Expert Systems with Applications; EJOR: European Journal of Operational Research; JCP: Journal of Cleaner Production; JMCDA: Journal of Multi-Criteria Decision Analysis; ASCJ: Applied Soft Computing Journal; IJPR: International Journal of Production Research; SUST (CHE): Sustainability (Switzerland); IJITDM: International Journal of Information Technology and Decision Making; JEM: Journal of Environmental Management; CIE: Computers and Industrial Engineering; IJAMT: International Journal of Advanced Manufacturing Technology; STE: Science of the Total Environment; ENER: Energy; ISORMS: International Series in Operations Research and Management Science; COR: Computers and Operations Research; RCR: Resources, Conservation and Recycling; DSS: Decision Support Systems; AOR: Annals of Operations Research; IJPE: International Journal of Production Economics; WRM: Water Resources Management; PLOS: PLoS ONE; OME: Omega; KBS: Knowledge-Based Systems; IJFS: International Journal of Fuzzy Systems; OME (UK): Omega (United Kingdom); IJMDM: International Journal of Multicriteria Decision Making; RSER: Renewable and Sustainable Energy Reviews; JIFS: Journal of Intelligent and Fuzzy Systems; EMS: Environmental Modelling and Software; EI: Ecological Indicators

Table 4 Top 30 papers in R-MCDA

R	Article	Authors	ST	TC	Year
1	A critical survey on the status of multiple criteria decision making theory and practice	Stewart [14]	OME	395	1992
2	Product-driven supply chain selection using integrated multi-criteria decision-making methodology	Wang et al. [13]	IJPE	380	2004
3	Choosing knowledge management strategies by using a combined ANP and DEMATEL approach	Wu [8]	ESA	296	2008
4	Analysis of the potentials of multi criteria decision analysis methods to conduct sustainability assessment	Cinelli et al. [16]	EI	226	2014
5	A GP-AHP method for solving group decision-making fuzzy AHP problems	Yu [9]	COR	200	2002
6	A combined methodology for supplier selection and performance evaluation	Zeydan et al. [10]	ESA	167	2011
7	A hybrid genetic algorithm for production and distribution	Chan et al. [15]	OME	155	2005
8	A fuzzy multi-criteria decision making model for supplier selection	Dalalah et al. [11]	ESA	151	2011
9	Urban flood hazard zoning in Tucumán Province, Argentina, using GIS and multicriteria decision analysis	Fernández and Lutz [25]	EG	134	2010
10	Application of fuzzy TOPSIS in evaluating sustainable transportation systems	Awasthi et al. [26]	ESA	111	2011
11	Robustness of additive value function methods in MCDM	Stewart [24]	JMCDA	108	1996
12	Optimal site selection of electric vehicle charging station by using fuzzy TOPSIS based on sustainability perspective	Guo and Zhao [17]	AE	107	2015
13	Designing an integrated AHP based decision support system for supplier selection in automotive industry	Dweiri et al. [27]	ESA	100	2016

(continued)

Table 4 (continued)

R	Article	Authors	ST	TC	Year
14	Distance-based fuzzy MCDM approach for evaluating flexible manufacturing system alternatives	Karsak [28]	IJPR	100	2002
15	Incremental analysis for MCDM with an application to group TOPSIS	Shih [29]	EJOR	99	2008
16	Personnel selection based on computing with words and fuzzy MULTIMOORA	Baležentis et al. [30]	ESA	99	2012
17	The analytic network process (ANP) approach to location selection: A shopping mall illustration	Cheng et al. [31]	CI	98	2005
18	Comparative analysis of MCDM methods for the assessment of sustainable housing affordability	Mulliner et al. [32]	OME (UK)	91	2016
19	A common weight MCDA-DEA approach to construct composite indicators	Hatefi and Torabi [18]	EE	91	2010
20	A robust hybrid multi-criteria decision making methodology for contractor evaluation and selection in third-party reverse logistics	Senthil et al. [33]	ESA	91	2014
21	A stochastic method for robustness analysis in sorting problems	Tervonen et al. [34]	EJOR	87	2009
22	The analytic hierarchy process (AHP) approach for assessment of urban renewal proposals	Lee and Chan [35]	SIR	86	2008
23	Combining scenario planning and multi-criteria decision analysis in practice	Montibeller G., Gummer H., Tumidei D	JMCDA	85	2006
24	Sustainability assessment of flooring systems in the city of Tehran: An AHP-based life cycle analysis	Reza et al. [36]	CBM	84	2011
25	Practical common weight multi-criteria decision-making approach with an improved discriminating power for technology selection	Karsak and Ahiska [37]	IJPR	83	2005

(continued)

Table 4 (continued)

R	Article	Authors	ST	TC	Year
26	Application of ANP and DEMATEL to evaluate the decision-making of municipal solid waste management in Metro Manila	Tseng [38]	EMA	82	2009
27	A decision model for information technology selection using AHP integrated TOPSIS-Grey: The case of content management systems	Oztaysi [39]	KBS	78	2014
28	Multi-criteria evaluation of alternative-fuel vehicles via a hierarchical hesitant fuzzy linguistic model	Yavuz et al. [40]	ESA	77	2015
29	A method for examining the spatial dimension of multi-criteria weight sensitivity	Feick and Hall [41]	IJGIS	77	2004
30	Incorporating uncertainty in the PROMETHEE MCDA method	Hyde et al. [42]	JMCDA	77	2003

Source own elaboration based on Scopus 2019. R: Ranking; ST: Source title; TC: Total citations; Y: Publication year. The abbreviation that not available in the previous table: EG: Engineering Geology; AE: Applied Energy; CI: Construction Innovation; EE: Ecological Economics; SIR: Social Indicators Research; CBM: Construction and Building Materials; EMA: Environmental Monitoring and Assessment; IJGIS: International Journal of Geographical Information Science

relates the characteristics of a product to the supply chain strategy by adopting specific performance metrics with supply chain operations reference (SCOR) model level I, as a decision criterion. An integrated analytic hierarchy process (AHP) and preemptive goal programming (PGP) based multi-criteria decision-making methodology is then developed to take into account both qualitative and quantitative factors in supplier selection, this allows a robust design of the supply chain. On the other hand, is [8], with his work entitled "Choosing knowledge management strategies by using a combined ANP and DEMATEL approach," mentions the importance of administrative knowledge as a competitive advantage. Therefore, it proposes a practical solution based on the combination of the analytic network process (ANP) and the Decision-Making Trial and Evaluation Laboratory (DEMATEL) to help companies that need to evaluate and select strategies of administrative knowledge. These previous studies established the principles and foundations of R-MCDA as a field of scientific interest and development for future studies.

It should be noted that Scopus and Wos have minimal variation concerning the main works. Indeed, Scopus has more documents cited than WoS. However, you can find jobs in WoS whose appointments are quite considerable such as Costa et al. [43] with 143 citations, [44] with 123, [45] with 115 quotes, and [46] with 113 total citations. These documents can easily be in the top 30 of the most cited papers in

R-MCDA, but because of the situation just described above, this work focuses on the main Scopus collection.

2.3 The 30 Most Influential Authors in R-MCDA

In this bibliometric, it can find a variety of authors who have contributed to the development of this field of research. Table 5 shows the most influential researchers in the field of R-MCDA, where thirty authors are listed. Kahraman and Greco, are the authors with more production in this field; however, Kahraman is one of the authors with the highest h index within the study and has a high amount of total citations than Greco, so this author is the leader in this top 30 and Greco in the second position. It should be noted that most of Kahraman's published articles are within the 1Q period (the highest in this study). In the third position is Karsak with a rather high production but not higher than Kahraman and Greco, and an index h less than or equal to these authors, however, he is the author with more total citations in this top 30.

Coles is probably the most prominent author in R-MCDA; With only 4 publications, it has a high enough number of total citations, with 286 citations (the fourth most cited in this table), with a citation by paper of 71.5 and a document within the 30 most cited articles in R-MCDA; Although there are authors like Karsak that has a total of 296 citations (the highest in the table), with a production of 9 documents and also presents two articles within the top 30 of the most cited articles, it could be an author in the future Stand out in this area.

2.4 The 30 Most Influential Institutions in R-MCDA

Universities are institutions that finance and support their researchers to develop research projects. Table 6 highlights the most important universities that have contributed to the development of R-MCDA. The Istanbul Technical University is the institution that leads this top with 15 publications, 378 total citations (The second best in the top 30 institutions), and most of its works are within the 1Q period. In the second position is the University of Tehran with 14 publications, 215 total citations, one publication in the top 30 of the most cited articles. It is followed by Poznan University of Technology and the University of Portsmouth with 12 publications.

It is essential to consider the rate of appointments within this set of universities. The Istanbul Technical University is the one with a high h index of 9, followed by the University of Tehran with an h index of 8 and the Poznan University of Technology, the Polish Academy of Sciences, the Higher Technical Institute, the University of Lisbon, each with an h index of 7. It is also essential to consider the "citation by paper" (CP) indicator in the leading institutions in R-MCDA. Therefore, The University of Warwick has the highest average citation by paper number of 49.6, followed by The University of Adelaide with an average citation by paper of 48.3, the University

Table 5 Top 30 authors in R-MCDA

R	Authors	Country	PR-MCDA	Percentage	HR-MCDA	TC	CP	T30	1Q	2Q	3Q	4Q	5Q	6Q
1	Kahraman, C	TUR	10	1.24	7	251	25.1	1	9	1	–	–	–	–
2	Greco, S	ITA	10	1.24	5	224	22.4	–	6	4	–	–	–	–
3	Karsak, E.E	TUR	9	1.12	5	296	32.9	2	2	–	4	3	–	–
4	Corrente, S	ITA	9	1.12	5	194	21.6	–	6	3	–	–	–	–
5	Linkov, I	USA	9	1.12	7	147	16.3	–	3	6	–	–	–	–
6	Słowiński, R	POL	8	0.99	6	221	27.6	–	5	3	–	–	–	–
7	Chung, E.S	KOR	8	0.99	6	116	14.5	–	8	–	–	–	–	–
8	Oztaysi, B	TUR	7	0.87	6	273	39.0	2	7	–	–	–	–	–
9	Barua, M.K	IND	7	0.87	6	163	23.3	–	7	–	–	–	–	–
10	Prakash, C	IND	6	0.74	5	131	21.8	–	6	–	–	–	–	–
11	Schultmann, F	DEU	6	0.74	4	37	6.2	–	1	5	–	–	–	–
12	Comes, T	DEU	6	0.74	4	36	6.0	–	–	6	–	–	–	–
13	Cinelli, M	GBR	5	0.62	5	292	58.4	1	5	–	–	–	–	–
14	Kirwan, K	GBR	5	0.62	5	292	58.4	1	5	–	–	–	–	–
15	Maier, H.R	AUS	5	0.62	5	283	56.6	1	1	–	3	1	–	–
16	Jain, V	NZL	5	0.62	5	163	32.6	1	5	–	–	–	–	–
17	Zavadskas, E.K	LTU	5	0.62	5	149	29.8	–	4	1	–	–	–	–
18	Lienert, J	CHE	5	0.62	4	135	27	–	3	2	–	–	–	–
19	Kim, Y	KOR	5	0.62	5	67	13.4	–	5	–	–	–	–	–
20	Wang, X	CHN	5	0.62	5	54	10.8	–	5	–	–	–	–	–
21	Zhou, F	CHN	5	0.62	5	54	10.8	–	5	–	–	–	–	–

(continued)

Table 5 (continued)

R	Authors	Country	PR-MCDA	Percentage	HR-MCDA	TC	CP	T30	1Q	2Q	3Q	4Q	5Q	6Q
22	Coles, S.R	GBR	4	0.50	4	286	71.5	1	4	–	–	–	–	–
23	Zarghami, M	IRN	4	0.50	4	201	50.3	–	–	3	1	–	–	–
24	Zhao, H	CHN	4	0.50	3	163	40.8	1	4	–	–	–	–	–
25	Lambert, J.H	USA	4	0.50	4	129	32.3	–	1	3	–	–	–	–
26	Figueira, J.R	PRT	4	0.50	4	128	32.0	1	2	2	-	–	–	–
27	Cevik Onar, S	TUR	4	0.50	3	125	31.3	1	4	–	–	–	–	–
28	Wang, J.q	CHN	4	0.50	4	99	24.8	–	4	–	–	–	–	–
29	Zhang, H.y	CHN	4	0.50	4	99	24.8	–	4	–	–	–	–	–
30	Chakraborty, S	IND	4	0.50	3	88	22.0	–	3	–	1	–	–	–

Source Own elaboration based on Scopus 2019; R: Ranking; PR-MCDA: Total number of documents published in R-MCDA; %: Percentage of articles published in R-MCDA: HR-MCDA: H-index only with R-MCDA; TC: Total Citations; CP: Citation by papers; N30: Number of publications within the 30 most cited papers; 1Q: 2014–2018; 2Q: 2009–2013; 3Q: 2004–2008; 4Q: 1999–2003; 5Q: 1994–1998; 6Q: 1989–1993

Table 6 Top institutions in R-MCDA

R	Institution	Country	PR-MCDA	Percentage	HR-MCDA	TC	CP	T30	1Q	2Q	3Q	4Q	5Q	6Q
1	Istanbul Technical University	TUR	15	1.86	9	378	25.2	2	14	1	–	–	–	–
2	University of Tehran	IRN	14	1.74	8	215	15.4	1	11	3	–	–	–	–
3	Poznan University of Technology	POL	12	1.49	7	244	20.3	–	8	4	–	–	–	–
4	University of Portsmouth	GBR	12	1.49	6	214	17.8	–	11	1	–	–	–	–
5	Galatasaray University	TUR	11	1.36	6	312	28.4	2	3	–	5	3	–	–
6	University of Catania	ITA	10	1.24	6	230	23.0	–	6	4	–	–	–	–
7	North China Electric Power University	CHN	10	1.24	5	207	20.7	1	10	–	–	–	–	–
8	Indian Institute of Technology Roorkee	IND	10	1.24	6	164	16.4	–	10	–	–	–	–	–
9	Polish Academy of Sciences	POL	9	1.12	7	236	26.2	–	6	3	–	–	–	–
10	Zhejiang University	CHN	8	0.99	6	181	22.6	–	6	2	–	–	–	–
11	Instituto Superior Técnico	PRT	8	0.99	7	221	27.6	1	5	2	1	–	–	–
12	Jadavpur University	IND	8	0.99	6	121	15.1	–	7	–	1	–	–	–
13	The University of Warwick	GBR	8	0.99	6	397	49.6	2	7	1	–	–	–	–
14	Seoul National University of Science and Technology	KOR	8	0.99	7	117	14.6	–	8	–	–	–	–	–

(continued)

Table 6 (continued)

R	Institution	Country	PR-MCDA	Percentage	HR-MCDA	TC	CP	T30	1Q	2Q	3Q	4Q	5Q	6Q
15	Vilnius Gediminas Technical University	LTU	8	0.99	6	228	28.5	1	7	1	–	–	–	–
16	University Of Lisbon	PRT	8	0.99	7	214	26.8	1	6	2	–	–	–	–
17	National Technical University of Athens	GRC	7	0.87	2	11	1.6	–	6	1	–	–	–	–
18	City University of Hong Kong	HKG	7	0.87	6	120	17.1	–	3	1	3	–	–	–
19	Central South University China	CHN	7	0.87	6	207	29.6	–	7	–	–	–	–	–
20	Islamic Azad University	IRN	7	0.87	3	17	2.4	–	5	2	–	–	–	–
21	Karlsruhe Institute of Technology	DEU	7	0.87	4	40	5.7	–	2	5	–	–	–	–
22	Aalto University	FIN	7	0.87	5	68	9.7	–	5	1	1	–	–	–
23	Ministry of Education China	CHN	6	0.74	4	161	26.8	–	5	1	–	–	–	–
24	Arizona State University	USA	6	0.74	4	82	13.7	–	3	2	–	1	–	–
25	University of Tabriz	IRN	6	0.74	5	220	36.7	–	2	3	1	–	–	–
26	The University of Adelaide	AUS	6	0.74	5	290	48.3	–	2	–	3	1	–	–
27	Panepistimion Pireos	GRC	6	0.74	3	28	4.7	–	5	1	–	–	–	–
28	National Cheng Kung University	TWN	6	0.74	4	127	21.2	–	2	2	2	–	–	–

(continued)

Table 6 (continued)

R	Institution	Country	PR-MCDA	Percentage	HR-MCDA	TC	CP	T30	1Q	2Q	3Q	4Q	5Q	6Q
29	Universidade de Coimbra	PRT	6	0.74	5	118	19.7	–	6	–	–	–	–	–
30	Chongqing University	CHN	6	0.74	5	65	10.8	–	6	–	–	–	–	–
31	The United States Army Corps of Engineers	USA	6	0.74	5	135	22.5	–	3	3	–	–	–	–

Source Own elaboration based on Scopus 2019; R: Ranking; PR-MCDA: Total number of documents published in R-MCDA; %: Percentage of articles published in R-MCDA; HR-MCDA: H-index only with R-MCDA; TC: Total citations; CP: citations by paper; T30: Number of publications within in 30 most citated papers; 1Q: 2014–2018; 2Q: 2009–2013; 3Q: 2004–2008; 4Q: 1999–2003; 5Q: 1994–1998; 6Q: 1989–1993

of Tabriz with 36.7, and Central South University China with a Citation by paper average of 29.6.

2.5 The 30 Most Influential Countries in R-MCDA

Scientific knowledge gives countries to position, grow, improve their economy and social welfare. This allows them to be at the forefront and allows them to reach a leading position in front of other countries in any field of research that they enter [47]. For this reason, countries seek to promote and support scientific knowledge through scientific research. Table 7 shows the most important countries that have contributed to the development of R-MCDA. The United States is the leading country in this field with a production of 115 documents, its total citations are the highest of all with 2118 citations, and its h index is 25 (the largest in this table), it presents a citation by paper of 18.4, it is lower than its production level. In second place is China, with 90 publications, its total citations are 1342 citations, its h index is 22, and it has a citation by paper of 14.9. Following this trend, India has important productivity, although it does not exceed the production of the United States and China. As with the most influential authors and institutions in R-MCDA, the "citation by paper" indicator is essential to consider to know the level of importance of the countries in R-MCDA. In this way, they find South Africa has the highest average number of citation by paper with a 60.7. Other countries, such as Hong Kong, Taiwan, Lithuania, Turkey, Portugal, and the United Kingdom, have a considerable citation by papers average: 60.7, 40.5, 38.2, 26.3, 22.0, 21.6, and 20.5, respectively.

3 Visualization with VOS Viewer

In this section, bibliographic information on R-MCDA will be analyzed graphically using the VOS viewer software. This software considers three analyzes: the bibliographic coupling, Co-citation, and Co-occurrence of author Keywords. The bibliographic coupling appears when two documents cite the same third document. Co-citation occurs when two documents receive an appointment from the same third document. The Co-occurrence of author Keywords measures the most common and most frequently occurring keywords in the same documents [47].

The first analysis to be performed is the bibliographic coupling in R-MCDA publications. Figure 2 shows the results of the bibliographic coupling of journals considering the most significant number of articles related to R-MCDA, and those most frequently cite these same documents. For this bibliographic coupling, a minimum threshold of 5 documents and the hundred strongest bibliographic coupling connections considered. The leading journals with the highest production of articles related to R-MCDA are Expert Systems with Applications, European Journal of Operational

Table 7 Top countries in R-MCDA

R	Country	PR-MCDA	Percentage	HR-MCDA	TC	CP	T30	1Q	2Q	3Q	4Q	5Q
1	United States	115	0.14	25	2118	18.4	4	60	35	13	4	3
2	China	90	0.11	22	1342	14.9	1	69	14	5	1	1
3	India	84	0.10	19	1193	14.2	2	68	9	5	2	–
4	United Kingdom	71	0.09	22	1459	20.5	3	50	11	8	2	–
5	Iran	54	0.07	14	723	13.4	1	42	11	1	–	–
6	Turkey	54	0.07	16	1190	22.0	5	38	6	7	3	–
7	Italy	50	0.06	17	729	14.6	–	35	10	3	2	–
8	France	36	0.04	15	470	13.1	–	21	13	2	–	–
9	Canada	34	0.04	10	611	18.0	3	17	13	2	2	
10	Australia	33	0.04	13	657	19.9	1	16	5	11	1	–
11	Brazil	32	0.04	9	223	7.0	–	25	6	1	–	–
12	Spain	31	0.04	11	448	14.5	–	17	10	1	3	–
13	Germany	23	0.03	8	246	10.7	–	12	10	1	–	–
14	Taiwan	23	0.03	9	879	38.2	4	9	4	7	2	1
15	Portugal	22	0.03	12	476	21.6	1	17	4	1	–	–
16	Switzerland	22	0.03	8	192	8.7	–	13	7	1	1	–
17	Greece	19	0.02	6	170	8.9	–	15	2	2	–	–
18	Malaysia	17	0.02	6	142	8.4	–	16	1	–	–	–
19	Poland	17	0.02	9	281	16.5	–	13	4	–	–	–
20	Belgium	16	0.02	5	247	15.4	–	9	5	2	–	–
21	South Korea	16	0.02	7	189	11.8	–	13	2	1	–	–
22	Netherlands	14	0.02	8	192	13.7	–	4	8	1	–	1
23	Finland	13	0.02	7	180	13.8	1	8	4	1	–	–
24	Hong Kong	13	0.02	10	526	40.5	3	5	1	6	1	–
25	Sweden	11	0.01	4	48	4.4	–	9	2	–	–	–
26	Austria	10	0.01	6	198	19.8	–	5	4	1	–	–
27	Lithuania	9	0.01	7	237	26.3	1	8	1	–	–	–
28	Saudi Arabia	9	0.01	3	64	7.1	–	8	–	–	1	–
29	South Africa	9	0.01	6	546	60.7	2	4	1	1	1	2

(continued)

Table 7 (continued)

R	Country	PR-MCDA	Percentage	HR-MCDA	TC	CP	T30	1Q	2Q	3Q	4Q	5Q
30	Denmark	8	0.01	5	84	10.5	–	5	2	1	–	–

Source Own elaboration based on Scoups 2019; R: Ranking; PR-MCDA: Total number of documents published in R-MCDA; %: Percentage of articles published in R-MCDA; HR-MCDA: H-index only with R-MCDA; TC: Total citations; CP: Citations by paper; T30: Number of publication within the 30 most cited papers; 1Q: 2014–2018; 2Q: 2009–2013; 3Q: 2004–2008; 4Q: 1999–2003; 5Q: 1977–1998

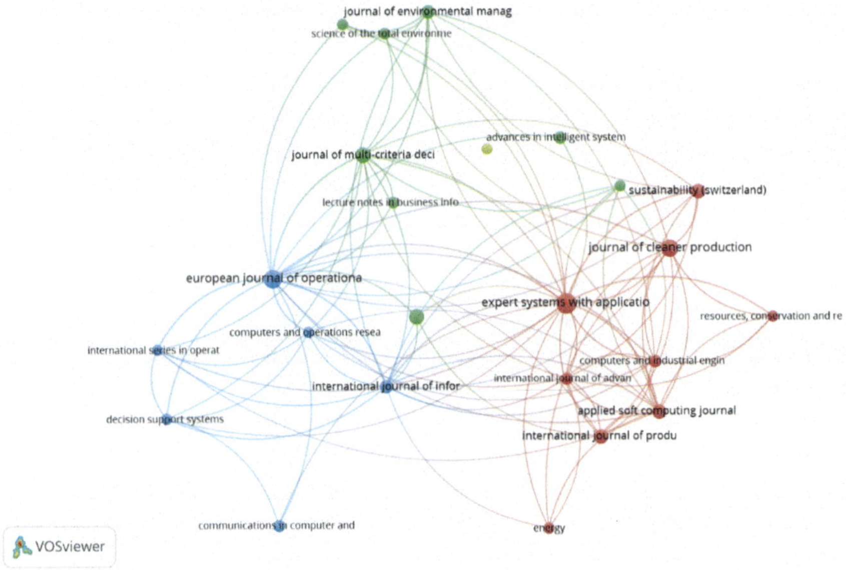

Fig. 2 Bibliographic coupling of journals in CSR-C publications

Research, and Journal of Cleaner Production. These journals are consistent with the results obtained in Table 3.

Another issue to consider is the bibliographic coupling of authors seen in Fig. 3; this map shows the most productive authors in R-MCDA and those who cite the same documents more frequently. This bibliographic coupling considers a threshold of 4 documents and one hundred stronger bibliographic coupling links. Cengız Kahraman and Salvatore Greco are the most productive authors in the field of R-MCDA, followed by Salvatore Corrente, E. Ertugrul Karsak, and Igor Linkov. Also, in this figure it's possible to see that the authors have worked together to develop scientific articles on this area. For example, in the blue cluster you can see that greco and slowinski are the authors who have collaborative works or that cite each other, the same happens in the red cluster with Kahraman and Karsak.

To provide a more general view on the future direction research will take in the R-MCDA field, we investigate the bibliographic coupling of countries. Which shows us which countries are more productive is this field. Figure 4 visualizes the results

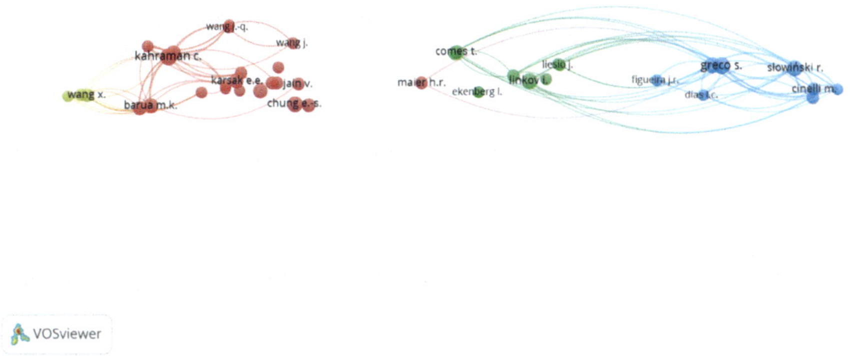

Fig. 3 Bibliographic coupling of authors in R-MCDA publications

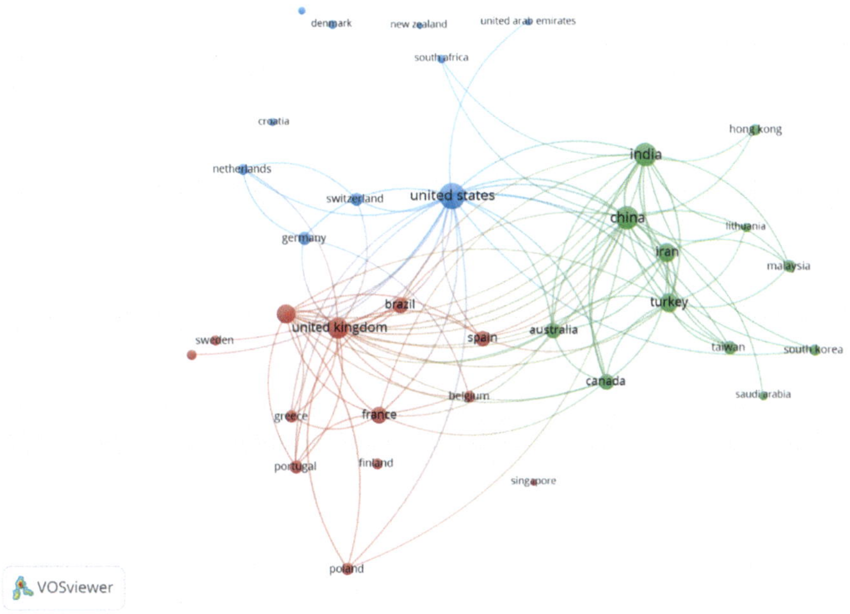

Fig. 4 Bibliographic coupling of countries in R-MCDA publications

considering a threshold of 5 documents and one hundred stronger bibliographic coupling links. The USA is the most productive country in the R-MCDA area; this implies that within its institutions, they produce a large number of documents focused on the R-MCDA, followed by China and the United Kingdom presenting a significant level of publications in this area.

Another analysis considered is the co-citation of journals and authors, which illustrates the most cited journals or authors and those who frequently receive citations from the same sources. Figure 5 presents the results of the co-citation of journals,

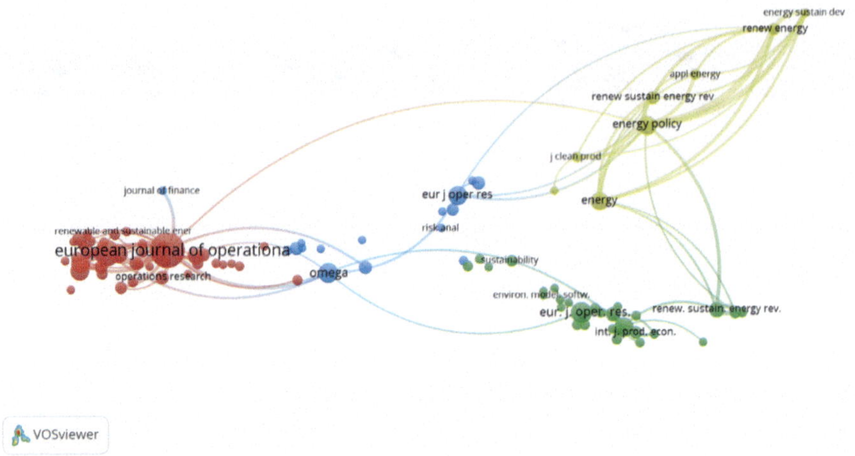

Fig. 5 Co-citation of journals among publications

considering a threshold of 30 citations and 100 stronger co-citations links. As it is possible to observe, the European Journal of Operational is the magazine that stands out. Also, it is seen two slopes within the journals a red colored cluster focused on operations research, while other yellow and green clusters are more focused on sustainability.

Figure 6 shows the co-citation of authors, where a threshold of 50 citations and 100 stronger co-citations links are considered. The most relevant and cited authors are T.L. Saaty, B. Roy, and Salvatore Greco. Furthermore, there are three principal slopes, the largest one is colored in green, presents authors such as Roy, Greco, and Slowinski, these areas focus on the development of new decision-making tools and their applications. The blue slope shows three relevant authors Linkov, Belton, and

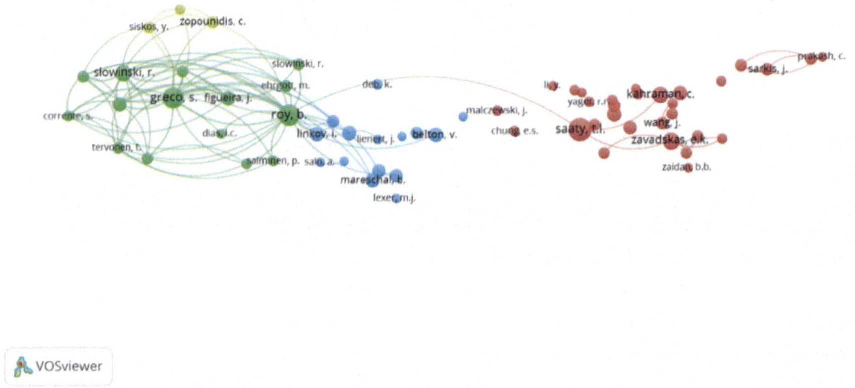

Fig. 6 Co-citation of authors among R-MCDA publications

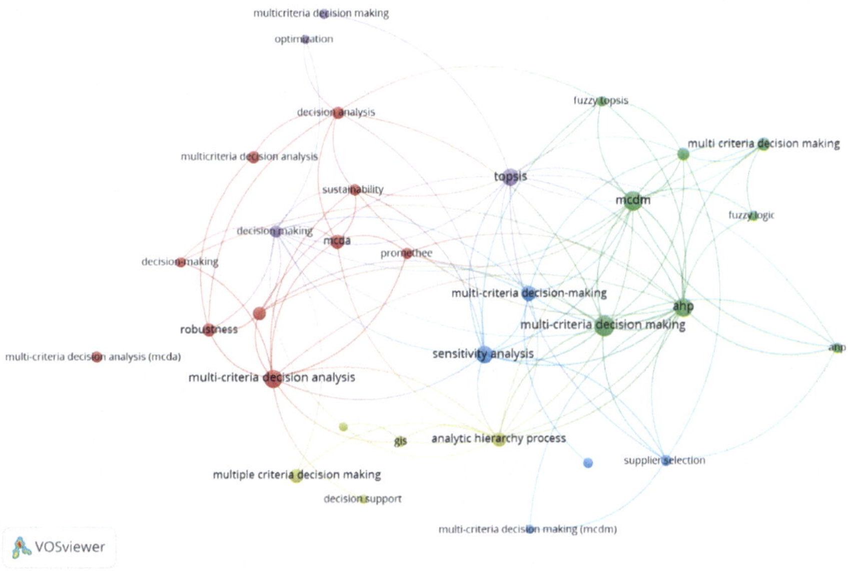

Fig. 7 Co-occurrence of author keywords in R-MCDA

Mareschal, here present works in environmental and sustainability decision making. The last red colored slope, reveal authors such as Saaty, and Zavadkas, which focuses mainly on Fuzzy logic and genetic algorithm.

Finally, a map is presented showing the most common keywords in the R-MCDA area; Fig. 7 shows the Co-occurrence of author keywords in R-MCDA publications. In this map, a minimum threshold of 10 occurrences and the 100 strongest co-occurrence links were considered. The keywords with the most occurrence are MCDA, Topsis, Analytic Hierarchy Process (AHP), sensitivity analysis, robustness, and fuzzy logic. It is evident in Fig. 7 that Topsis keyword presents a great relevance as a methodology used in the decision-making process and to provide a robust solution. On the other hand, the sensitivity analysis shows a large scope in the field of study. These results display the diversity of topics that exist and demonstrate the wide variety of applications of R-MCDA. Also, it's possible to observe that the keywords related to the multicriteria concept appear on many occasions limiting other keywords that show more areas of research.

4 Conclusions

This study presents an overview of R-MCDA from 1977 to 2018, using information from the Scopus database as the main source with a sample size of 807 documents and supplemented with the WoS database. It is important to note that there is a significant increase in the production of documents since 2004 and has been increasing until

2018. In the 2014–2018 period, it represents 48% of the total production of documents in the R-MCDA field; this allows concluding that this topic is a crucial area of research today. The analysis of journals reveals that the leading journals published in R-MCDA are Expert Systems with Applications with a production of 22 documents, followed by the European Journal of Operational Research with a production of 17 documents and the Journal of Cleaner Production with a production of 15 documents.

The type of document that stands out in this investigation is the articles that makeup 89% of the total documents found. The most cited document is "A critical survey on the status of multiple criteria decision-making theory and practice," published by Stewart T. in 1992. The most notable authors come from Italy, Turkey, and the USA. The 10 most important universities in terms of production 2 are from Turkey, China, and Poland, respectively. The countries with the highest production in terms of R-MCDA are the USA, China, and India. The maps that were made in this research show us the interrelations that exist within the R-MCDA research field, such as computer science, engineering, economics, decision sciences, among others.

With the results of this research, it can be determined that this area is of interest to researchers due to the versatility and effectiveness in its results; therefore, it is expected that R-MCDA production will increase in the future.

References

1. McKenna R, Bertsch V, Mainzer K, Fichtner W (2018) Combining local preferences with multi-criteria decision analysis and linear optimization to develop feasible energy concepts in small communities. Eur J Oper Res 268(3):1092–1110. https://doi.org/10.1016/j.ejor.2018.01.036
2. Duarte BPM, Reis A (2006) Developing a projects evaluation system based on multiple attribute value theory. Comput Oper Res 33(5):1488–1504. https://doi.org/10.1016/j.cor.2004.11.003
3. Roy B (2016) Paradigms and challenges. In: Multiple criteria decision analysis: state of the art surveys, pp 19–39. https://doi.org/10.1007/978-1-4939-3094-4_2
4. Figueira J, Greco S, Ehrogott M (2016) Multiple criteria decision analysis. In: Greco S, Ehrgott M, Figueira JR (eds), Multiple criteria decision analysis: state of the art surveys, vol 233. Springer, New York, NY. https://doi.org/10.1007/978-1-4939-3094-4
5. Aissi H, Roy B (2010) Robustness in multi-criteria decision aiding, pp 87–121. https://doi.org/10.1007/978-1-4419-5904-1_4
6. Sniedovich M (2016) From statistical decision theory to robust optimization: a maximin perspective on robust decision-making. In: International series in operations research & management science, vol 241, pp 59–87. https://doi.org/10.1007/978-3-319-33121-8_4
7. Roy B (2010) Robustness in operational research and decision aiding: a multi-faceted issue. Eur J Oper Res 200(3):629–638. https://doi.org/10.1016/j.ejor.2008.12.036
8. Wu W-W (2008) Choosing knowledge management strategies by using a combined ANP and DEMATEL approach. Expert Syst Appl 35(3):828–835. https://doi.org/10.1016/j.eswa.2007.07.025
9. Yu C-S (2002) A GP-AHP method for solving group decision-making fuzzy AHP problems. Comput Oper Res 29(14):1969–2001. https://doi.org/10.1016/S0305-0548(01)00068-5
10. Zeydan M, Çolpan C, Çobanoğlu C (2011) A combined methodology for supplier selection and performance evaluation. Expert Syst Appl 38(3):2741–2751. https://doi.org/10.1016/j.eswa.2010.08.064

11. Dalalah D, Hayajneh M, Batieha F (2011) A fuzzy multi-criteria decision making model for supplier selection. Expert Syst Appl 38(7):8384–8391. https://doi.org/10.1016/j.eswa.2011.01.031
12. Tervonen T (2014) JSMAA: open source software for SMAA computations. Int J Syst Sci 45(1):69–81. https://doi.org/10.1080/00207721.2012.659706
13. Wang G, Huang SH, Dismukes JP (2004) Product-driven supply chain selection using integrated multi-criteria decision-making methodology. Int J Prod Econ 91(1):1–15. https://doi.org/10.1016/S0925-5273(03)00221-4
14. Stewart T (1992) A critical survey on the status of multiple criteria decision making theory and practice. Omega 20(5–6):569–586. https://doi.org/10.1016/0305-0483(92)90003-P
15. Chan F, Chung S, Wadhwa S (2005) A hybrid genetic algorithm for production and distribution. Omega 33(4):345–355. https://doi.org/10.1016/j.omega.2004.05.004
16. Cinelli M, Coles SR, Kirwan K (2014) Analysis of the potentials of multi criteria decision analysis methods to conduct sustainability assessment. Ecol Ind 46:138–148. https://doi.org/10.1016/j.ecolind.2014.06.011
17. Guo S, Zhao H (2015) Optimal site selection of electric vehicle charging station by using fuzzy TOPSIS based on sustainability perspective. Appl Energy 158:390–402. https://doi.org/10.1016/j.apenergy.2015.08.082
18. Hatefi SM, Torabi SA (2010) A common weight MCDA–DEA approach to construct composite indicators. Ecol Econ 70(1):114–120. https://doi.org/10.1016/j.ecolecon.2010.08.014
19. Diaby V, Campbell K, Goeree R (2013) Multi-criteria decision analysis (MCDA) in health care: a bibliometric analysis. Oper Res Health Care 2(1–2):20–24. https://doi.org/10.1016/j.orhc.2013.03.001
20. Guerrero-Baena MD, Gómez-Limón JA, Fruet Cardozo JV (2014) Are multi-criteria decision making techniques useful for solving corporate finance problems? A bibliometric analysis. Revista de Métodos Cuantitativos Para La Economía y La Empresa 17:60–79
21. Blanco-Mesa F, Merigó JM, Gil-Lafuente AM (2017) Fuzzy decision making: a bibliometric-based review. J Intell Fuzzy Syst 32(3):2033–2050. https://doi.org/10.3233/JIFS-161640
22. Bonilla CA, Merigó JM, Torres-Abad C (2015) Economics in Latin America: a bibliometric analysis. Scientometrics 105(2):1239–1252. https://doi.org/10.1007/s11192-015-1747-7
23. Doumpos M, Zopounidis C, Grigoroudis E (2016) Robustness analysis in decision aiding, optimization, and analytics, vol 241. In: Doumpos M, Zopounidis C, Grigoroudis E (eds). Springer International Publishing, Cham. https://doi.org/10.1007/978-3-319-33121-8
24. Stewart TJ (1996) Robustness of additive value function methods in MCDM. J Multi-Criteria Decis Anal 5(4):301–309. https://doi.org/10.1002/(SICI)1099-1360(199612)5:4%3c301::AID-MCDA120%3e3.0.CO;2-Q
25. Fernández DS, Lutz MA (2010) Urban flood hazard zoning in Tucumán Province, Argentina, using GIS and multicriteria decision analysis. Eng Geol 111(1–4):90–98. https://doi.org/10.1016/j.enggeo.2009.12.006
26. Awasthi A, Chauhan SS, Omrani H (2011) Application of fuzzy TOPSIS in evaluating sustainable transportation systems. Expert Syst Appl 38(10):12270–12280. https://doi.org/10.1016/j.eswa.2011.04.005
27. Dweiri F, Kumar S, Khan SA, Jain V (2016) Designing an integrated AHP based decision support system for supplier selection in automotive industry. Expert Syst Appl 62:273–283. https://doi.org/10.1016/j.eswa.2016.06.030
28. Karsak EE (2002) Distance-based fuzzy MCDM approach for evaluating flexible manufacturing system alternatives. Int J Prod Res 40(13):3167–3181. https://doi.org/10.1080/00207540210146062
29. Shih H-S (2008) Incremental analysis for MCDM with an application to group TOPSIS. Eur J Oper Res 186(2):720–734. https://doi.org/10.1016/j.ejor.2007.02.012
30. Baležentis A, Baležentis T, Brauers WKM (2012) Personnel selection based on computing with words and fuzzy MULTIMOORA. Expert Syst Appl 39(9):7961–7967. https://doi.org/10.1016/j.eswa.2012.01.100

31. Cheng EWL, Li H, Yu L (2005) The analytic network process (ANP) approach to location selection: a shopping mall illustration. Constr Innov 5(2):83–97. https://doi.org/10.1108/14714170510815195
32. Mulliner E, Malys N, Maliene V (2016) Comparative analysis of MCDM methods for the assessment of sustainable housing affordability. Omega 59:146–156. https://doi.org/10.1016/j.omega.2015.05.013
33. Senthil S, Srirangacharyulu B, Ramesh A (2014) A robust hybrid multi-criteria decision making methodology for contractor evaluation and selection in third-party reverse logistics. Expert Syst Appl 41(1):50–58. https://doi.org/10.1016/j.eswa.2013.07.010
34. Tervonen T, Figueira JR, Lahdelma R, Dias JA, Salminen P (2009) A stochastic method for robustness analysis in sorting problems. Eur J Oper Res 192(1):236–242. https://doi.org/10.1016/j.ejor.2007.09.008
35. Lee GKL, Chan EHW (2008) The analytic hierarchy process (AHP) approach for assessment of urban renewal proposals. Soc Indic Res 89(1):155–168. https://doi.org/10.1007/s11205-007-9228-x
36. Reza B, Sadiq R, Hewage K (2011) Sustainability assessment of flooring systems in the city of Tehran: an AHP-based life cycle analysis. Constr Build Mater 25(4):2053–2066. https://doi.org/10.1016/j.conbuildmat.2010.11.041
37. Karsak EE, Ahiska SS (2005) Practical common weight multi-criteria decision-making approach with an improved discriminating power for technology selection. Int J Prod Res 43(8):1537–1554. https://doi.org/10.1080/13528160412331326478
38. Tseng M-L (2009) Application of ANP and DEMATEL to evaluate the decision-making of municipal solid waste management in Metro Manila. Environ Monit Assess 156(1–4):181–197. https://doi.org/10.1007/s10661-008-0477-1
39. Oztaysi B (2014) A decision model for information technology selection using AHP integrated TOPSIS-grey: the case of content management systems. Knowl-Based Syst 70:44–54. https://doi.org/10.1016/j.knosys.2014.02.010
40. Yavuz M, Oztaysi B, Cevik Onar S, Kahraman C (2015) Multi-criteria evaluation of alternative-fuel vehicles via a hierarchical hesitant fuzzy linguistic model. Expert Syst Appl 42(5):2835–2848. https://doi.org/10.1016/j.eswa.2014.11.010
41. Feick R, Hall B (2004) A method for examining the spatial dimension of multi-criteria weight sensitivity. Int J Geogr Inf Sci 18(8):815–840. https://doi.org/10.1080/13658810412331280185
42. Hyde K, Maier HR, Colby C (2003) Incorporating uncertainty in the PROMETHEE MCDA method. J Multi-Criteria Decis Anal 12(4–5):245–259. https://doi.org/10.1002/mcda.361
43. Costa BE, C. A., De Corte, J.-M., & Vansnick, J.-C. (2012) MACBETH. Int J Inf Technol Decis Mak 11(02):359–387. https://doi.org/10.1142/S0219622012400068
44. Verderame PM, Elia JA, Li J, Floudas CA (2010) Planning and scheduling under uncertainty: a review across multiple sectors. Ind Eng Chem Res 49(9):3993–4017. https://doi.org/10.1021/ie902009k
45. Peng Y, Kou G, Wang G, Wu W, Shi Y (2011) Ensemble of software defect predictors: an AHP-based evaluation method. Int J Inf Technol Decis Mak 10(01):187–206. https://doi.org/10.1142/S0219622011004282
46. Buchholz T, Rametsteiner E, Volk TA, Luzadis VA (2009) Multi criteria analysis for bioenergy systems assessments. Energy Policy 37(2):484–495. https://doi.org/10.1016/j.enpol.2008.09.054
47. Blanco-Mesa F, León-Castro E, Merigó JM (2019) A bibliometric analysis of aggregation operators. Appl Soft Comput 81(May). https://doi.org/10.1016/j.asoc.2019.105488

A Bibliometric Analysis on Innovation Management Research

Denisse Ballardo-Cárdenas, Ernesto León-Castro, Fabio Blanco-Mesa, and Ramón Martínez-Huerta

Abstract This article aims to provide an analysis of the information on scientific research publications related to Innovation Management available on the Web of Science. For this, a bibliometric study of the publications of the Web of Science Core Collection database on Innovation Management is carried out. The study considers all the documents published during the period from 1982 to 2018. In the search, the type of document selected is by article, review, and note for researchers interested in the subject. The study graphically maps the bibliographic material by using the VOS viewer software, to provide a more in-depth analysis of data obtained from the search. The present work shows the analysis of data of the publications on Innovation management. Identify the leading publications, the most productive authors, the most productive countries, the most productive institutions, and analyze their evolution over time. The above are some of the data obtained in the analysis. Therefore, the results of the bibliometric study show that Innovation Management remains a topic of interest in research, presents significant development of the journal over time, and provides a synthesis of research for new researchers.

Keywords Innovation management · Bibliometric analysis · VOS viewer visualization

---

D. Ballardo-Cárdenas (✉)
Universidad Autónoma de Occidente, Blvd. Lola Beltrán, Blvd. Rolando Arjona Amabilis y, 4 de Marzo, 80020 Culiacán Rosales, Sin., Mexico
e-mail: denisse.ballardo@uadeo.mx

E. León-Castro
Universidad Católica de la Santísima Concepción, Alonso de Ribera 2850, Concepción, Chile

F. Blanco-Mesa
Escuela de Administración de Empresas, Universidad Pedagógica y Tecnológica de Colombia, Av. Central del Norte, 39-115, 150001 Tunja, Colombia

R. Martínez-Huerta
Doctorado en Ciencias Administrativas, Universidad Autónoma de Occidente, Blvd. Rolando Arjona Amabilis y, Blvd. Lola Beltrán4 de Marzo, 80020 Culiacán Rosales, Sin., Mexico

E. León-Castro et al. (eds.), *Soft Computing and Fuzzy Methodologies in Innovation Management and Sustainability*, Lecture Notes in Networks and Systems 337,
https://doi.org/10.1007/978-3-030-96150-3_17

1 Introduction

The competitiveness of companies in a globalized world has highlighted the need for our organizations to recognize the strategic value of innovation and incorporate techniques and tools for its management (Hidalgo, 2011). An essential instrument for the transformation of companies in the management of innovation when the entire organization is willing to update and reform [1]. According to the literature, innovation management (MI) is the application of new ideas to improve organizational structures and systems, and processes related to the social structure of an organization [2–4]. Innovation management should be aimed at products, markets, production skills and administrative skills [5].

Innovation management can be seen as a form of innovation capacity, which seeks to stimulate innovation for the organization successfully [6, 7] by continually transforming knowledge and ideas into new products, processes and systems for the benefit of the company and its stakeholders. The capacity for innovation could be defined as a group of *characteristics* of organizations, understood as capacities or abilities through which better competitive performance can be achieved, if adequately managed [8].

The literature on innovation management and innovation capability is extensive, has several approaches and a large number of researchers involved. In recent years, researchers and academics have paid attention to the field of innovation management studies in organizations receiving different names. For example, it had called administrative innovation [9–11], and then this was changed to organizational innovation [12], as time goes by the name changed to managerial innovation [13]. Nowadays, it is called innovation management [14–19], Hollen et al. (2013), and research in this field has increased.

Hence, the main aim is to use a bibliometric approach to map authors, institutions, countries, journals and the evolution of the publication also to provide useful information on scientific research publications in the area of Innovation Management and Innovation capability, through bibliometric analysis and the use of software for bibliographic mapping [20]. This study can be a support for researchers who need an appropriate way to analyze new developments in the area and provide the necessary path to build a profoundly theoretical framework on the subject.

Through bibliometric analysis, it seeks to provide useful information on Innovation Management scientific research publications. A bibliometric review is the quantitative study of bibliographic material that can show a general picture of a research field [21]. Also, it can be based on the counting of citations [22], co-citations [23, 24], bibliographic coupling [25], co-authors [26] Glänzel [27], similarly in its application in institutions, countries, authors, magazines, others. The study graphically maps the bibliographic material through the use of VOS visualization software to provide a more in-depth analysis of the data obtained from the search. The VOS visualization program can be used to build and view graphic maps through the use of bibliographic coupling [28]. For example, the bibliographic coupling of the authors occurs when the authors of the two documents cite the same third document [25]. Likewise, the

VOS viewer allows to collect the data and creating maps in terms of bibliographic coupling, analysis of citations and citations, co-authorship and co-occurrence of author keywords [21], Blanco et al. [29], Ortiz-de Urbina-Created, Nájera-Sánchez and Mora-Valentín [30], [31], Blanco-Mesa et al. [32]. Findings indicate an emerging field of study and a large number of articles and citations concentrated in different journals. It is noteworthy that the principal authors are Bessant, J, Verganti R, and Frattini F, and the leading institutions are The Polytechnic University of Milan is the most productive institution is located in Italy, Lappeenranta University of Technology and Aalto University, both located in Finland. Likewise, the main countries that publish innovation management studies are the USA, Germany, and China. Finally, this document is organized as follows. Section 2 reviews the bibliometric methods used in this document. Section 3 presents the results; Sect. 4 develops a visual analysis of the Innovation Management bibliographic data with the VOS viewer software. Section 5 summarizes the main findings and concludes the document.

2 Bibliometric Methods

This analysis is performed with the bibliographic search, where the database of the Web of Science collection is used. The WoS database has been used in different academic research as the primary source [33–35]. The research methodology process starts with the selection of the keywords that help to determine the study sample. The keywords in the topics section are "Innovation management" or "Innovation capacity*," for this study. It is labeled as Innovation management (IM). The initial search was performed in August 2019, and the analysis considers all the documents published by the innovation management in a time range from 1982 to 2018. Note that the field is considered from the first publication of Management of the innovation from the base of Web of Science data. Web of Science has published 2448 documents during this period. A refined search is performed for the type of document considering articles, reviews and notes. Therefore, the search is reduced to 2288. Until 2018, innovation management has received 42,612 citations from other documents available in the Web of Science database. This makes an average number of citations per document of 18.62. The h index is 92. That is, of the 2288 documents, 92 have received 92 citations or more. In addition to 30,455 articles in which it is cited.

Based on data obtained in the search, a detailed analysis of the information provided by the WoS carried out, considering different bibliometric indicators such as paper citations, citations per year, thresholds for citations, the articles they cite and the evolution throughout the weather. For countries, the analysis uses the number of documents and citations per million inhabitants. At the university level, the work complements the data with the ranking of universities in the Academic Ranking of World Universities (ARWU) and the Quacquarelli and Symonds University Ranking (QS), in others. Likewise, a general evaluation of the most productive and influential variables in Innovation Management is presented. The above, to show the information and facilitate the understanding of the subject to the reader. Besides, the study graphically maps the bibliographic material using VOS visualization software [28]. The VOS viewer collects data and creates maps in terms of bibliographic coupling, quotation, joint appointment, co-authorship, and concurrence of author keywords [21, 31]. It is essential to mention that the bibliographic coupling of the authors occurs when the authors of the two documents cite the same third document [25].

3 Results

The search conducted in the topic of Innovation Management provides as a result of what is presented in the following tables and figures illustrated by The VOS viewer.

3.1 Publication and Citation Structure

The topic of Innovation Management has been studied over the years and has shown increasing evolution. This evolution is observed significantly in the last quartile of Fig. 1.

Table 1 presents the growth in the number of citations over time; mainly, it is observed that the study of innovation management is increasing with a total of 2288 productions and 38,083 citations. It is identified that innovation management has been mostly studied in 2017 and 2018. This last year, it shows a total of 7062 citations as the most significant number of citations over time. It is observed that from the second decade, the growth is substantial.

Table 2 presents the fifty most cited papers of the journal of all-time. Note that account that since 1982 the first publications of Innovation Management are presented. In this table, the three most cited papers are positioned in the years

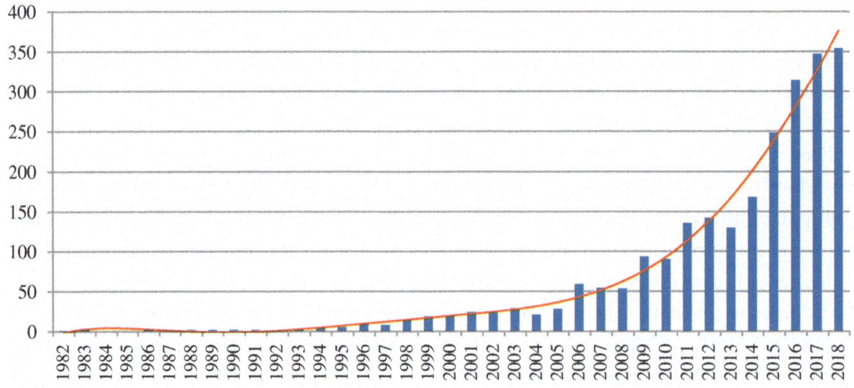

Fig. 1 The annual number of papers published in Innovation Management

Table 1 Annual citation structure of Innovation Management

Year	TP	TC	Thresholds						
			≥200	≥100	≥50	≥20	≥10	≥5	≥1
1982	1	1	0	0	0	0	0	0	1
1983	1	1	0	0	0	0	0	0	1
1984	0	0	0	0	0	0	0	0	0
1985	0	0	0	0	0	0	0	0	0
1986	2	1	0	0	0	0	0	1	0
1987	1	1	0	0	0	0	0	0	1
1988	0	0	0	0	0	0	0	0	0
1989	1	0	0	0	0	0	0	0	0
1990	2	2	0	0	0	0	1	0	0
1991	3	0	0	0	0	0	0	0	0
1992	3	2	0	0	1	0	1	1	0
1993	3	2	0	0	0	0	0	2	0
1994	6	9	0	0	0	1	2	2	1
1995	5	8	0	0	1	1	0	0	2
1996	11	15	1	0	1	3	2	0	4
1997	8	15	0	0	3	0	2	1	2
1998	13	29	2	0	2	3	2	2	2
1999	16	43	0	0	3	2	3	3	4
2000	20	57	3	1	4	4	4	3	0
2001	21	58	1	3	3	6	2	4	0
2002	24	93	3	2	0	7	2	4	5
2003	27	129	1	2	3	10	4	2	5
2004	19	162	0	3	8	2	1	1	2
2005	29	327	0	3	7	6	4	2	2
2006	53	444	2	8	9	11	5	9	6
2007	50	570	1	5	1	12	12	4	7
2008	47	809	1	4	6	12	6	8	6
2009	84	935	1	10	8	21	11	10	12
2010	85	1196	4	2	9	28	13	5	13
2011	123	1861	2	6	8	29	22	26	17
2012	137	2253	1	4	10	29	29	17	26
2013	119	2756	0	3	7	22	19	34	16
2014	162	3441	2	2	7	32	42	36	28
2015	235	4034	0	2	4	24	37	48	71
2016	310	5475	0	0	0	13	37	60	123

(continued)

Table 1 (continued)

Year	TP	TC	Thresholds						
			≥200	≥100	≥50	≥20	≥10	≥5	≥1
2017	341	6292	0	0	2	8	18	46	142
2018	326	7062	0	0	0	1	7	15	136
Total	2288	38,083	25	60	107	287	288	346	635
	100.00%		1.09%	2.62%	4.68%	12.54%	12.59%	15.12%	27.75%

Abbreviation TP and TC = Total papers and citations; ≥ 200, ≥ 100, ≥ 50, ≥ 20, ≥ 10, ≥ 5, ≥ 1 = Number of papers with equal or more than 200, 100, 50, 20, 10, 5 and 1 citations

2010, 2002 and 1998, being Business Models, Business Strategy and Innovation the most cited article and corresponds to the author Teece, DJ occupying the first place. The purpose of this article is to understand the significance of business models and explore their connections with business strategy, innovation management, and economic theory. Followed by the paper Learning orientation, firm innovation capability, and firm performance managing. That article investigates the relationships among learning orientation, firm innovativeness, and firm performance, through a model drawn from organizational learning theory and the new product literature [36]. In position three, is presented the paper Does entrepreneurial self-efficacy distinguish entrepreneurs from managers? With a total of 826 citations by the authors Chen, CC; Greene, PG; Crick, A.

3.2 *Leading Authors, Institutions and Countries*

In this section, let us analyze those authors, institutions, and countries that have most publications in Innovation Management. First, let us look into the 50 most productive authors (see Table 3). Note that in the case of a tie, the ranking is according to the number of citations. Bessant, J is the author that has most publications in Innovation Management. Followed by Verganti R., Note that both of them are among the most productive authors. Furthermore, it is worth mentioning the author Frattini Fother of Polytechnic University of Milan, Italy, that he is the third position of the ranking of 50 most productive authors.

Table 4 shows the temporal evolution of the most productive authors in Innovation Management. The author Verganti R. shows to be the most productive in the third decade. Also, during the third and fourth decade, is observed significant influence with the author Bessant, J. This author occupies the second important position in recent six years after the author Parida V. who has been in the first position in the fourth decade.

Next, let us analyze the most productive institutions to know which universities produce the most documents and know the universities that are publishing the most science in the area of innovation management. The Polytechnic University of Milan is

Table 2 The 50 most cited papers in Innovation Management between 1982 and 2018

R	TC	Title	Author/s	Year	C/Y
1	1502	Business models, business strategy and innovation	Teece [37]	2010	187.8
2	951	Learning orientation, firm innovation capability, and firm performance	Calantone et al. [36]	2002	59.44
3	826	Does entrepreneurial self-efficacy distinguish entrepreneurs from managers?	Chen et al. [38]	1998	41.3
4	555	Open innovation: state of the art and future perspectives	Huizingh [39]	2011	79.29
5	498	Innovation and creativity in organizations: a state-of-the-science review, prospective commentary, and guiding framework	Anderson et al. [40]	2014	124.5
6	482	An examination of organizational factors influencing new product success in internal and alliance-based processes	Sivadas and Dwyer [41]	2000	26.78
7	418	Product complexity, innovation and industrial organization	Hobday [42]	1998	20.9
8	399	Innovation management measurement: a review	Adams [43]	2006	33.25
9	395	Knowledge sharing and firm innovation capability: an empirical study	Lin [44]	2007	35.91
10	373	Determinants of innovation capability in small electronics and software firms in southeast England	Romijn and Albaladejo [45]	2002	23.31
11	371	Performance assessment of the lead user idea-generation process for new product development	Lilien et al. [46]	2002	23.19
12	357	Leveraging external sources of innovation: a review of research on open innovation	West and Bogers [47]	2014	89.25
13	306	A collective action model of institutional innovation	Hargrave and Van de Ven [48]	2006	25.5
14	273	Innovation management in context: environment, organization and performance	Tidd [49]	2001	16.06
15	271	Development of a technical innovation audit	Chiesa et al. [50]	1996	12.32

(continued)

Table 2 (continued)

R	TC	Title	Author/s	Year	C/Y
16	261	Communities of creation: managing distributed innovation in turbulent markets	Sawhney and Prandelli [51]	2000	14.5
17	254	Knowledge collaboration in online communities	Faraj et al. [52]	2011	36.29
18	254	The past and the future of international entrepreneurship: a review and suggestions for developing the field	Keupp and Gassmann [53]	2009	28.22
19	247	Design, meanings, and radical innovation: A metamodel and a research agenda	Verganti [54]	2008	24.7
20	246	Adaptive management in agricultural innovation systems: the interactions between innovation networks and their environment	Klerkx et al. [55]	2010	30.75
21	241	Innovative capability and export performance of Chinese firms	Guan and Ma [56]	2003	16.07
22	233	Capabilities for managing service innovation: towards a conceptual framework	den Hertog et al. [57]	2010	29.13
23	230	Innovation practice and its performance implications in small and medium enterprises (SMES) in the manufacturing sector: a resource-based view	Terziovski [58]	2010	28.75
24	226	Organizing for Innovation in the Digitized World	Yoo et al. [59]	2012	37.67
25	222	Successful execution of product development projects: balancing firmness and flexibility in the innovation process	Tatikonda and Rosenthal [60]	2000	12.33
26	199	Legitimizing research, technology and innovation policies for transformative change combining insights from innovation systems and multi-level perspective in a comprehensive 'failures' framework	Weber and Rohracher [61]	2012	33.17
27	197	Selling university technology: Patterns from MIT	Shane, S	2002	12.31

(continued)

Table 2 (continued)

R	TC	Title	Author/s	Year	C/Y
28	195	An audit of technological innovation capabilities in Chinese firms: some empirical findings in Beijing, China	Yam, RCM; Guan, JC; Pun, KF; Tang, EPY	2004	13.93
29	192	Organizational innovation as an enabler of technological innovation capabilities and firm performance	Camison, C; Villar-Lopez, A	2014	48
30	189	Linking routines to operations capabilities: a new perspective	Peng et al. [62]	2008	18.9
31	187	Conceptualizing innovation orientation: a framework for study and integration of innovation research	Siguaw et al. [63]	2006	15.58
32	179	Emergence of innovations in services	Toivonen and Tuominen [64]	2009	19.89
33	178	Targeting innovation and implications for capability development	Francis and Bessant [65]	2005	13.69
34	176	Knowledge sharing, absorptive capacity, and innovation capability: an empirical study of Taiwan's knowledge-intensive industries	Liao et al. [66]	2007	16
35	172	Customer empowerment in new product development	Fuchs and Schreier [67]	2011	24.57
36	171	Relationships between innovation stimulus, innovation capacity, and innovation performance	Prajogo and Ahmed [68]	2006	14.25
37	170	Small-firm performance: modeling the role of product and process improvements	Wolff and Pett [69]	2006	14.17
38	164	A measure of absorptive capacity: scale development and validation	Flatten et al. [70]	2011	23.43
39	163	The lead user method: an outline of empirical findings and issues for future research	Luthje, C; Herstatt, C	2004	11.64
40	157	Brand communities embedded in social networks	Zaglia [71]	2013	31.4
41	156	Green innovation in technology and innovation management—an exploratory literature review	Schiederig et al. [72]	2012	26

(continued)

Table 2 (continued)

R	TC	Title	Author/s	Year	C/Y
42	156	A Dialectic Perspective on Innovation: conflicting demands, multiple pathways, and ambidexterity	Bledow [73]	2009	17.33
43	156	Breakthrough? China's and India's transition from production to innovation	Altenburg et al. [74]	2008	15.6
44	156	Innovation types and innovation management practices in service companies	Oke [75]	2007	14.18
45	155	The open innovation journey: how firms dynamically implement the emerging innovation management paradigm	Chiaroni et al. [76]	2011	22.14
46	150	Integration of market pull and technology push in the corporate front end and innovation management-Insights from the German software industry	Brem and Voigt [77]	2009	16.67
47	149	Market orientation and the new product paradox	Baker and Sinkula [78]	2005	11.46
48	148	Community engineering for innovations: the ideas competition as a method to nurture a virtual community for innovations	Ebner et al. [79]	2009	16.44
49	148	Auditing best practice for effective product innovation management	Cormican and O'Sullivan [80]	2004	10.57
50	141	Strategic knowledge management, innovation and performance	Lopez-Nicolas, C; et al	2011	20.14

Abbreviations R = Rank; TC = Total citations; C/Y = Citations per year

the most productive institution, in addition to being the institution with the highest H index and appointments. The institutions that follow are Lappeenranta University of Technology and Aalto University, both located in Finland. Likewise, it is observed in this top 50 that the country with the most significant institutions' number is England; this country has six institutions: The University of Sussex, University of Manchester, University of London, The University of Cambridge, The University of Nottingham and The University of Exeter. In this last institution, the author Bessant carried out his studies and is considered as the most productive author of Innovation Management (Table 5).

Table 3 Top 50 most productive authors in innovation management

R	Author/s	University	Country	TP	TC	H	C/P	Threshold		
								≥100	≥50	≥1
1	Bessant, J	University of Exeter	England	12	744	7	62	2	1	7
2	Verganti R	Polytechnic University of Milan	Italy	10	430	7	43	1	1	8
3	Frattini F	Polytechnic University of Milan	Italy	9	485	8	53.89	1	3	4
4	Brem A	University of Erlangen Nuremberg	Germany	8	166	3	20.75	1	0	5
5	Pantanoe	University of Calabria	Italy	8	99	5	12.38	0	0	7
6	Parida V	Lulea University of Technology	Sweden	8	153	7	19.13	0	0	8
7	Kodama M	Nihon University	Japan	7	11	1	1.571	0	0	2
8	Meissner D	National Research University - Higher School of Economics	Russia	7	80	5	11.43	0	0	7
9	Akgun AE	Gebze Technical University	Turkey	6	237	5	39.5	1	0	4
10	Chen CC	Tzu Chi University	Taiwan	6	1040	4	173.3	2	1	4
11	Corso M	Polytechnic University of Milan	Italy	6	171	5	28.5	0	1	5
12	Keskin H	Gebze Technical University	Turkey	6	237	5	39.5	1	1	4
13	Lee H	Seoul National University of Science & Technology	South Korea	6	173	4	28.83	1	0	5
14	Rohrbeck R	Technical University of Berlin	Germany	6	330	6	55	1	3	2
15	Saunila M	Lappeenranta University of Technology	Finland	6	91	4	15.17	0	0	6
16	Sneck T	VTT Technical Research Center Finland	Finland	6	0	0	0	0	0	0
17	Tuominen M	Lappeenranta University of Technology	Finland	6	69	3	11.5	0	1	4
18	Yang J	University of Houston	USA	6	70	4	11.67	0	0	6
19	Chiesa V	University of London	England	5	503	4	100.6	2	1	1

(continued)

Table 3 (continued)

R	Author/s	University	Country	TP	TC	H	C/P	Threshold		
								≥100	≥50	≥1
20	De Massis A	Lancaster University	England	5	259	5	51.8	0	2	3
21	Fan PL	Massachusetts Institute of Technology (MIT)	USA	5	141	4	28.2	1	0	4
22	Figueiredo PN	Getulio Vargas Foundation	Brazil	5	103	4	20.6	0	0	5
23	Herstatt C	Hamburg University of Technology	Germany	5	350	5	70	2	0	3
24	Hoegl M	Aalto University	Finland	5	43	3	8.6	0	0	5
25	Hu MC	National Chung Hsing University	Taiwan	5	109	4	21.8	0	0	4
26	Karkkainen H	Lappeenranta University of Technology	Finland	5	96	3	19.2	0	1	3
27	Liao SH	Tamkang University	Taiwan	5	227	4	45.4	1	0	4
28	Lichtenthaler U	University of Mannheim	Germany	5	59	4	11.8	0	0	4
29	Luoma T	VTT Technical Research Center Finland	Finland	5	0	0	0	0	0	0
30	Nagano MS	Universidad de Sao Paulo	Brazil	5	41	3	8.2	0	0	4
31	Ngo LV	University of New South Wales Sydney	Australia	5	264	4	52.8	1	2	2
32	Paasi J	VTT Technical Research Center Finland	Finland	5	0	0	0	0	0	0
33	Piippo P	Lappeenranta University of Technology	Finland	5	64	2	12.8	0	3	1
34	Ratten V	La Trobe University	Australia	5	9	2	1.8	0	0	3
35	Saenz J	University of Deusto	Spain	5	72	3	14.4	0	0	4
36	Salerno MS	Universidade de Sao Paulo	Brazil	5	70	3	14	0	0	4
37	Schultz C	Technical University of Berlin	Germany	5	53	5	10.6	0	0	5
38	Song M	University of Washington	USA	5	173	4	34.6	0	1	4

(continued)

Table 3 (continued)

R	Author/s	University	Country	TP	TC	H	C/P	Threshold		
								≥100	≥50	≥1
39	Zhao XD	China Europe International Business School	China	5	76	4	15.2	0	0	5
40	Aramburu N	University of Deusto	Spain	4	69	3	17.25	0	0	3
41	Blindenbach-Driessen F	Erasmus University Rotterdam	Netherlands	4	181	4	45.25	0	1	3
42	Busse C	Technical University of Berlin	Berlin	4	82	3	20.5	0	0	4
43	Camison C	Universitat Jaume I	Spain	4	276	3	69	1	1	2
44	Chen J	Zhejiang University	China	4	32	2	8	0	0	2
45	Cormican K	National University of Ireland (NUI) Galway	Ireland	4	187	3	46.75	1	0	2
46	Dahlgaard JJ	Aarhus University	Denmark	4	57	4	14.25	0	0	4
47	Dooley L	University College Cork	Ireland	4	35	3	8.75	0	0	4
48	Elmquist M	Chalmers University of Technology	Sweden	4	56	3	14	0	0	4
49	Frishammar J	Lulea University of Technology	Sweden	4	90	3	22.5	0	0	3
50	Griffin A	University of Utah	USA	4	89	3	22.25	0	0	4

Abbreviations available in Tables 1 and 3 except: C/P = Cites per paper; H = h -index

Table 4 Temporal evolution of the most productive authors

R	1982–1991			2002–2011		
	Author	TP	TC	Author	TP	TC
1	Chakrabarti, Ak	1	66	Verganti R	5	357
2	Sanchez, Am	1	21	Paasi J	5	0
3	Thom, N	1	14	Luoma T	5	0
4	Rasler, K	1	8	Bessant J	4	640
5	Rothwell, R	1	8	Yam RCM	4	508
6	Martin, MJC	1	7	Guan JC	4	389
7	Mensch, Go	1	2	Cormican K	4	187
8	Kasper, H	1	1	Wonglimpiyarat J	4	60
9	Moenaert, R	1	0	Karkkainen H	4	44
10	Strebel, H	1	0	Tuominen M	4	8
R	*1992–2001*			*2012–2018*		
	Author	*TP*	*TC*	*Author*	*TP*	*TC*
1	Hobday M	3	481	Parida V	8	153
2	Boer H	3	161	Bessant J	8	104
3	Corso M	3	118	Pantano E	8	99
4	Coughlan P	2	308	Meissner D	7	80
5	Chiesa V	2	285	Brem A	7	16
6	Rush, H	2	97	Kodama M	7	11
7	Piippo P	2	61	Frattini F	6	229
8	Jones O	2	54	Saunila M	6	91
9	Dahlgaard JJ	2	23	Sneck T	6	0
10	Dooley L	2	14	De Massis A	5	259

Abbreviations are available in the previous tables. Note that each five-year period indicates the number of papers and citations in that period.

In the same way, these results can be observed over time in the following Table 6, which shows the most productive institutions in the subject classified by four periods: 1982–1991, 1992–2001, 2002–2011 and 2012–2018. This table shows the evolution of the innovation management per university. The Polytechnic University of Milan from the second decade begins to take relevance as one of the most productive institutions occupying the most critical positions in the last thirty years.

Table 7 presents the fifty most productive countries in Innovation Management. The USA is already the most prolific and influential country in Innovation Management, although still very close to Germany. Under this perspective, China and England obtain the most remarkable results being in the third and fourth positions, respectively, in absolute terms. However, keep in mind that the most productive author is from England and the institutions that publish the most are from Italy. The country

Table 5 Top 50 the most productive and influential institutions in Innovation Management

R	Institution	Country	TP	TC	H	C/P	Thresholds		ARWU		QS
							≥100	≥50	≥1		
1	Polytechnic University of Milan	Italy	35	1397	17	39,91	3	4	26	151–200	170
2	Lappeenranta University of Technology	Finlad	29	396	11	13,66	0	2	25	801–900	501–550
3	Aalto University	Finlad	27	697	11	25,81	2	0	23	301–400	137
4	Universidade De Sao Paulo	Brazil	24	188	6	7,83	0	0	19	151–200	121
5	Chalmers University of Technology	Sweden	20	402	11	20,10	1	1	18	201–300	133
6	University of Sussex	England	19	1103	11	58,05	3	2	13	201–300	158
7	Universitat Politecnica de Valencia	Spain	18	356	8	19,78	1	1	15	401–500	373
8	Zhejiang University	China	18	176	6	9,78	1	0	13	67	87
9	Copenhagen Business School	Denmark	17	487	10	28,65	2	2	13	29	73
10	University of Manchester	England	17	250	9	14,71	0	1	14	34	34
11	Lund University	Sweden	16	197	7	12,31	0	1	13	101–150	78
12	Chinese Academy of Sciences	China	15	186	7	12,40	0	0	14	101–150	97
13	Delft University of Technology	Netherlands	15	303	10	20,20	0	0	15	151–200	54
14	Erasmus University Rotterdam	Netherlands	15	485	8	32,33	1	1	12	79	147
15	Technical University of Berlin	Germany	15	330	8	22,00	1	1	11	301–400	144
16	University of Southern Denmark	Denmark	15	430	5	28,67	1	0	12	301–400	384
17	Vtt Technical Research Center Finland	Finland	15	21	3	1,40	0	0	4	–	–
18	Eth Zurich	Switzerland	14	298	9	21,29	0	2	11	54	10
19	Fraunhofer Gesellschaft	Germany	14	131	4	9,36	0	1	6	–	–
20	Islamic Azad University	Iran	14	41	5	2,93	0	0	8	–	–

(continued)

Table 5 (continued)

R	Institution	Country	TP	TC	H	C/P	Thresholds		ARWU		QS
							≥100	≥50	≥1		
21	Lulea University of Technology	Sweden	14	249	10	17,79	0	0	13	801–900	–
22	National Research University Higher	Russia	14	109	7	7,79	0	0	11	901–1000	382
23	Tsinghua University	China	14	144	5	10,29	0	1	11	45	25
24	University of London	England	14	519	7	37,07	2	1	10	17	23
25	Aalborg University	Denmark	13	236	5	18,15	1	0	11	201–300	379
26	Massachusetts Institute of Technology Mit	USA	13	942	11	72,46	3	4	6	4	1
27	Bocconi University	Italy	12	346	7	28,83	1	0	11	–	–
28	European Business School Ebs University	Germany	12	218	7	18,17	0	2	7	–	–
29	Georgia Institute of Technology	USA	12	287	9	23,92	0	1	10	79	70
30	Linkoping University	Sweden	12	206	7	17,17	0	1	11	301–400	287
31	National Kaohsiung University of Science Tech	Taiwan	12	143	6	11,92	0	0	12	–	–
32	National University of Ireland Nui Galway	Ireland	12	228	6	19,00	1	0	9	401–500	243
33	Penn State University	USA	12	772	8	64,33	2	1	8	74	19
34	University of Cambridge	England	12	136	5	11,33	0	1	8	3	5
35	University of Exeter	England	12	202	5	16,83	0	2	8	151–200	158
36	University of Nottingham	England	12	183	7	15,25	0	1	9	101–150	84
37	University of St Gallen	Switzerland	12	414	8	34,50	1	1	8	–	372
38	Wageningen University Research	Netherlands	12	568	7	47,33	2	2	7	101–150	124
39	City University of Hong Kong	China	11	687	11	62,45	3	2	6	201–300	49

(continued)

Table 5 (continued)

R	Institution	Country	TP	TC	H	C/P	Thresholds		ARWU		QS
							≥100	≥50	≥1		
40	Michigan State University	USA	11	1208	7	109,82	1	1	9	101–150	21
41	South China University of Technology	China	11	210	6	19,09	0	1	10	201–300	551–600
42	Universidade Da Beira Interior	Portugal	11	60	5	5,45	0	0	11	–	–
43	Federal University of Santa Catarina Ufsc	Brazil	11	26	3	2,36	0	0	5	601–700	751–800
44	University of Deusto	Spain	11	94	5	8,55	0	0	7	–	–
45	University of Groningen	Netherlands	11	739	8	67,18	1	0	10	66	113
46	University of Ljubljana	Slovenia	11	165	6	15,00	1	0	9	401–500	651–700
47	University of New South Wales Sydney	Australia	11	701	8	63,73	1	2	6	101–150	45
48	University of Twente	Netherlands	11	292	7	26,55	0	1	9	501–600	179
49	University of Vaasa	Finland	11	139	6	12,64	0	0	9	–	–
50	Whu Otto Beisheim Sch Management	Germany	11	180	7	16,36	0	1	9	–	–

Abbreviations are in Tables 1 and 3 except ARWU and QS = Ranking in the general ARWU and QS university rankings

Table 6 Temporal evolution of the most productive institutions

R	*1982–1991*			*2002–2011*		
	University	*TP*	*TC*	*University*	*TP*	*TC*
1	Universidad Drexel	1	66	Polytechnic University Of Milan	12	670
2	University of Kiel	1	66	Aalto University	9	590
3	Universidad De Friburgo	1	14	Cranfield University	7	727
4	Indiana University Bloomington	1	8	Massachusetts Institute of Technology MIT	7	682
5	University of Sussex	1	8	City University Of Hong Kong	7	556
6	Universidad Dalhousie	1	7	Erasmus University Rotterdam	7	368
7	Caso Reserva Occidental Universidad	1	2	Lappeenranta University of Technology	7	147
8	Carl Von Ossietzky Universitat Oldenburg	1	0	Michigan State University	6	1157
9	Universidad de Alabama Huntsville	1	0	Eindhoven University of Technology	6	492
10	Universidad Ghent	1	0	Bocconi University	6	284
R	*1992–2001*			*2012–2018*		
	University	*TP*	*TC*	*University*	*TP*	*TC*
1	University of Sussex	5	827	Universidade De Sao Paulo	22	179
2	Polytechnic University Of Milan	5	404	Lappeenranta University Of Technology	20	188
3	University of Brighton	4	147	Polytechnic University of Milan	18	323
4	University of London	3	349	Aalto University	17	98
5	Aalborg University	3	186	Universitat Politecnica De Valencia	15	220
6	University of Quebec	3	86	Chalmers University of Technology	14	193
7	Aarhus University	3	25	Lulea University of Technology	14	246
8	Fraunhofer Gesellschaft	2	70	Lund University	14	84
9	City University of Hong Kong	2	65	University of Southern Denmark	14	429

(continued)

Table 6 (continued)

10	Chinese University of Hong Kong	2	58	Islamic Azad University	13	39

Abbreviations are available in the previous tables

of Italy occupies the sixth position of the most productive countries in Innovation Management.

Next, let us analyze these results through time. To do so, Table 8 presents the number of papers published annually by the Top 50 countries. Note that the table focuses on the last seven years providing a total number for each of the first three decades. During the first two decades, the USA published more than twenty papers. This number started to grow significantly in the third decade with over a hundred publications and today the USA participates with almost four hundred documents of Innovation Management. On the other hand, Germany in the third decade also presents an important share with consistent growth. It is observed that most of the countries are growing their numbers of papers.

4 Graphical Analysis of IM or IC with VOS Viewer

The previous section provides a general description of the most productive and influential variables in the journal. Hence, to give a more in-depth analysis of the data obtained from the search allows visualizing the bibliographic connections between the leading sources. For doing so, the work uses the VOS viewer software. This program can be used to build and view graphic maps through the use of bibliographic coupling [28]. Likewise, analysis of citations and co-citations, co-authorship and co-occurrence of author keywords [21, 30, 31].

First, let us analyze co-citation that occurs when two documents published in different journals receive a citation from the third document of another journal [21]. Figure 2 presents the general visualization between 1982 and 2018 with a minimum threshold of two hundred ninety cites and the fifty most representative connections. Note that the colors of the circles of the journals indicate the cluster to which the journals belong.

To see how citations evolve, Figs. 3, 4, 5 and 6 presents the co-appointment of journals between 1982–1991, 1992–2001, 2002–2011 and 2012–2018. Note that the thresholds for these figures are one, ten, fifty-two, two hundred and twenty-eight, respectively.

Next, let us analyze the bibliographic coupling of authors that publish of Innovation Management. Figure 7 presents the most productive authors with a minimum threshold of four published documents and the fifty most robust bibliographic coupling links between authors. The advantage of this figure is the graphical mapping of authors that connects or clusters together those that have similar research profiles.

Table 7 Top 50 the most productive and influential country in innovation management

R	Country	TP	TC	H	C/P	Thresholds			P/Po	C/Po
						≥200	≥100	≥50		
1	USA	373	13,969	54	37,45	11	15	293	1,14	42,69
2	Germany	255	4121	33	16,16	0	13	177	3,07	49,65
3	China	248	2780	24	11,21	1	4	189	0,18	2,00
4	England	213	5883	37	27,62	6	6	171	3,83	105,81
5	Spain	142	2611	29	18,39	0	5	109	3,04	55,91
6	Italy	122	2737	30	22,43	2	3	97	2,02	45,39
7	Taiwan	108	2096	24	19,41	1	3	85	4,60	89,19
8	Netherlands	105	3888	33	37,03	4	4	85	6,18	228,71
9	Australia	102	2151	21	21,09	2	3	81	4,06	85,70
10	Sweden	101	1427	22	14,13	0	1	94	9,90	139,90
11	Brazil	100	611	13	6,11	0	0	58	0,48	2,92
12	Finland	100	1462	19	14,62	1	1	80	18,18	265,82
13	France	72	766	16	10,64	0	1	59	1,08	11,47
14	Denmark	66	1713	20	25,95	1	5	57	11,38	295,34
15	South Korea	66	498	11	7,55	0	1	56	1,28	9,65
16	Canada	63	1501	17	23,83	1	1	51	1,70	40,57
17	Switzerland	56	1091	16	19,48	1	0	47	6,59	128,35
18	Japan	43	348	10	8,09	0	1	29	0,34	2,75
19	India	39	144	7	3,69	0	0	25	0,03	0,11
20	Turkey	38	489	11	12,87	0	1	28	0,46	5,96
21	Austria	34	748	9	22,00	0	3	21	1,35	29,80
22	Belgium	34	604	12	17,76	0	2	20	2,98	52,98
23	Portugal	34	385	11	11,32	0	0	28	3,33	37,75
24	Russia	34	194	9	5,71	0	0	19	0,24	1,34
25	Norway	33	270	9	8,18	0	0	26	6,35	51,92
26	Iran	30	129	6	4,30	0	0	20	0,37	1,58
27	Ireland	26	601	9	23,12	1	1	27	5,31	122,65
28	Malaysia	25	168	7	6,72	0	0	16	0,77	5,20
29	Colombia	24	128	6	5,33	0	0	11	0,48	2,58
30	Poland	24	59	4	2,46	0	0	13	0,63	1,56
31	Slovenia	19	257	7	13,53	0	1	16	9,50	128,50
32	Ukraine	18	4	1	0,22	0	0	3	0,43	0,10
33	South Africa	17	190	6	11,18	0	0	10	0,29	3,29
34	Thailand	17	95	5	5,59	0	0	10	0,24	1,37
35	Mexico	15	28	4	1,87	0	0	5	0,12	0,22

(continued)

Table 7 (continued)

R	Country	TP	TC	H	C/P	Thresholds			P/Po	C/Po
						≥200	≥100	≥50		
36	New Zealand	15	89	5	5,93	0	0	10	3,13	18,54
37	Scotland	15	674	8	44,93	1	0	13	2,83	127,17
38	Saudi Arabia	13	52	3	4,00	0	0	7	0,39	1,55
39	Chile	12	82	5	6,83	0	0	8	0,64	4,39
40	Greece	12	310	7	25,83	0	1	9	1,12	28,97
41	Pakistan	12	37	3	3,08	0	0	8	0,06	0,18
42	Wales	12	192	6	16,00	0	1	9	40,00	640,00
43	Czech Republic	10	32	4	3,20	0	0	4	0,94	3,02
44	Lithuania	10	39	4	3,90	0	0	8	3,70	14,44
45	Singapore	10	437	6	43,70	0	2	6	1,79	78,04
46	Hungary	9	19	3	2,11	0	0	6	0,93	1,96
47	Indonesia	9	42	4	4,67	0	0	7	0,03	0,16
48	Romania	9	29	4	3,22	0	0	6	0,46	1,48
49	Serbia	9	9	2	1,00	0	0	4	1,30	1,30
50	U Arab Emirates	8	53	3	6,63	0	0	6	0,83	5,52

Abbreviations available in previous tables except: P/Po and C/Po = Papers and cites per million inhabitants

That is, those that cite the same bibliographic material. It is worth mentioning that the results of this figure are under Table 3.

Next, let us look into the bibliographic coupling of institutions publishing in the journal. Figure 8 visualizes the data considering a minimum threshold of nine documents and fifty connections.

The leading institutions are consistent with those of Table 5. Observe that the bibliographic data of institutions depends on the authors working at these organizations. To deepen in the analysis, let us present a graphical visualization of the citations between institutions. To do so, Fig. 9 shows the results with the same threshold and connections as Fig. 8. Besides, it can be seen that the Polytechnic University of Milan it the institution with the highest citations and registered documents, has a strong relationship with various institutions.

Figure 10 presents the co-authorship of institutions with the same threshold and connections as Figs. 8 and 9. The results are similar to those of the previous figures. However, here the links between institutions of the same country are stronger, which indicates that authors from the same country tend to collaborate more than with foreign authors.

To summarize the results at the country level, Fig. 11 presents bibliographic coupling of countries with a threshold of five documents and one hundred connections. The results are under Table 7, where the USA is the most productive country, followed by Germany. It is observed that the countries in the first positions are

Table 8 Temporal evolution of the country publications

R	Country	D1	D2	D3	2012	2013	2014	2015	2016	2017	2018	Total
1	USA	5	24	116	20	23	36	25	45	35	44	373
2	Germany	1	7	71	11	19	16	21	34	38	37	255
3	China	0	4	44	10	13	21	30	36	44	46	248
4	England	1	17	41	12	10	12	26	30	30	34	213
5	Spain	1	0	36	14	12	12	17	19	11	20	142
6	Italy	0	8	29	7	8	6	12	16	23	13	122
7	Taiwan	0	0	27	10	5	11	14	16	11	14	108
8	Netherlands	0	5	43	4	11	14	5	8	8	7	105
9	Australia	0	4	23	5	4	11	10	19	13	13	102
10	Sweden	0	3	14	5	9	14	15	14	14	13	101
11	Brazil	0	1	10	5	3	7	14	22	23	15	100
12	Finland	0	3	23	11	1	11	12	17	10	12	100
13	France	0	6	15	4	5	10	1	10	11	10	72
14	Denmark	0	7	10	5	5	4	12	8	7	8	66
15	South Korea	0	2	11	4	3	4	6	14	9	13	66
16	Canada	1	5	14	3	1	5	8	8	11	7	63
17	Switzerland	1	2	17	6	4	5	2	4	7	8	56
18	Japan	0	1	12	2	3	6	2	7	5	5	43
19	India	0	3	1	0	0	2	1	5	13	14	39
20	Turkey	0	0	11	5	3	2	2	9	3	3	38
21	Austria	1	0	4	3	3	3	2	3	3	12	34
22	Belgium	1	2	4	3	4	1	4	4	4	7	34
23	Portugal	0	0	3	0	0	3	8	3	9	8	34
24	Russia	0	1	4	0	0	2	6	4	9	8	34
25	Norway	0	1	5	0	1	1	6	4	7	8	33
26	Iran	0	0	3	1	1	1	3	7	9	5	30
27	Ireland	0	4	7	1	0	0	2	2	5	5	26
28	Malaysia	0	0	3	2	1	0	5	4	3	7	25
29	Colombia	0	0	2	1	0	0	5	8	3	5	24
30	Poland	0	0	1	1	1	1	4	6	5	5	24
31	Slovenia	0	0	7	3	0	5	2	1	1	0	19
32	Ukraine	0	0	1	0	0	0	2	2	9	4	18
33	South Africa	0	1	2	2	0	0	1	2	3	6	17
34	Thailand	0	1	4	1	0	2	0	3	3	3	17
35	Mexico	0	0	1	0	0	2	2	3	5	2	15
36	New Zealand	0	0	2	1	0	3	1	1	4	3	15

(continued)

Table 8 (continued)

R	Country	D1	D2	D3	2012	2013	2014	2015	2016	2017	2018	Total
37	Scotland	0	0	4	1	1	3	0	4	2	0	15
38	Saudi Arabia	0	0	1	0	0	1	2	1	6	2	13
39	Chile	0	0	2	1	0	1	1	5	2	0	12
40	Greece	0	1	5	0	1	0	2	0	2	1	12
41	Pakistan	0	0	1	0	0	0	1	1	3	6	12
42	Wales	0	0	3	1	0	2	2	1	2	1	12
43	Czech Republic	0	0	0	0	0	1	0	1	5	3	10
44	Lithuania	0	0	2	0	1	0	2	2	1	2	10
45	Singapore	0	0	5	2	0	2	0	1	0	0	10
46	Hungary	0	1	0	1	1	1	3	1	1	0	9
47	Indonesia	0	0	0	0	0	0	2	2	2	3	9
48	Romania	0	0	1	0	1	0	2	1	4	0	9
49	Serbia	0	0	0	2	1	0	2	0	2	2	9
50	U Arab Emirates	0	0	1	0	1	0	0	1	1	4	8

Abbreviations D1, D2 and D3 = Number of publications in 1982–1991, 1992–2001, 2002–2011. The rest of the years indicate the number of publications in that year

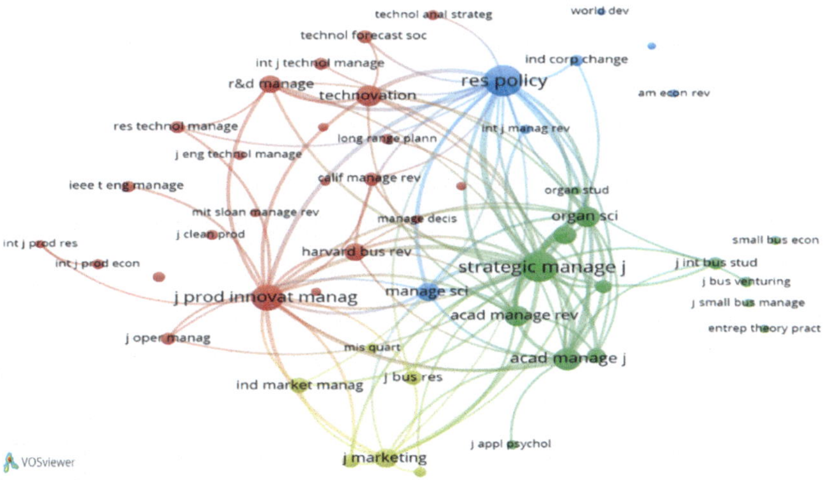

Fig. 2 Co-citation of journals cited in Innovation Management

related to most other countries. On the other hand, countries with lower weights are connected only to countries that are in the first places of production and data. Figure 11 presents the data is quite diverse with countries from different regions linked together.

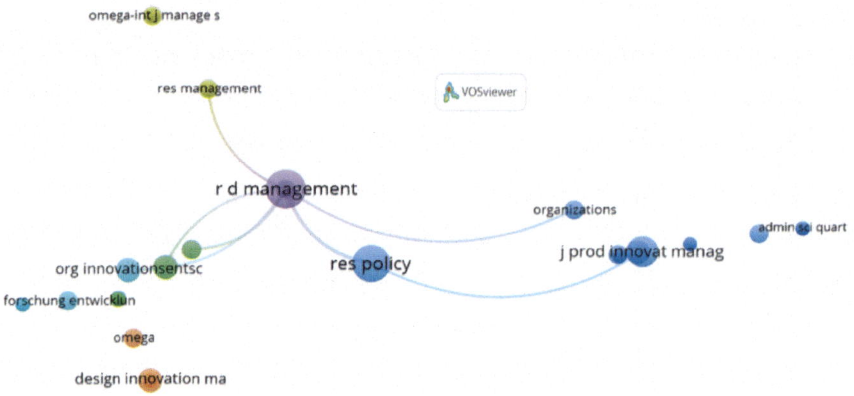

Fig. 3 Co-citation of journals cited in Innovation Management: 1982–1991

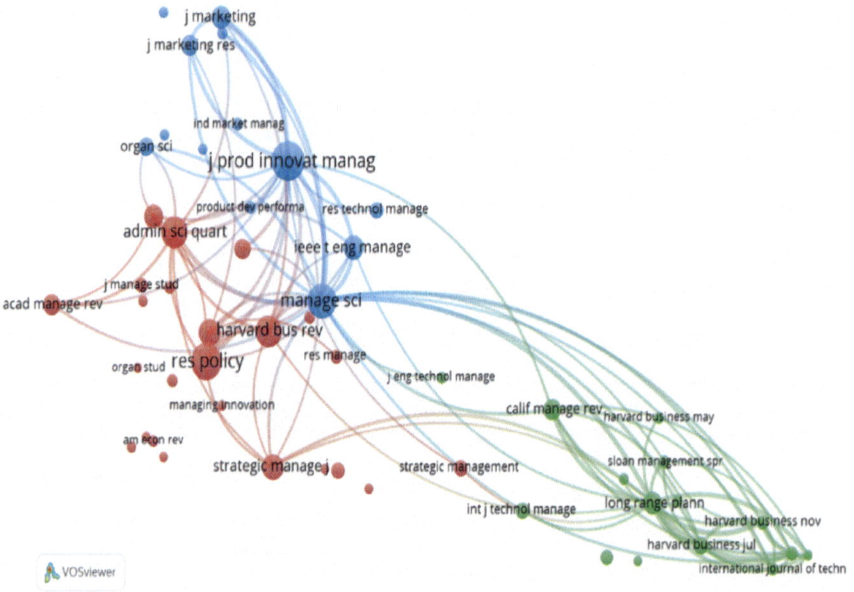

Fig. 4 Co-citation of journals cited in Innovation Management: 1992–2001

The present study considers the coincidence of the author's keywords by analyzing the main keywords of the journal. Figure 12 visualizes the map considering a threshold of seventeen occurrences and the fifty most representative connections. It is noted that the larger node refers to the keyword Innovation Management, as a central theme that has been carried out for the development of the research field. The above, followed by the nodes of innovation, innovation capacity, open innovation, and knowledge management as the most common keywords in the magazine. To

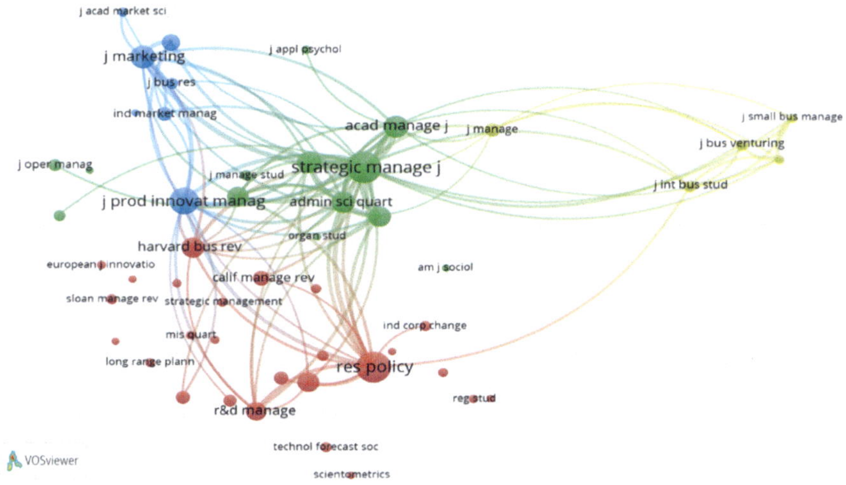

Fig. 5 Co-citation of journals cited in Innovation Management: 2002–2011

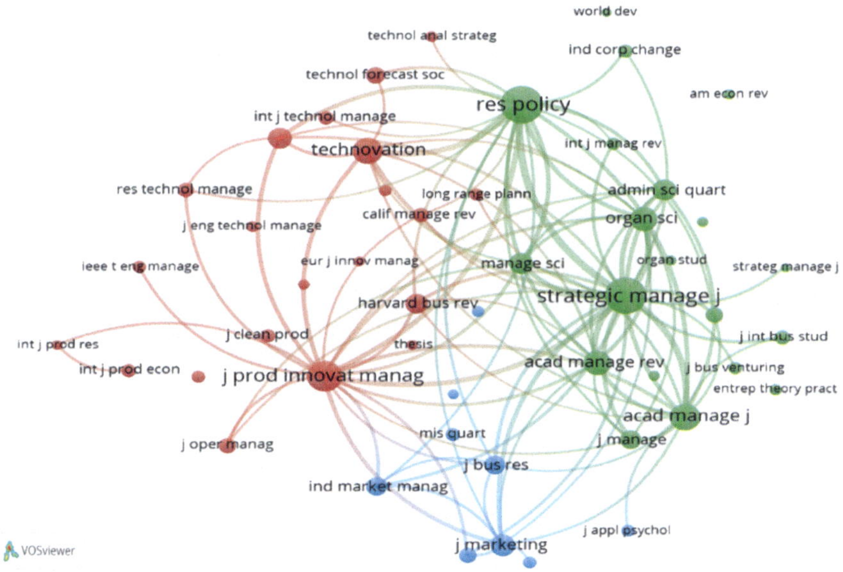

Fig. 6 Co-citation of journals cited in Innovation Management: 2012–2018

see how the use of these keywords is evolving through time, Table 9 presents the Top 10 occurrences of the author keywords divided into four periods: 1982–1991, 1992–2001, 2002–2011 and 2012–2018.

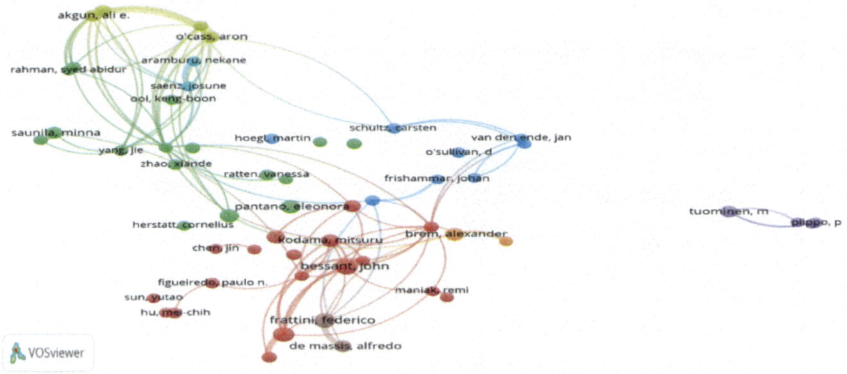

Fig. 7 Bibliographic coupling of authors that publish in innovation management

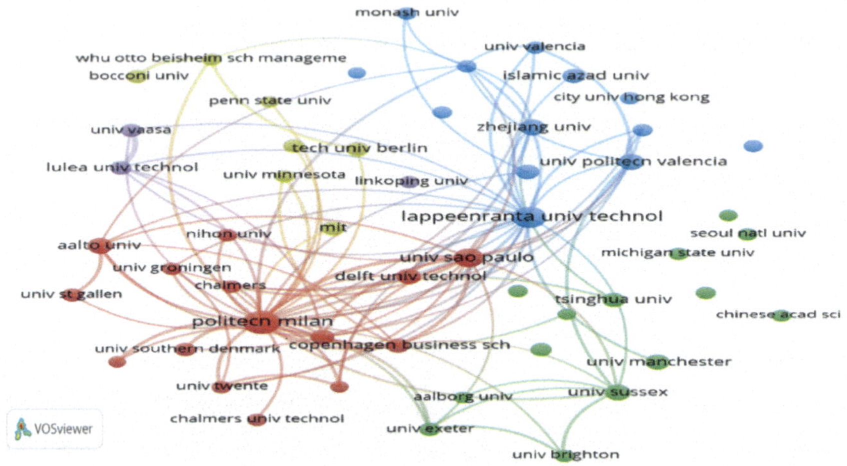

Fig. 8 Bibliographic coupling of institutions that publish in Innovation Management

5 Conclusions

The study presents an overview of innovation management from 1982 to 2018 based on the Web of Science database, where the sample size is 2288. It is noteworthy that innovation management has been mostly studied in 2017 and 2018, and the last year has got the most significant number of citations over time with a total of 7062. Among the papers about innovation management, the most cited paper is Business Models, Business Strategy and Innovation, the most cited article and corresponds to the author Teece, DJ, which is published in 2010 and has 1502. This article is highlighted because it addresses a comprehensive understanding of business models and their relations with business strategy, innovation management, and economic

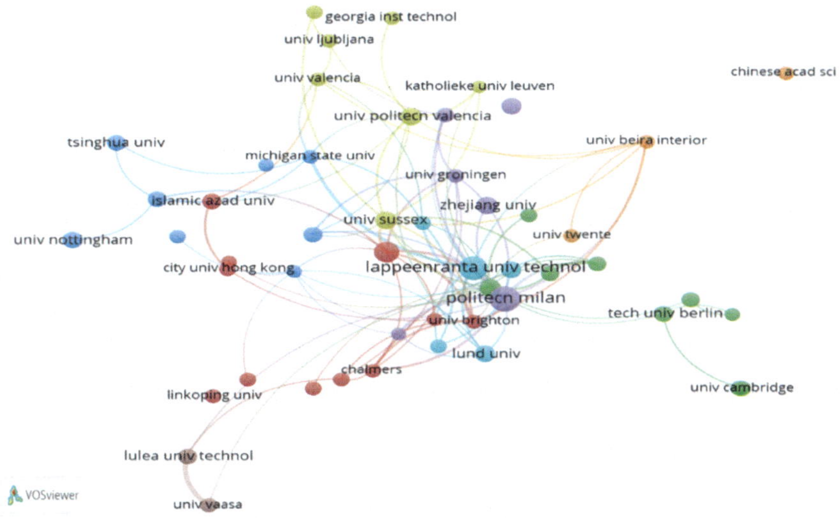

Fig. 9 Citation analysis of institutions publishing

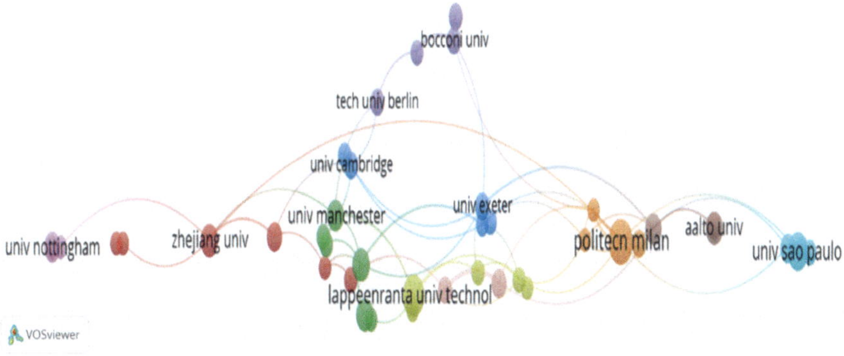

Fig. 10 Co-authorship of institutions that publish in innovation management

theory. The second position stands out the article published by [36] titled Learning orientation, firm innovation capability, and firm performance managing, which is focused on the relationships among learning orientation, firm innovativeness, and firm performance.

About authors that have most publications in Innovation Management is highlighted that Bessant, J is the author that has most publications in Innovation Management, followed by Verganti R. It is noted that both of them are among the most productive authors. Additionally, in the temporal evolution of the most productive authors, Verganti shows to be the most productive in the third decade. Also, during the third and fourth decades, it is observed significant influence with the author Bessant, J This author occupies the second important position in recent six years

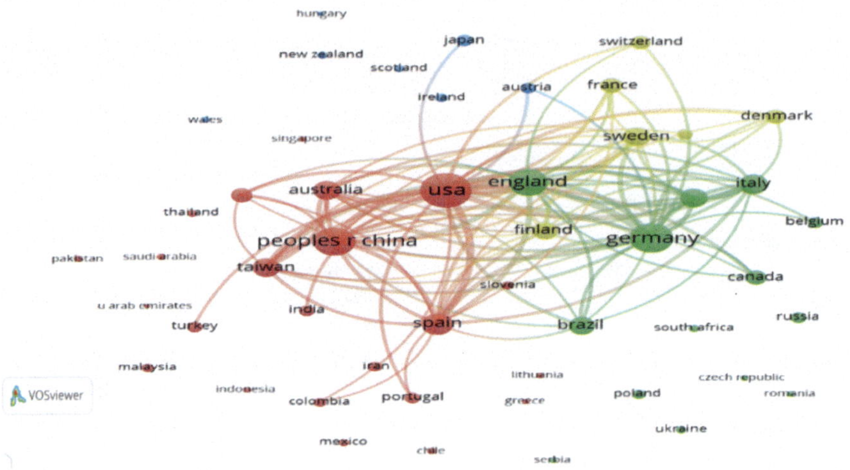

Fig. 11 Bibliographic coupling of countries that publish in innovation management

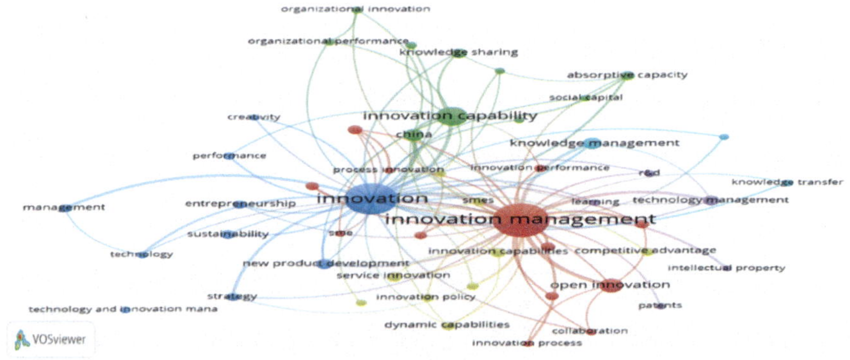

Fig. 12 Co-occurrence of author keywords of documents published in IM

after author Parida V who has been in the first position in the fourth decade. For institutions, the most productive universities are highlighted The Polytechnic University of Milan, following to Lappeenranta University of Technology and Aalto University, both located in Finland. Likewise, it is observed in this top 50 that the country with the most significant institutions' number is England with six institutions. For countries, the USA is already the most productive and influential country in Innovation Management, although still very close to Germany. Besides, China and England obtain the most remarkable results being in the third and fourth positions. However, it is essential to take into account that the most productive author is from England and the institutions that publish the most are from Italy.

Table 9 Most common author keyword occurrences in innovation management

R	*1982–1991*			*2002–2011*		
	Keyword	*Oc*	*Co*	*Keyword*	*Oc*	*Co*
1	Model	2	5	Innovation	104	31
2	Innovation	1	3	Innovation management	92	39
3	Personnel	1	3	Knowledge management	21	17
4	Product	1	3	China	20	18
5	Granger causality	1	2	Innovation capability	19	7
6	Power	1	2	New product development	12	8
7				Product development	11	11
8				R&D	9	11
9				Patents	9	6
10				Technology management	8	10
R	*1992–2001*			*2012–2018*		
	Keyword	*Oc*	*Co*	*Keyword*	*Oc*	*Co*
1	Innovation management	22	24	Innovation management	358	128
2	Innovation	13	12	Innovation	283	95
3	Technology management	7	8	Innovation capability	122	34
4	Product development	5	7	Open innovation	75	47
5	Project management	4	6	China	44	26
6	New product development	3	8	Knowledge sharing	33	14
7	Competitive advantage	3	7	Innovation capabilities	33	13
8	Learning	3	7	Knowledge management	32	26
9	Technological innovation	3	1	Service innovation	32	11
10	National system of innovation	2	6	Case study	31	33
R	*Global*					
	Keyword	*Oc*	*Co*			

(continued)

Table 9 (continued)

1	Innovation management	472	384				
2	Innovation	400	348				
3	Innovation capability	142	104				
4	Open innovation	82	93				
5	China	65	71				
6	Knowledge management	55	71				
7	Technology management	46	63				
8	New product development	45	50				
9	Knowledge sharing	38	37				
10	Innovation capabilities	38	32				

Abbreviations Oc = Author keyword occurrences; Co = Author keyword co-occurrences links

Finally, graphical analysis shows citations and co-citations, co-authorship and co-occurrence of author keywords, which allows different observer points of relationships and understand the origin and demystification about Innovation Management. From these graphical representations is noteworthy the common keywords, which can aid in observing the themes in which this field of research is oriented. In this case, the familiar words are innovation management, innovation, innovation capabilities and open innovation.

Although the study presents an exciting overview of innovation management, it has several limitations due to the methodology that was employed to obtain the main results. Most of these limitations are attributed to the notion that the Web of Science database was the only database applied to the study and as most of the journals are published in English, which explains why some notable publications in other journals. Thus, it is convenient in the next approximations to use the Scopus database to extend the results and include conference proceedings and books. Future research will focus on the more specific issues of innovation management, such as widely studied service innovation, innovation capabilities and so on.

References

1. Bilton C, Cummings S (2010) Creative strategy: reconnecting business and innovation. John Wiley & Sons, Chichester
2. Damanpour F (1987) The adoption of technological, administrative, and ancillary innovations: impact of organizational factors. J Manage 13(4):675–688
3. Kim D-Y, Kumar V, Kumar U (2012) Relationship between quality management practices and innovation. J Oper Manage 30(4):295–315

4. Weerawardena J (2003) The role of marketing capability in innovation based competitive strategy. J Strateg Mark 11(1):15–36
5. Tomala.F, Séméchal. O. (2004) Innovation management: a synthesis of academic and industrial points of view. Int J Project Manage 22(2):281–287
6. Lawson B, Samson D (2001) Developing innovation capability in organisations: a dynamic capabilities approach. Int J Innovation Manage 5(03):377–400
7. Santos J, Zawislak P, Franzoni G, y Vieira, H. (2015) Buscando un camino: un estudio bibliométrico sobre innovación y capacidades tecnológicas. Revista Internacional de Innovación - IJI 3(2):54–66. https://doi.org/10.5585/iji.v3i2.58
8. Yam RC, Guan JC, Pun KF, Tang EP (2004) An audit of technological innovation capabilities in Chinese firms: some empirical findings in Beijing, China. Res Policy 33(8):1123–1140
9. Damanpour F (1991) Organizational innovation: A meta-analysis of effects of determinants and moderators. Acad Manage J 34:555–590
10. Damanpour F, Evan WM (1984) Organizational innovation and performance: the problem of organizational lag. Adm Sci Q 29:392–409
11. Damanpour F, Szabat KA, Evan WM (1989) The relationship between types of innovation and organizational performance. J Manage Stud 26(6):587–601
12. Battisti G, Stoneman P (2010) How innovative are UK firms? Evidence from the fourth UK community innovation survey on synergies between technological and organizational innovations. Br J Manage 21:187–206
13. Damanpour F, Aravind D (2011) Managerial innovation: conceptions, processes, and antecedents. Manage Org Rev
14. Damanpour F (2014) Footnotes to research on management innovation. Organ Stud 35:1265–1285
15. Hervas-Oliver JL, Sempere-Ripoll F, Boronat-Moll C, Rojas R (2015) Technological innovation without R&D: unfolding the extra gains of management innovations on technological performance. Technol Anal Strat Manage 27(1):19–38
16. Hervas-Oliver J, Ripoll-Sempere F, y Moll, C. (2016) ¿Beneficia la innovación de la gestión en las PYME? Evidencia empírica para las PYMES españolas. Small Bus Econ 47(2):507–533
17. Hervas-Oliver F-R, Boronat-Moll C, Rojas-Alvarado R (2018) On the joint effect of technological and management innovations on performance: increasing or diminishing returns? Technol Anal Strat Manage 30(5):569–581
18. Volberda HW, Frans VD, Bosch AJ, Mihalache OR (2014) Advancing management innovation: synthesizing processes. Levels Anal Change Agents Org Stud 2014(35):1245
19. Volberda HW, Bosch D, Van F, A. J., & Heij, Cornelis V. (2013) Management innovation: management as fertile ground for innovation. Eur Manage Rev 10(1):1–15
20. Wang W, Laengle S, Merigó JM, Yu D, Herrera-Viedma E, Cobo MJ, Bouchon-Meunier B (2018) A bibliometric analysis of the first twenty-five years of the international journal of uncertainty, fuzziness and knowledge based systems. Int J Uncertain Fuzz Knowl Based Syst 26:169–193
21. Merigo JM, Yang JB (2017) A bibliometric analysis of operations research and management science. Omega-Int J Manage Sci 73:37–48
22. de Solla Price DJ (1965) Networks of science papers. Science 149(3683):510–155
23. Small H (1973) Co-citation in the relationship between two documents. J Am Soc Inf Sci 24:256–269
24. Small H (2010) Emerging research fronts in science and technology: patterns of new knowledge development. Scientometrics 83:15–38
25. Kessler MM (1963) Bibliographic coupling between scientific papers. Am Doc 14(1):10–25
26. White HD, Griffith B (1981) Author co-citation: a literature Measure of Intellectual structure. J Am Soc Inf Sci 32(3):163–172
27. Glänzel W (2001) National characteristics in international scientific co-authorship relations. Scientometrics 51:69–115
28. Van Eck NJ, Waltman L (2010) Software survey: VOSviewer, a computer program for bibliometric mapping. Scientometrics 84(2):523–538. https://doi.org/10.1007/s11192-009-0146-3

29. Blanco-Mesa F, Merigó JM, Gil-Lafuente AM (2017) Fuzzy decision making: a bibliometric-based review. J Intell Fuzzy Syst 32(3):2033–2050. https://doi.org/10.3233/JIFS-161640
30. Ortiz-de-Urbina-Criado M, Najera-Sanchez JJ, Mora-Valentin EM (2018) A research agenda on open innovation and entrepreneurship: a co-word analysis. Admin Sci 8(3):34
31. Caputo A, Marzi G, Pellegrini MM, Rialti R (2018) Conflict management in family businesses: a bibliometric analysis and systematic literature review. Int J Confl Manage 29(4):519–542
32. Blanco-Mesa F, León-Castro E, Merigó JM (2019) A bibliometric analysis of aggregation operators. Appl Soft Comput 105488. https://doi.org/10.1016/j.asoc.2019.105488
33. Rey-Martí A, Ribeiro-Soriano D, Palacios-Marqués D (2016) A bibliometric analysis of social entrepreneurship. J Bus Res 69(5):1651–1655. https://doi.org/10.1016/j.jbusres.2015.10.033
34. Soosaraei M, Khasseh AA, Fakhar M, Hezarjaribi HZ (2018) A decade bibliometric analysis of global research on leishmaniasis in Web of Science database. Ann Med Surg 26:30–37. https://doi.org/10.1016/j.amsu.2017.12.014
35. Van Nunen K, Li J, Reniers G, Ponnet K (2018) Bibliometric analysis of safety culture research. Saf Sci 108:248–258. https://doi.org/10.1016/j.ssci.2017.08.011
36. Calantone RJ, Cavusgil ST, Zhao Y (2002) Learning orientation, firm innovation capability, and firm performance. Ind Mark Manage 31(6):515–524
37. Teece DJ (2010) Business models, business strategy and innovation. Long Range Plan 43(2–3):172–194
38. Chen CC, Greene PG, Crick A (1998) Does entrepreneurial self-efficacy distinguish entrepreneurs from managers? J Bus Ventur 13(4):295–316
39. Huizingh EK (2011) Open innovation: state of the art and future perspectives. Technovation 31(1):2–9
40. Anderson N, Potočnik K, Zhou J (2014) Innovation and creativity in organizations: a state-of-the-science review, prospective commentary, and guiding framework. J Manage 40(5):1297–1333
41. Sivadas E, Dwyer FR (2000) An examination of organizational factors influencing new product success in internal and alliance-based processes. J Mark 64(1):31–49
42. Hobday M (1998) Product complexity, innovation and industrial organisation. Res Policy 26(6):689–710
43. Adams R, Bessant J, Phelps R (2006) Innovation management measurement: a review. Int J Manag Rev 8(1):21–47
44. Lin HF (2007) Knowledge sharing and firm innovation capability: an empirical study. Int J Manpow 28(3/4):315–332
45. Romijn H, Albaladejo M (2002) Determinants of innovation capability in small electronics and software firms in southeast England. Res Policy 31(7):1053–1067
46. Lilien GL, Morrison PD, Searls K, Sonnack M, Hippel EV (2002) Performance assessment of the lead user idea-generation process for new product development. Manage Sci 48(8):1042–1059
47. West J, Bogers M (2014) Leveraging external sources of innovation: a review of research on open innovation. J Prod Innov Manage 31(4):814–831
48. Hargrave TJ, Van de Ven AH (2006) A collective action model of institutional innovation. Acad Manage Rev 31(4):864–888
49. Tidd J (2001) Innovation management in context: environment, organization and performance. Int J Manage Rev 3(3):169–183
50. Chiesa V, Coughlan P, Voss CA (1996) Development of a technical innovation audit. J Prod Innov Manage Int Publ Prod Dev Manage Assoc 13(2):105–136
51. Sawhney M, Prandelli E (2000) Communities of creation: managing distributed innovation in turbulent markets. Calif Manage Rev 42(4):24–54
52. Faraj S, Jarvenpaa SL, Majchrzak A (2011) Knowledge collaboration in online communities. Organ Sci 22(5):1224–1239
53. Keupp MM, Gassmann O (2009) The past and the future of international entrepreneurship: a review and suggestions for developing the field. J Manage 35(3):600–633

54. Verganti R (2008) Design, meanings, and radical innovation: a metamodel and a research agenda. J Prod Innov Manage 25(5):436–456
55. Klerkx L, Aarts N, Leeuwis C (2010) Adaptive management in agricultural innovation systems: the interactions between innovation networks and their environment. Agric Syst 103(6):390–400
56. Guan J, Ma N (2003) Innovative capability and export performance of Chinese firms. Technovation 23(9):737–747
57. Den Hertog P, Van der Aa W, De Jong MW (2010) Capabilities for managing service innovation: towards a conceptual framework. J Serv Manage 21(4):490–514
58. Terziovski M (2010) Innovation practice and its performance implications in small and medium enterprises (SMEs) in the manufacturing sector: a resource-based view. Strateg Manage J 31(8):892–902
59. Yoo Y, Boland RJ Jr, Lyytinen K, Majchrzak A (2012) Organizing for innovation in the digitized world. Organ Sci 23(5):1398–1408
60. Tatikonda MV, Rosenthal SR (2000) Successful execution of product development projects: balancing firmness and flexibility in the innovation process. J Oper Manage 18(4):401–425
61. Weber KM, Rohracher H (2012) Legitimizing research, technology and innovation policies for transformative change: combining insights from innovation systems and multi-level perspective in a comprehensive 'failures' framework. Res Policy 41(6):1037–1047
62. Peng DX, Schroeder RG, Shah R (2008) Linking routines to operations capabilities: a new perspective. J Oper Manag 26(6):730–748
63. Siguaw JA, Simpson PM, Enz CA (2006) Conceptualizing innovation orientation: a framework for study and integration of innovation research. J Prod Innov Manage 23(6):556–574
64. Toivonen M, Tuominen T (2009) Emergence of innovations in services. Serv Ind J 29(7):887–902
65. Francis D, Bessant J (2005) Targeting innovation and implications for capability development. Technovation 25(3):171–183
66. Liao SH, Fei WC, Chen CC (2007) Knowledge sharing, absorptive capacity, and innovation capability: an empirical study of Taiwan's knowledge-intensive industries. J Inf Sci 33(3):340–359
67. Fuchs C, Schreier M (2011) Customer empowerment in new product development. J Prod Innov Manage 28(1):17–32
68. Prajogo DI, Ahmed PK (2006) Relationships between innovation stimulus, innovation capacity, and innovation performance. R&D Manage 36(5):499–515
69. Wolff JA, Pett TL (2006) Small-firm performance: modeling the role of product and process improvements. J Small Bus Manage 44(2):268–284
70. Flatten TC, Engelen A, Zahra SA, Brettel M (2011) A measure of absorptive capacity: scale development and validation. Eur Manage J 29(2):98–116
71. Zaglia ME (2013) Brand communities embedded in social networks. J Bus Res 66(2):216–223
72. Schiederig T, Tietze F, Herstatt C (2012) Green innovation in technology and innovation management–an exploratory literature review. R&D Manage 42(2):180–192
73. Bledow R, Frese M, Anderson N, Erez M, Farr J (2009) A dialectic perspective on innovation: conflicting demands, multiple pathways, and ambidexterity. Ind Organ Psychol 2(3):305–337
74. Altenburg T, Schmitz H, Stamm A (2008) Breakthrough? China's and India's transition from production to innovation. World Dev 36(2):325–344
75. Oke A (2007) Innovation types and innovation management practices in service companies. Int J Oper Prod Manage 27(6):564–587
76. Chiaroni D, Chiesa V, Frattini F (2011) The open innovation journey: how firms dynamically implement the emerging innovation management paradigm. Technovation 31(1):34–43
77. Brem A, Voigt KI (2009) Integration of market pull and technology push in the corporate front end and innovation management—insights from the German software industry. Technovation 29(5):351–367
78. Baker WE, Sinkula JM (2005) Market orientation and the new product paradox. J Prod Innov Manage 22(6):483–502

79. Ebner W, Leimeister JM, Krcmar H (2009) Community engineering for innovations: the ideas competition as a method to nurture a virtual community for innovations. R&D Manage 39(4):342–356
80. Cormican K, O'Sullivan D (2004) Auditing best practice for effective product innovation management. Technovation 24(10):819–829

Environmental Sustainability: A 10-Year Bibliometric Analysis of the Developments in Management, Business, Finance and Economics

Rodrigo Gómez Monge, Víctor G. Alfaro-García, Irma C. Espitia-Moreno, Dalia García-Orozco, and Manuel Ricardo Romo de Vivar Mercadillo

Abstract The current needs in production of goods and services required to sustain the fast-growing population reveal unprecedented challenges bound to the way humankind relates to its surroundings. Developments in sustainability, understood as the field of knowledge focused on satisfying the needs of people without jeopardizing the resources of future generations play a significant role in this relation, moreover the systems designed for the transformation, consumption, strategic coordination, and allocation of resources are key to the achievement of sustainable goals. The present study conducts a bibliometric analysis designed to find core academic developments indexed in the Web of Science scientific database from 2010 to 2019. Results reveal leading authors, journals, organizations, countries, documents, and categories in the field of sustainable management, business, finance, and economics. The aim is to present trends, connections and networks using analytical and statistical tools based on a systematic detailed methodology. These bibliometric techniques have proven to be effective when analysing large numbers of scientific material, moreover the use of the VoSviewer bibliographic network mapping allows a fast and visual representation of the retrieved core documents. This paper tries to shed light in the representation of the actual landscape of sustainability in business economics and the possible conformation of forthcoming contributions, collaborations, and the creation of synergies.

Keywords Environmental sustainability · Business · Finance · Management · Bibliometrics

R. G. Monge · V. G. Alfaro-García (✉) · I. C. Espitia-Moreno · D. García-Orozco · M. R. R. de Vivar Mercadillo
Universidad Michoacana de San Nicolás de Hidalgo, Gral. Francisco J. Múgica S/N, C.U, 58030 Morelia, México
e-mail: victor.alfaro@umich.mx

E. León-Castro et al. (eds.), *Soft Computing and Fuzzy Methodologies in Innovation Management and Sustainability*, Lecture Notes in Networks and Systems 337,
https://doi.org/10.1007/978-3-030-96150-3_18

1 Introduction

Human production and consumption systems are subject to the natural resources and ecological limits that the planet can sustain [1]. The level of development and the quality of life that citizens can expect answer to these planetary limits too [2]. In the last decades, several nations have established multiple pacts, laws, and regulations, seeking to preserve the natural resources of the planet [3], nonetheless, current evidence shows considerable environmental deterioration [4–7].

The accelerated population growth and the economic policies adopted until today imply great challenges for humanity including degradation of cultivable soils, shortages of food supply, loss of biodiversity, global warming, marine pollution, air pollution, potable water supply, energy supply, etc. [8–13]. These complex and profound problems require the participation and commitment of key stakeholders, such as, governments, society, public and private organizations [7].

Sustainability is currently considered a priority issue in most of the political and research agendas of the world [6], this has fostered the proliferation of this research field in wide ranging areas of knowledge [14]. However, despite the rapid and diverse proliferation of the topic, there is no single definition of sustainability [15]. One of the most accepted and used concepts of sustainability in literature is the satisfaction of humanities' current needs without compromising the ability of future generations to satisfy their own [16]. Ever since the adoption of this concept by the World Commission on Environment and Development in 1987, the evolution of scientific publications regarding sustainability increased annually. This led to the creation of specialized research journals that compile and disseminate scientific documents in the field of sustainability [4].

The economic and business models rely on the means of production and consumption, these affect the extraction, production, transfer, consumption, and waste disposal systems [17]. The relation of business and economic sciences, and environmental sustainability is the consideration of production and consumption systems with respect to environmental limits and the regeneration capacity of the planetary resources [2]. This requires a paradigm shift that implies the direct contribution of society and organizations [18]. Several authors agree on the impossibility of sustainable development without the integration of organizations [19]. Here, the business, financial, managerial and economic (BFME) sciences play a vital role, as a profound analyzes is needed to understand how sustainable business models can maintain or improve economic performance minimizing environmental and social impacts [19].

As the science of sustainability becomes a categorical research field [20], given its wide ranging lines of research, the accumulation of contributions can become both enriching and risky [21]. Diluting the conceptual definitions of sustainability can limit its effectiveness [22]. The emergence of independent and isolated views on sustainability jeopardize assertive research lines and scientific synergies, hence an overview of the evolution of BFME sciences in sustainability is appealing [15, 23].

Bibliometric techniques are systematic reviews of information designed to evaluate and monitor the evolution of a particular field of study [24–31]. Bibliometric studies use meta-analytical data sets of scientific publications to generate statistics and ordered information of authors, citations, keywords, co-authorship, affiliations, production, influence, etc., [32]. These are commonly applied to specific topics, during a pre-established period of time, thus allowing the visualization of links and patterns of the retrieved documents (see e.g. [33]), and visualize the evolution of publications [34], detecting components in the theoretical construction of knowledge [22, 27].

Some bibliometric studies focusing on BFME and environmental sustainability (ES-BFME) include e.g. the structure of sustainability research in marketing [35], systematic reviews on innovation in hospitality and tourism [36]; project management and sustainability [37]; the landscape of ecological economics [38], and social entrepreneurship [39]. However, to the best of our knowledge, little effort has been put in the description of the scientific structure of environmental sustainability in the field of BFME. The objective of this paper is applying bibliometric techniques and network maps to describe the evolution of ES-BFME in the past 10 years.

The structure of the study is presented as follows. Section 2 presents the data collection and treatment methodology followed in this paper. Section 3 shows the results of the methodological approach. Finally, Sect. 4 presents the concluding comments.

2 Methodology

Perhaps the foremost important section to detail in any bibliometric study is the methodology applied for the extraction and treatment of information. A clear descriptive path of the retrieval and further processing of information is key for the replicability and validity of the obtained data. The present study uses the Web of Science core collection as the main database for the obtention of information. It is conventionally accepted that this scientific database has evolved into one of the most robust academic library research tools and a world-leading source for large-scale data-driven scientometric studies [40]. Today, Eugene Garfield's input to the scientific community compiles around 1.9 billion cited references and 161 million records across 254 subject areas. The robust selection and indexation of contributions to the WoS core collection makes it a dependable source of scientific information [41]. The WoS core collection includes 10 sub-databases classified in eight citation indexes and two chemical indexes [42]. The sub-databases used in the present paper and its coverage is as follows: science citation index expanded (SCI-EXPANDED), 1900—present; social sciences citation index (SSCI), 1900—present; arts & humanities citation index (A&HCI), 1975—present; conference proceedings citation index-science (CPCI-S) 1990—present; conference proceedings citation index—social science & humanities (CPCI-SSH), 1990—present; book citation index—science (BKCI-S), 2005—present, book citation index—social sciences & humanities (BKCI-SSH), 2005—present; emerging sources citation index (ESCI), 2015—present. The present study mainly focuses on WoS core collection and the specific sub-databases before

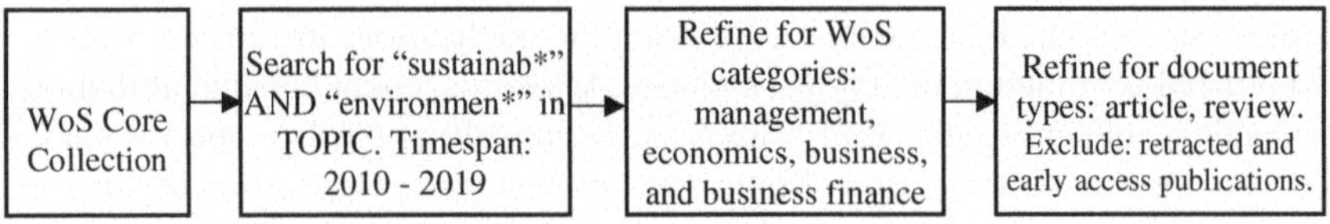

Fig. 1 Flowchart of the information retrieval process

mentioned, however, there are other scientific repositories that are also relevant in the academic community e.g. Google Scholar or SCOPUS, for a comparison of these databases please see [43].

The search process combines the keywords "sustainab*" AND "environment*" in the "topic" section from the timespan 2010–2019. The search delivers a total of 116,359 documents that include the specified keywords in the title, abstract, author keywords and keywords plus. From here, we refine for the WoS categories "management", "business", "economics" and "business finance", segmenting the results to 13,389 documents. Aiming to address the core proposals in the segmented fields, we perform a further refinement, this time for "article" and "review" and exclude "early access" and "retracted publications". In August 2020, a total of 9,735 papers addressing ES-BFME were retrieved using the proposed method. Please see Fig. 1 for a flowchart of the general procedure of information extraction followed in the present study.

Please observe that, in this document we have established the search body on the environmental sustainability concept, as an improvement to the refinement of the data, for two reasons: (1) the compatibility of the concept with economics-business and environmental sciences. Environmental sustainability studies the systems of sustainable production and consumption with respect to environmental limits and the regeneration capacity of planetary resources [2] and (2) It improves the selection, the term sustainable in business is widely used by other topics that may not be related to the field of sustainability, such as sustainability of competitive advantage or sustainable finance, making references to sustaining a condition over time [44]. In this way, when adding the prefix environmental, we select those documents that, in addition to including in the sustainable theme, address environmental issues. This does not guarantee that all the documents are of full relevance to sustainability, given the broad consultation methodology used for data collection, detailed above.

Along with presenting the retrieved information in tables using conventionally accepted metrics for the representation of influence and productivity of authors, journals, universities, etc. [33], a bibliometric network analyses has also been performed using VoS viewer technology [45]. Bibliometric network analysis have proven to be effective when dealing with large amount of data [46, 47] and represent it in visual landscapes of co-authorship, co-occurrence, citation, bibliographic coupling and co-citation see e.g. [24, 48].

3 Results

The 9,735 core documents in ES-BFME are catalogued as articles (9,412), book chapters (914), reviews (324) and proceedings papers (115). Figure 2 shows the annual production of papers along the selected years including the total citations accumulated per year. The graph shows an increasing rate in productivity and influence over the years.

A general citation structure of the core papers shows that 94.47% of the retrieved documents have less than 50 citations, 6.89% reach more than 50 citations, 1.41% obtained over 150 citations and 0.07% received more than 500 cites.

This section introduces a series of data, statistics and network maps trying to show the reader a clear picture of the ES scope over the last 10 years. Information about most cited papers, leading authors, journals, countries, universities, categories, and co-word analyses are displayed in Tables 1, 2, 3, 4, 5, 6 and 7 and Figs. 3, 4, 5, 6 and 7.

3.1 *Most Cited Papers*

Table 1 presents the 25 most influential articles in the field of environmental sustainability in ES-BFME, of the last 10 years. The results show that ten of the twenty-five documents analyzed were published in 2010 and are predominantly topics on innovation and entrepreneurship, for the years 2011 and 2012, six and five articles were published respectively, which are mostly on supply chains and the four remaining in 2014, were topics about social responsibility and the human factor.

The most cited article is *sustainability transitions: an emerging field of research and its prospects* by Markard et al. [18], which contains 946 citations and was published in Research Policy, the document makes a bibliometric review of 540 articles from 1998 to 2011, on sustainability transitions. Its most relevant results

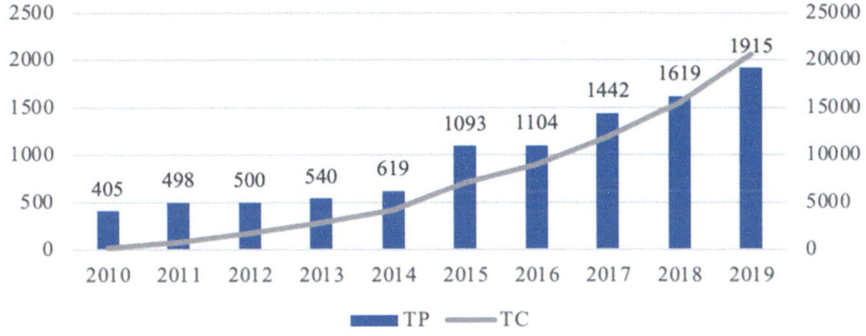

Fig. 2 Sum of papers and papers citing in ES-BFME

Table 1 Most influential research titles in ES-BFME

R	J	TC	TI	AT	PY	C/PY	DT
1	RPO	946	Sustainability transitions: an emerging field of research and its prospects	Markard, J; Raven, R; Truffer, B	2012	118	A
2	IJPDLM	586	Sustainable supply chain management: evolution and future directions	Carter, CR; Easton, PL	2011	65	A
3	RPO	569	Ontologies, socio-technical transitions (to sustainability), and the multi-level perspective	Geels, FW	2010	57	A
4	SMJ	534	Corporate social responsibility and access to finance	Cheng, BT; Ioannou, I; Serafeim, G	2014	89	R
5	AER	509	The environment and directed technical change	Acemoglu, D; Aghion, P; Bursztyn, L; Hemous, D	2012	64	A
6	EJOR	495	Quantitative models for sustainable supply chain management: developments and directions	Brandenburg, M; Govindan, K; Sarkis, J; Seuring, S	2014	83	A
7	BSE	476	Sustainable entrepreneurship and sustainability innovation: categories and interactions	Schaltegger, S; Wagner, M	2011	53	A
8	SCMIJ	417	Green supply chain management practices: impact on performance	Green, KW; Zelbst, PJ; Meacham, J; Bhadauria, VS	2012	52	A
9	FPO	400	Where are the best opportunities for reducing greenhouse gas emissions in the food system (including the food chain)?	Garnett, T	2011	44	A
10	IJMR	384	Social enterprises as hybrid organizations: a review and research agenda	Doherty, B; Haugh, H; Lyon, F	2014	64	A
11	EJOR	376	Operations research for green logistics - an overview of aspects, issues, contributions and challenges	Dekker, R; Bloemhof, J; Mallidis, I	2012	47	A

(continued)

Table 1 (continued)

R	J	TC	TI	AT	PY	C/PY	DT
12	AOS	368	Is accounting for sustainability actually accounting for sustainability … and how would we know? an exploration of narratives of organisations and the planet	Gray, R	2010	37	A
13	JBV	362	Sustainable development and entrepreneurship: past contributions and future directions	Hall, JK; Daneke, GA; Lenox, MJ	2010	36	A
14	JBV	362	Greening goliaths versus emerging davids - theorizing about the role of incumbents and new entrants in sustainable entrepreneurship	Hockerts, K; Wustenhagen, R	2010	36	A
15	CSREM	360	Sustainable supply chain management and inter-organizational resources: a literature review	Gold, S; Seuring, S; Beske, P	2010	36	A
16	JBE	356	Mainstreaming green product innovation: why and how companies integrate environmental sustainability	Dangelico, R; Pujari, D	2010	36	A
17	JMK	346	The sustainability liability: potential negative effects of ethicality on product preference	Luchs, MG; Naylor, RW; Irwin, JR; Raghunathan, R	2010	35	A
18	MISQ	339	Information systems innovation for environmental sustainability	Melville, NP	2010	34	A
19	JOM	330	Balancing priorities: decision-making in sustainable supply chain management	Wu, ZH; Pagell, M	2011	37	A
20	JAMS	329	Mindful consumption: a customer-centric approach to sustainability	Sheth, JN; Sethia, NK; Srinivas, S	2011	37	A

(continued)

Table 1 (continued)

R	J	TC	TI	AT	PY	C/PY	DT
21	SCMIJ	317	Making connections: a review of supply chain management and sustainability literature	Ashby, A; Leat, M; Hudson-Smith, M	2012	40	R
22	FPO	316	Sustainability labels on food products: consumer motivation, understanding and use	Grunert, KG; Hieke, S; Wills, J	2014	53	A
23	AMP	312	Building sustainable organizations: the human factor	Pfeffer, J	2010	31	A
24	EE	309	Payments for ecosystem services: from local to global	Farley, J; Costanza, R	2010	31	A
25	EE	308	In defence of degrowth	Kallis, G	2011	34	A

Acronyms R, Ranking; J, Journal; TC, Sum of citations in WoS; TI, Title; AT, Author; PY, Published Year; C/PY, Sum of times cited per published year; DT, Document Type; RPO, Research Policy; IJPDLM, International Journal of Physical Distribution & Logistics Management; RPO, Research Policy; SMJ, Strategic Management Journal; AER, American Economic Review; EJOR, European Journal of Operational Research; BSE, Business Strategy and the Environment; SCMIJ, Supply Chain Management an International Journal; FPO, Food Policy; IJMR, International Journal of Management Reviews; EJOR, European Journal of Operational Research; AOS, Accounting Organizations and Society; JBV, Journal of Business Venturing; JBV, Journal of Business Venturing; CSREM, Corporate Social Responsibility and Environmental Management; JBE, Journal of Business Ethics; JMK, Journal of Marketing; MISQ, Management Information Systems Quarterly; JOM, Journal of Operations Management; JAMS, Journal of the Academy of Marketing Science; SCMIJ, Supply Chain Management an International Journal; FPO, Food Policy; AMP, Academy of Management Perspectives; EE, Ecological Economics; EE, Ecological Economics

mentioned a new but growing interest in sustainability issues in the fields of governance, production, consumption, and social sciences, identifying four theoretical frameworks commonly used in these transition studies: transition management, niche management strategic, level perspective on socio-technical transitions and technological innovation systems.

The second most influential document is *sustainable supply chain management: evolution and future directions* by Carter and Easton [49], published in the International Journal of Physical Distribution and Logistics Management with 586 citations, which analyzes 80 articles on sustainability in 7 specialized logistics journals for the period 1991–2010, finding in the evolution of the literature the absence of theoretical references In 55% of the documents, the change of the paradigm from the environmental to the sustainable is visualized between the period from 1991 to 2000 where 87% of the articles haven't sustainability theories references or them only refers some environmental theories, and from 2001 to 2010 where the 33% of the investigations begins from theories of corporate responsibility and start to create

Table 2 Most productive authors in ES-BFME

R	AT	C	%BFME	TP-BFME	TC-BFME	H-BFME	TC/TP-BFME	TP	TC	TC/TP	H
1	Schaltegger S	DEU	39	28	1561	16	56	72	3884	54	31
2	Garcia-Sanchez IM	ESP	21	26	874	14	34	124	3253	26	32
3	Sarkis J	USA	8	25	1773	16	71	305	20,506	67	72
4	Sueyoshi T	USA	18	25	886	18	35	136	4598	34	42
5	Jabbour CJC	FRA	15	23	632	12	27	158	3861	24	35
6	Boiral O	CAN	22	20	631	12	32	89	3135	35	31
7	Wagner M	DEU	24	16	1111	12	69	67	2686	40	25
8	Svensson G	NOR	8	15	216	8	14	187	1836	10	24
9	Maroun W	ZAF	27	15	238	7	16	55	568	10	13
10	Frey M	DEU	17	15	147	7	10	88	1782	20	22
11	Longoni A	ESP	56	14	333	9	24	25	640	26	13
12	Searcy C	CAN	19	14	453	10	32	72	2883	40	25
13	Streimikiene D	LTU	5	14	81	6	6	306	2892	9	25
14	Martinez-Ferrero J	ESP	23	13	340	11	26	56	804	14	14
15	Seuring S	DEU	17	13	1445	9	111	78	7255	93	34
16	Valentinov V	DEU	14	13	205	8	16	91	885	10	18
17	Goto M	JPN	17	12	508	11	42	70	3169	45	37
18	Horisch J	DEU	50	12	332	7	28	24	490	20	12
19	Rezaee Z	USA	24	12	102	5	9	49	655	13	13
20	Sovacool BK	DNK	4	12	347	8	29	342	10,691	31	56
21	Figge F	FRA	25	11	789	9	72	44	2116	48	21

(continued)

Table 2 (continued)

R	AT	C	%BFME	TP-BFME	TC-BFME	H-BFME	TC/TP-BFME	TP	TC	TC/TP	H
22	Hahn T	ESP	42	11	776	9	71	26	1666	64	19
23	Managi S	JPN	4	11	256	7	23	308	3776	12	29
24	Shmelev SE	CHE	38	11	26	2	2	29	243	8	7
25	Toppinen A	FIN	10	11	234	9	21	109	1549	14	21

Acronyms C, Country; CAN, Canada; CHE, Switzerland; DEU, Germany; DNK, Denmark; ESP, Spain; FIN, Finland; FRA, France; H, H index in WoS; H-BFME, H index in BFME; JPN, Japan; LTU, Lituania; NOR, Norway; TC-BFME, Sum of citations in BFME; TP, Sum of papers in WoS; TP-BFME, Sum of papers in BFME; USA, United States of America; ZAF, South Africa; (en blanco); TC/TP-BFME, Citations per paper in BFME; TC/TP, Citation per paper in WoS; %BFME, Sum of papers in BFME over sum of papers published in WoS

Table 3 Most productive journals in ES-BFME

R	J	TP-BFME	H-BFME	TC-BFME	TC/TP-BFME	TP	H	TC	TC/TP	%BFME
1	BSE	455	59	12,841	28	653	65	16,658	26	70
2	EP	434	56	12,629	29	7,085	141	197,530	28	6
3	EE	417	50	10,630	25	2,678	102	67,294	25	16
4	CSREM	287	43	7,077	25	540	52	11,354	21	53
5	JBE	267	56	9,373	35	3,354	108	82,164	24	8
6	FPE	129	23	1,735	13	1,253	45	16,355	13	10
7	TFSC	125	27	2,412	19	2,395	77	44,626	19	5
8	QAS	.110	5	130	1	1,079	11	1,262	1	10
9	AECON	108	12	696	6	626	20	3,005	5	17
10	IJCS	96	22	1,442	15	784	41	9,448	12	12
11	JTG	88	27	2,141	24	1,392	75	3,1301	22	6
12	TM	85	33	3,808	45	1,781	111	71,751	40	5
13	EJOR	83	31	4,008	48	6,002	118	132,726	22	1
14	OE	83	25	2,130	26	208	36	4,789	23	40
15	AAAJ	75	25	1,644	22	580	44	9,784	17	13
16	TRPOL	74	18	1,148	16	1,176	57	19,641	17	6
17	SCMIJ	71	32	4,272	60	440	61	14,347	33	16
18	ENECON	68	29	2,815	41	2,625	101	64,252	24	3
19	WD	68	19	1,049	15	2,285	86	48,608	21	3
20	BIJ	66	14	587	9	602	20	3,101	5	11
21	JBR	66	22	1,643	25	3,575	98	77,475	22	2

(continued)

Table 3 (continued)

R	J	TP-BFME	H-BFME	TC-BFME	TC/TP-BFME	TP	H	TC	TC/TP	%BFME
22	FUT	61	16	826	14	921	43	11,439	12	7
23	SAMPJ	59	12	447	8	146	18	1,085	7	40
24	TRPP	59	21	1,161	20	1,702	71	33,688	20	3
25	BJECON	57	2	21	0	729	5	253	0	8

Acronyms EP, Energy Policy; FPE, Forest Policy and Economics; TFSC, Technological Forecasting and Social Change; QAS, Quality Access To Success; AECON, Amfiteatru Economic; IJCS, International Journal of Consumer Studies; JTG, Journal of Transport Geography; TM, Tourism Management; OE, Organization Environment; AAAJ, Accounting Auditing Accountability Journal; TRPOL, Transport Policy; ENECON, Energy Economics; WD, World Development; BIJ, Benchmarking An International Journal; JBR, Journal of Business Research; FUT, Futures; SAMPJ, Sustainability Accounting Management and Policy Journal; TRPP, Transportation Research Part A Policy and Practice; BJECON, Baltic Journal of Economic Studies; EJOR, European Journal of Operational Research

Table 4 Most productive countries in ES-BFME

R	C	TP-BFME	H-BFME	TC-BFME	CA-BFME	>500	>250	>100	>50	>25	TP/I	TC/I
1	USA	1,862	100	47,358	33,145	4	20	78	140	211	5.68	144.38
2	UK	1,199	83	30,971	24,112	3	9	55	88	183	21.42	553.25
3	AUS	750	62	14,953	12,511	0	1	25	58	81	30.01	598.36
4	CHN	683	57	12,740	10,661	0	0	18	54	83	0.49	9.15
5	ESP	617	52	11,278	8,761	0	3	13	45	67	13.14	240.26
6	DEU	590	64	15,262	11,526	0	4	29	50	75	7.11	183.84
7	ITA	536	48	9,150	7,759	0	1	11	34	54	8.88	151.59
8	CAN	514	63	14,431	11,772	1	4	26	50	61	13.67	383.91
9	FRA	461	45	8,693	7,581	0	3	11	27	52	6.88	129.77
10	BRA	403	23	2,206	1,996	0	0	4	5	13	1.92	10.53
11	IND	389	26	3,201	2,827	0	1	1	10	15	0.29	2.37
12	NLD	377	54	11,587	10,177	2	3	15	41	53	21.82	670.54
13	SWE	248	37	5,592	5,221	1	2	7	15	28	24.24	546.63
14	ROU	221	14	964	865	0	0	0	3	2	11.39	49.67
15	MYS	192	21	2,350	2,181	0	1	6	5	4	6.09	74.53
16	ZAF	184	22	1,703	1,519	0	0	1	3	14	3.18	29.47
17	DNK	173	34	4,756	4,441	0	3	4	14	25	29.80	819.15
18	NZL	171	29	3,170	2,637	0	1	3	15	14	35.00	648.79
19	CHE	161	33	4,256	4,011	1	1	5	8	26	18.79	496.62
20	UKR	160	6	157	149	0	0	0	0	0	3.81	3.74
21	FIN	158	29	2,931	2,717	0	1	2	9	22	28.63	531.17
22	TWN	155	26	2,504	2,343	0	0	3	8	19	6.52	105.30
23	KOR	141	24	2,198	2,031	0	0	3	8	12	2.73	42.56
24	TUR	139	19	1,532	1,505	0	0	5	2	8	1.70	18.68
25	PRT	136	23	1,525	1,477	0	0	4	17	115	13.23	148.35

Acronyms UK, United Kingdom; AUS, Australia; CHN, PR China; ITA, Italy; FRA, France; BRA, Brazil; IND, India; NLD, Netherlands; SWE, Sweden; ROU, Romania; MYS, Malaysia; NZL, New Zealand; UKR, Ukraine; TWN, Taiwan; KOR, South Korea; TUR, Turkey; PRT, Portugal; CA-BFME, Sum of citing articles in BFME; >500, TP with more than 500 citations; >250, TP with more than 250 citations; >100, TP with more than 100 citations; >50, TP with more than 50 citations; >25, TP with more than 25 citations; TP/I, Sum of papers in BFME per capita; TC/I, Sum of citations in BFME per capita

sustainability frameworks using to the 3 dimensions of this. The theories of the parts interested are identified [50], the resource-based view and the resource view based on natural resources [51], as the most recurrent.

In the third position is the article *ontologies, socio-technical transitions (towards sustainability) and the multilevel perspective* by Geels [52], published in Research Policy, which accumulates 569 citations, this research analyzes seven ontologies

Table 5 Most productive organizations in ES-BFME

R	Organization	C	TP-BFME	H-BFME	TC-BFME	CA-BFME	TC/TP - BFME
1	Bucharest U of Economic Studies	ROU	90	9	258	245	2.87
2	U Leeds	UK	75	27	2257	2441	30.09
3	Hong Kong Polytech U	HKG	67	20	1312	1257	19.58
4	U Manchester	UK	57	24	1939	1851	34.02
5	U Queensland	AUS	57	24	2138	1959	37.51
6	Vrije U Amsterdam	NLD	57	24	1857	1803	32.58
7	Griffith U	AUS	54	21	1390	1329	25.74
8	U Sao Paulo	BRA	51	10	592	513	11.61
9	Monash U	AUS	48	17	858	838	17.88
10	U Cambridge	UK	48	19	1747	1718	36.40
11	U Sussex	UK	47	19	2465	2175	52.45
12	Wageningen U	NLD	46	17	1211	1195	26.33
13	Lund U	SWE	45	16	1016	984	22.58
14	U Oxford	UK	45	14	578	567	12.84
15	Deakin U	AUS	44	14	781	758	17.75
16	U Sydney	AUS	44	16	1360	1283	30.91
17	Vilnius Gediminas Tech U	LTU	44	15	767	688	17.43
18	U Granada	ESP	41	17	975	891	23.78
19	Copenhagen Business Sch	DNK	40	16	1094	1053	27.35
20	Macquarie U	AUS	39	15	835	797	21.41
21	Queensland U Technol	AUS	39	16	818	794	20.97
22	U Kassel	DEU	39	16	1942	1633	49.79
23	U South Australia	AUS	39	9	331	317	8.49
24	U Ghent	BEL	38	17	795	787	20.92
25	Aarhus U	DNK	37	16	1043	1003	28.19

Acronyms U, University; HKG, Hong Kong; BEL, Belgium

Table 6 Frequently introduced keywords in ES-BFME

R	Keywords	T	Keywords plus	T	Titles	T	Abstracts	T
1	sustainability	1,920	performance	1,387	sustainable	1,863	environmental	12,853
2	sustainable development	781	management	1,203	sustainability	1,801	sustainability	10,152
3	corporate social responsibility	446	sustainability	1,168	environmental	1,524	sustainable	8,848
4	environment	325	impact	825	development	970	study	7,727
5	environmental	292	corporate social-responsibility	712	management	913	social	7,598
6	sustainable	259	model	599	social	907	development	7,034
7	innovation	231	framework	561	corporate	825	paper	6,554
8	management	216	green	504	green	802	economic	5,611
9	development	198	innovation	458	performance	670	management	5,292
10	climate change	188	determinants	399	case	632	performance	4,896
11	environmental sustainability	182	governance	395	analysis	606	environment	4,290
12	performance	181	behavior	348	study	567	analysis	4,247
13	china	158	policy	342	supply	551	green	4,106
14	supply chain management	156	consumption	337	energy	518	energy	4,026
15	social	148	business	325	responsibility	446	business	3,902
16	environmental management	144	strategy	313	business	418	findings	3,771
17	corporate sustainability	138	systems	311	economic	411	model	3,649
18	responsibility	122	perspective	306	chain	403	companies	3,616
19	environmental policy	118	strategies	284	role	390	firms	3,356
20	energy	116	industry	283	industry	377	corporate	3,343
21	green	114	financial performance	278	innovation	377	practices	3,164

(continued)

Table 6 (continued)

R	Keywords	T	Keywords plus	T	Titles	T	Abstracts	T
22	CSR	113	CSR	274	approach	353	based	3,131
23	social responsibility	113	firm	264	evidence	347	purpose	3,105
24	environmental performance	110	climate-change	263	practices	335	data	2,973
25	stakeholders	107	organizations	261	impact	329	supply	2,818

Acronyms T, Total

Table 7 Top WoS categories in ES-BFME

R	Keywords	TP	H	TC	CA	TC/TP
1	Management	4,288	114	82,482	42,732	19.24
2	Economics	3,555	88	53,717	42,411	15.11
3	Business	3,383	108	69,541	35,288	20.56
4	Environmental studies	2,330	94	54,357	36,565	23.33
5	Environmental sciences	892	69	23,474	19,743	26.32
6	Business finance	606	43	8,015	4,255	13.23
7	Energy fuels	449	56	12,775	11,034	28.45
8	Ecology	424	50	10,655	9,133	25.13
9	Transportation	334	42	6,521	5,494	19.52
10	Ethics	320	56	10,188	7,613	31.84
11	Operations research management science	308	53	9,605	7,207	31.19
12	Regional urban planning	273	32	4,075	3,792	14.93
13	Green sustainable science technology	269	13	663	562	2.46
14	Hospitality leisure sport tourism	258	44	6,426	4,827	24.91
15	Agricultural economics policy	235	28	3,738	3,302	15.91
16	Forestry	137	23	1,802	1,570	13.15
17	Geography	133	31	3,076	2,817	23.13
18	Engineering industrial	128	24	1,804	1,584	14.09
19	Development studies	124	23	1,968	1,898	15.87
20	Transportation science technology	104	30	2,605	2,340	25.05
21	Law	71	13	400	384	5.63
22	Psychology applied	71	23	1,806	1,465	25.44
23	Social sciences interdisciplinary	65	17	737	666	11.34
24	Engineering civil	63	23	1,600	1,395	25.4
25	Information science library science	60	22	2,301	1,619	38.35

from the social sciences, the structuralism, relationships, power struggle, rational choice, interpretivism, theory of evolution, and conflict, in relation with the transitions to sustainability. Deducing from this investigation that the complexity of the interested parties, given the beliefs and values ingrained, become conflicting when private parts show little incentive to develop sustainable activities. They suggest as a change mechanism a regulatory transition from the governmental and social part, which through the consumption and economy, stimulates the industry to reorient its activities.

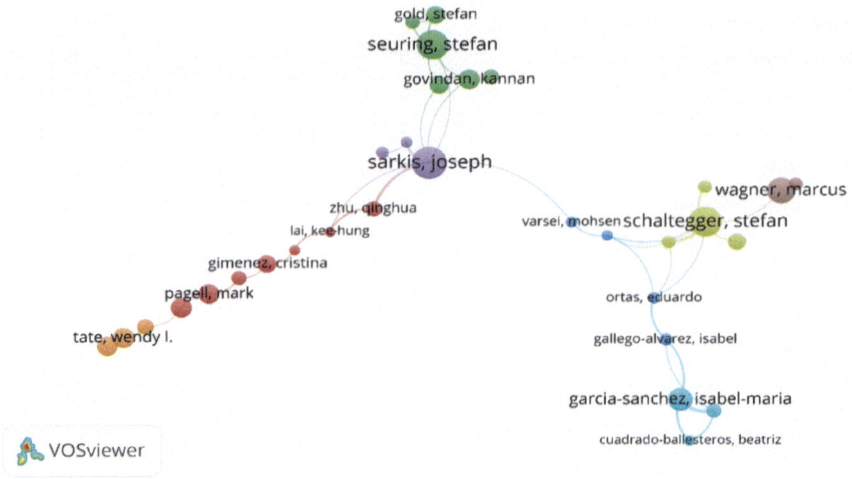

Fig. 3 Co-authorship leading authors

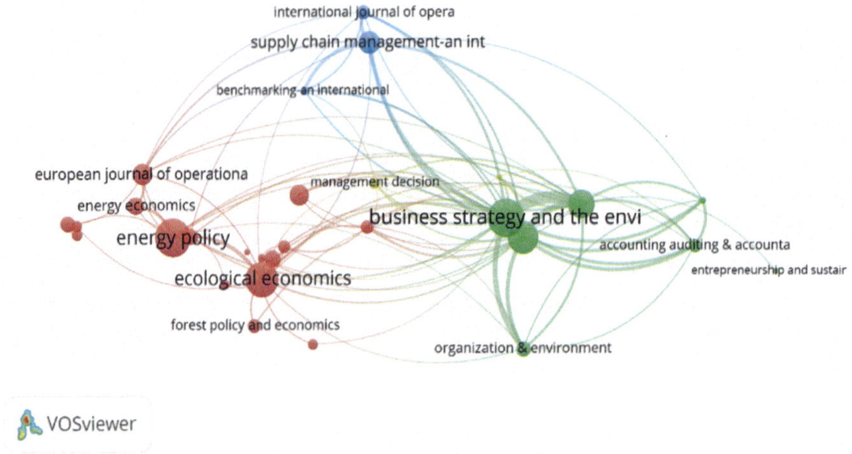

Fig. 4 Journal's citation analysis

3.2 *Authors*

Table 2 identifies the most productive authors with respect to their publications of articles on sustainability in the areas of business, economics, administration, and finance in the last 10 years. This ranking is made up of researchers enrolled in a research center in one of 12 different countries found in the table, in order from highest to lowest recurrence the countries are respectively Germany, Spain, United States of America, France, Japan, Canada, Norway, South Africa, Lithuania, Denmark, Switzerland, and Finland.

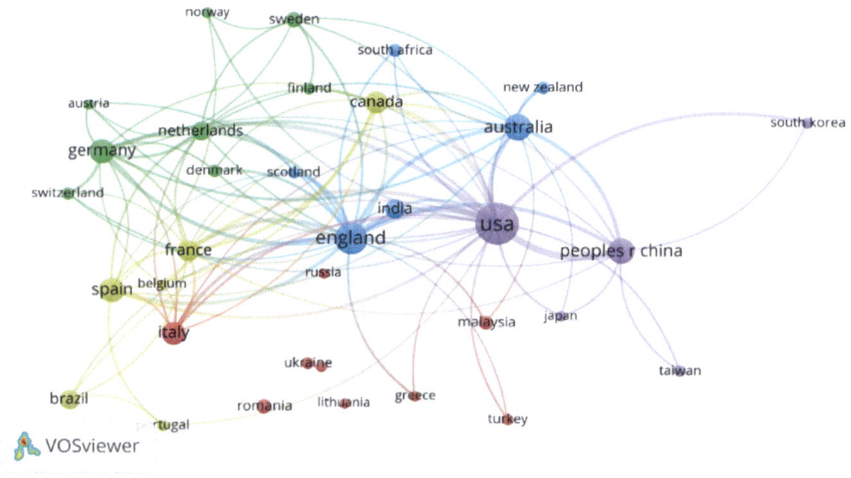

Fig. 5 Country co-authorship network

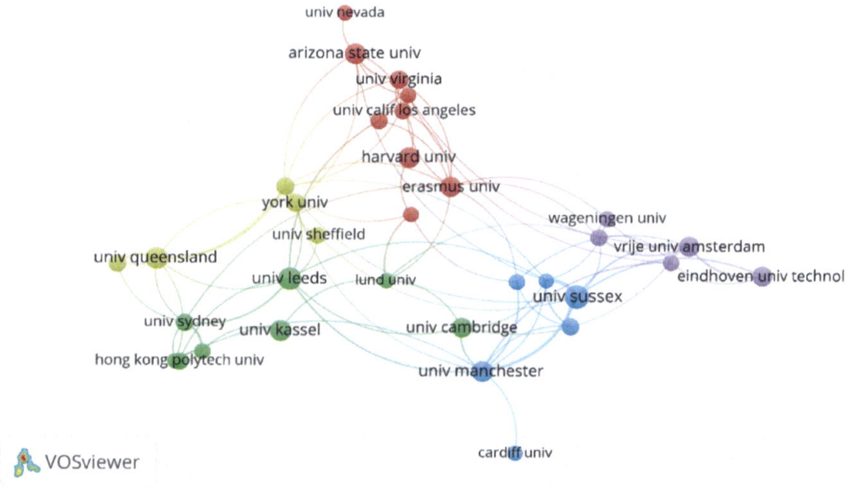

Fig. 6 Co-authorship network of organizations

The most productive author is Schaltegger who has published 28 documents on ES-BFME, predominantly in the areas of Business & Economics and Environmental Sciences & Ecology, accumulating a total of 1561 citations and an H index of 16 for this segment. His most cited article is "Sustainable Entrepreneurship and Sustainability Innovation: Categories and Interactions" [53], which is in the 7th position in the ranking of the 25 most cited documents in this investigation (See Table 1), in general, the author holds a H index of 31, of which 51% is due to his specialization in ES-BFME, has published 72 articles and has 3884 global citations.

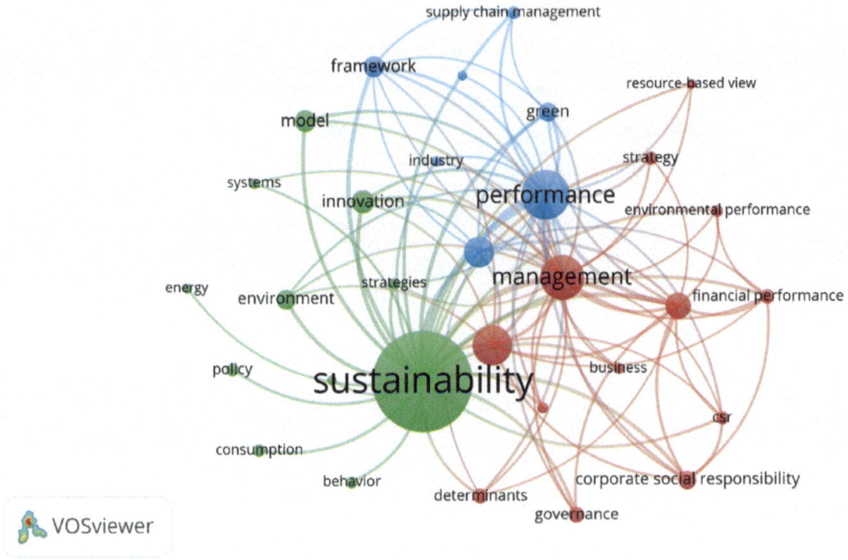

Fig. 7 Co-ocurrence words

In the second position of this ranking is the author García-Sánchez with 26 publications in ES-BFME, whose documents are mostly in the areas of business & economics and environmental sciences & ecology, accumulating a total of 874 citations and an H index of 14 for this segment, her most cited publication in ES-BFME is *the role of the board in the dissemination of integrated corporate social reporting* [54] with 192 citations. Overall, the author has published 124 documents, of which she has accumulated 3,253 citations and reached a general H index of 32.

In the third position by numbers of articles published in ES-BFME, is Sarkis with 25 publications, however, in the degree of influence this is the main author in the table based on the number of citations, he has been referenced a total of 1,773 times and he has reaching an H index for this category of 16, his most cited publication in ES-BFME is *quantitative models for sustainable supply chain management: developments and directions* [55] and has the 6th position in the ranking of the most cited articles in this documents (see Table 1). Overall, the author has the highest H index in the table with 72, has published 305 documents, and accumulates more than 20,500 citations.

Figure 3 shows a co-authorship bibliographic network analyses of the leading authors in ES-BFME. The size of the spheres depends on the authors' number of citations. It is observable that the top 3 most productive authors, Schaltegger, García-Sanchez, and Sarkis share common links. The network is produced with at least one paper sharing common connections of co-authorship. In this case Mohsen establishes the link between the three most productive authors with the article *Framing sustainability performance of supply chains with multidimensional indicators* [56]. It can also be seen that other leading authors share connections, e.g., Sarkis with

Seuring and Schaltegger with Wagner. Please note that Fig. 3 is constructed on papers sharing authorship, therefore, there might be some top productive authors missing in the graph. This is interesting as it might reveal some room for synergy and further collaboration.

3.3 Journals

The most influential journal on ES-BFME of the last 10 years is Business Strategy and the Environment (see Table 3), in which 455 have been published documents in this category, accumulating 12,841 citations and an H-BFME of 59. This journal holds a high content of documents in ES-BFME based on its thematic objectives, which search answers for commercial objectives, obtaining the lowest repercussions for environmental development. In general, this journal published 653 documents and a total of 16,658 citations, which represents in ES-BFME 70% of its composition. Registered in the Journal Citation Report of Clarivate Analytics since 2011, it obtained a JCR Impact Factor of 5.48 in 2019 and its publications are classified in three categories: “Business” which is ranked 19 out of 152 journals evaluated by JCR for that category, "Environmental Studies" at position 10 out of 123 and "Management" at position 21 out of 226 for the last year of analysis. All the journal categories are valued within Quartile 1, for the last 3 years ago and have an average percentile for the last 5 years of 86, 92, and 89 respectively, their highest Impact Factor was 6,381 in 2018.

In the second position in the ranking is Energy Policy magazine, with 343 publications on ES-BFME, an H-BFME of 56, accumulating for this category of 12,629 citations, is the one with the highest number of publications with 7,085 in this table, as well as the highest H-Index of 141 and the one with the highest number of general citations 197,530. However, given the thematic diversity of the journal, it only contains 6% of articles in the ES-BFME area. With registration in the Journal Citation Report of Clarivate Analytics, it obtained a JCR Impact Factor of 5.69 in 2019 and its publications are classified in two categories: “Economic” which is in position 13 of 371 journals evaluated by JCR for that category, and "Environmental Studies" at 14th out of 123 for the last year of analysis. All these journal categories are rated within Quartile 1 and hold an average percentile of 96.26 and 88.63 respectively in the last 3 years.

The third position in the ranking is occupied by the Journal "Ecological Economics" with 417 publications in ES-BFME, accumulating 10,630 citations for this category and holds a 50 H-BFME, in general, the journal has published 2,678 articles, of which 16% belong to the ES-BFME category, in general, has 67,294 citations and a 102 H-index. With registration in the Journal Citation Report of Clarivate Analytics, this journal obtained a JCR Impact Factor of 5.23 in 2019 and its publications are classified in two categories: "Economic" which is ranked 26th out of 371 journals evaluated by JCR, and "Environmental Studies" which is ranked 20 out of 123, both in the last year of analysis. All journal categories are rated within

Quartile 1 and have an average percentile of 93.79 and 86.90 respectively from the last 5 years.

When analyzing the sources in WoS where most of the documents in ES-BFME are published, we find the journals BSE, EP and EE as leading publishers in both, productivity, and influence in ES-BFME. However, there is a clear distinction in the specialization of the journals, when analyzing the ratio %BFME i.e., ES-BFME papers published over the total production of the journal indexed in WoS, we can see that BSE maintains the leadership in the ranking with 7 out of 10 documents falling in the category of ES-BFME, following, CSREM with 53%, and OE and SAMPJ with 40% each. In general terms, EP, EJOR and JBR, show the largest amount of TP and TC, this is consistent with the characteristics of these traditional and firmly established journals.

Figure 4 shows the results of a citation bibliographic network analysis of the found journals, the size of the spheres corresponds to the number of citations received. It is observable that BSE, EP, EE share most of the citations. It is also observable that the networks generated between the journals create at least 4 clusters of relation. In green we can see BSE, JBE and CSR as the most representative, this cluster is represented by the words business and sustainability included in most of the journals. A red cluster where EP and EE show preponderance, here the word economics leads the count of journals names. A blue cluster led by SCMIJ and the word international as representative of the array. Finally, a yellow cluster in which none of the top 25 journals appear.

3.4 Countries

Table 4 shows a picture of the leading countries in ES-BFME. In the case of countries, the USA leads the list in both productivity and influence, followed by the UK (England, Scotland, Wales, and North Ireland) and AUS. It is interesting to note that a dissection on highly cited papers is also performed, here, the reader can observe that USA, UK, CAN, NLD, SWE and CHE share the characteristic of at least one paper exceeding the 500 citations. When observing the H-BFME index, we can also see some trends concerning the influence and productivity of the countries, in this case, after the already mentioned counties, Germany and P.R. China also holds a leading role. The last two columns of the table show productivity and influence considering the total population of the country, in this case led by DNK, NLD and NZL.

Figure 5 shows a country co-authorship bibliographic network analysis. The size of the spheres also corresponds to the number of times cited. In general, we can observe that most of the leading countries previously mentioned share some common connections in terms of co-authored papers, however some trends and links appear, e.g. USA shares more papers with CHN than with other European countries, on the other hand a clear connection and distribution of the links between European nations is also observed. England stays at the center of the graph showing a big network of connections and collaboration.

3.5 Organizations

Table 5 shows the distribution of documents and citations by organizations. This ranking is led by Bucharest University of Economic Studies in Romania, followed by the University of Leeds in the United Kingdom and the Hong Kong Polytechnic University in Hong Kong. It is worth to note that 32% of the organizations in the list are in AUS and 20% of them in the UK. In terms of influence, the U of Sussex, Leeds and Queensland share the common characteristic of having received more than 200 citations in their published papers indexed in WoS concerning ES-BFME. Another ratio worth being considered is the TC/TP – BFME, here, the U of Sussex, Kassel and Queensland lead the list correspondingly.

3.6 Most Frequent Keywords

Table 6 presents information regarding co-occurrence of words i.e., words frequently appearing in key sections of the indexed articles published in ES-BFME. It can be observed that the most frequent words used by authors keywords are sustainability, sustainable development, and corporate social responsibility. In keywords plus, which is a special feature of the WoS Clarivate Analytics database, performance, management, and sustainability lead the list. As per the titles of the documents published in ES-BFME, the most co-occurring words are sustainable, sustainability and environmental. Finally, the most used words in the abstract section of the indexed publications are environmental, sustainability and sustainable. It is interesting to view these top 3 words in each category, moreover, the rest of the top co-occurrence word analyses allows the verification of papers including sustainability with relation to the environment and the selected research areas in BFME.

Figure 7 shows the co-occurrence of words network bibliographic analysis for the retrieved ES-BFME documents. The size of the spheres corresponds to the amount of co-occurrence of a word in the selection. As it is expected, the word sustainability is the largest element with the highest number of connections in the graphic. Following, we can also observe the words, performance, management, and sustainable development. In general, three clusters of words are built, in green we can find more environmental related topics, in blue, performance and industry related words and in red management, determinants and CSR topics.

3.7 WoS Categories

Finally, Table 7 presents the distribution of the ES-BFME found documents in the WoS categories. Please observe that these categories correspond to WoS classifications. Once a journal is indexed in WoS at least one category is assigned to the periodical. Even though our methods are specifically designed to find the core BFME papers published and indexed in the WoS regarding sustainability, it results interesting to depict the secondary categories in which these studies are categorized. Here we can observe Management, economics and Business as top 3 categories considering productivity; however, business finance is the 6th category in the list. Please also note that there is a significant difference of around 1400 papers in TP-BFME from the 4th to the 5th place in the ranking. Environmental studies and sciences are the top 4th and 5th category. In terms of influence, and besides management and economics, environmental studies have received much attention by ES-BFME papers.

4 Conclusion

Bibliometric studies have proven to be effective in the description of the advancements of research areas and fields of knowledge using analytical and statistical tools. The objective of the present study is to collect and describe in a systematized manner, the contributions that have shaped the landscape of environmental sustainability, moreover, the proposals that have appeared in the fields of business, finance, management and economics. This document seeks to contribute to two main purposes, (1) to the knowledge of the sustainability field, trough brings a general vision focalized to the importance of the economic sphere, necessary for the adoption of long-term sustainable changes, and (2) to provide to researchers and general readers, a current panorama of the structure of environmental sustainability science through disciplines like the administration, business, finance, and economics as a specific area. Seeking that it be useful for the investigator's community like beginning point to allow them to trace the recent lines of research of the economic-administrative sciences adapted to the new paradigm of sustainability.

The benefits in the use of bibliometrics as a methodology, become visible in the advances prior to this research and those reflected in this document too, because it allowing appreciating the evolution, growth, and trends of an environmental sustainability, a recently studied field, even more in the economic-administrative area [18, 49, 52]. As previous of this investigation examples, through the use of the bibliometrics method, Schoolman et al. [57], identified to the economic dimension as the most integrated of the three, and Kajikawa et al. [14, 44], registered an exponential growth for the publications on sustainability in the Economic and Business group from the period 2007–2013. These kinds of studies provide elements to measure the expansion of the topic in the economic and administrative sciences and allow infer the time between periods where the changes in the scientific communication present.

Standing the utility to the methodology used, in the results of the present investigation, it can be seen out as a common factor among the most cited documents (see Table 1), the common use of bibliometric tools for the construction of literary, theoretical, and conceptual reviews addressed in the documents of ES-BFME most influential to the last decade. As an example, the reader can see the description of the three most cited articles in this document *sustainability transitions: an emerging field of research and its prospects* by Markard et al. [18]; *sustainable supply chain management: evolution and future directions* from Carter and Easton [49] and *ontologies, socio-technical transitions (to sustainability), and the multi-level perspective* by Geels [52], all the mentioned papers, carry out a detailed literary analysis of their subject, and the first two start from a bibliometric analysis for this.

The recent expansion of sustainability in scientific research has been highlighted throughout this study, so that theoretical constructions are not completely delimited and constructs are still being redefined by the sciences that adopt them, is the same case for the environmental sustainability concept, which the definition is not finally established for all the community (environmental, economic and social), however this science has emerged as an important field of research dedicated to the study of sustainable production and consumption systems with respect to environmental limits and the regeneration capacity of planetary resources [2].

Is important to say that sustainability nor planetary limits are new concepts, today the studies based on them are in the full expansion, being that only two decades ago the Nobel laureate Crutzen distinguished the current period such as the *anthropocene*, here the humanity is the dominant force of change in the Earth [58, 59], this theory between other implications, advise that the current pace of social and economic development exerts great pressure on the planet, which leads us to the destabilization of biophysical systems and abrupt environmental changes in many irreversible scenarios, a catastrophic scenario for human well-being [2].

From the economic and managerial standpoint, this is highly relevant given that the current economic system, especially the industries, has been pointed responsible for most frequent environmental problems in the world [19, 60, 61], but also, in their ability to act, to achieve a true change towards sustainability [2, 51, 62], since current technologies, lifestyles, business models and value chains are strongly determined by predisposed the social and economic development [18].

This study includes information retrieved from the WoS scientific database. This repository includes worldwide peer-reviewed journals indexed with robust criteria, thus allowing a transparent and reliable source of information. The search process is designed to find core documents addressing ES-BFME published in the last 10 years. A total of 9,735 documents are collected following the methodological approach detailed in the present paper. Together, these papers portray in a global way the latest advancements of the scientific community in these matters.

Specific information of the most cited documents reveals that two out of three most cited documents in environmental sustainability are available in the journal RPO. It also shows that three documents have consistently influenced the scientific community, acquiring more than 80 citations per year since their publication. One of those documents belongs to Sarkis (USA), author that along to Schaltegger (DEU), and Garcia-Sanchez (ESP), share the top three positions in documents publication productivity. The predominant theme of these investigations can be grouped into 5 areas, ordered from highest to lowest, are the investigations about the management of the supply chain and logistics; innovation and entrepreneurship; economics, accounting, and finance; consumer behavior and corporate social responsibility. On the other hand, further statistics show that the most productive and specialized authors in these types of publications are Longoni and Horisch with 56% and 50% of their total production focused on core ES-BFME.

A network analysis is also performed on authors co-citation, exposing a connection between most of the leaders in the field. Some other results are also interesting to mention, e.g., the periodical BSE is the leading journal in productivity, moreover its specialization in ES-BFME is calculated in 70%. As per countries, the USA leads in the ES-BFME ranking with 1,862 papers published and the highest citation with 47,358 total citations. This country has also produced 50% of papers receiving more than 500 citations worldwide and some of its strongest connections in co-authored papers involve the UK, AUS and CHN. The most productive university is the Romanian Bucharest University of Economic Studies; however, the most influential organization in ES-BFME is the University of Sussex in the UK. A co-word analysis shows that sustainability, performance, sustainable and environment are most co-occurrent or frequent words used by the authors in the field and a network analysis shows the connections between them. Finally, as it is expected, management, economics and business are the top 3 categories in which the retrieved core ES-BFME documents lie, however the most cited papers per document are in the categories of information science, library science, ethics, operations research management science, and energy fuels.

The design of this paper tries to share some light in the depiction of leaders in some specific categories of the core ES-BFME retrieved documents. The aim is to find some trends, connections and reveal space for collaboration and creation of synergies. Most cited papers, authors, journals, countries, words, and categories are described using several statistical calculations, moreover network analyses using the software VoSviewer also reveals key information of the last 10 years shape of the ES-BFME field. Nonetheless some limitations are found in the creation of this study, these are mainly observed in the selection of the scientific database, as other well-known scientific repositories are also important to study e.g., SCOPUS or Google Scholar, this required analysis is suggested for future research. Sustainability in business, finance, management, and economics is a key research field that needs to be nourished and promoted. Today, the unprecedented challenges that humanity faces reveal the need for prompt, science-based solutions specifically designed from the core of human production, consumption, strategic coordination, allocation, and transfer of wealth.

References

1. Sala S, Crenna E, Secchi M, Sanyé-Mengual E (2020) Environmental sustainability of European production and consumption assessed against planetary boundaries. J Environ Manage 269:110686
2. Rockstrom J, Steffen W, Noone K, Persson A, Chapin F, Stuart I, Lambin E, Foley J (2009) Planetary boundaries: exploring the safe operating space for humanity. Ecol Soc 14(2):32
3. Torregrosa CG (2005) An analytical study of the relation between international business and sustainable development in the context of free trade agreements: practical experience.

Innovaciones de Negocios 1–14
4. Buter RK, Van Raan AFJ (2013) Identification and analysis of the highly cited knowledge base of sustainability science. Sustain Sci 8(2):253–267
5. Gómez C (2006) El desarrollo sostenible: conceptos básicos, alcance y criterios para su evaluación. Cuestiones de Sociología 5–6:295–312
6. ONU (2015) Objetivos de Desarrollo Sustentable 2015–2030
7. PNUD (2019) Medio ambiente y desarrollo sustentable
8. García-Orozco D, Alfaro-García VG, Espitia-Moreno IC, Gil-Lafuente AM (2020) Forgotten effects analysis of the consumer behavior of sustainable food products in Mexico. J Intell Fuzzy Syst 1–10. https://doi.org/10.3233/JIFS-189194
9. Godfray HCJ, Beddington JR, Crute IR, Haddad L, Lawrence D, Muir JF, Toulmin C (2010) Food security: the challenge of feeding 9 billion people. Science 327(5967):812–818
10. Jackson JBC (2001) Historical overfishing and the recent collapse of coastal ecosystems. Science 293(5530):629–637
11. Pauly D, Christensen V, Guénette S, Pitcher TJ, Sumaila UR, Walters CJ, Zeller D (2002) Towards sustainability in world fisheries. Nature 418(6898):689–695
12. Rosenzweig C, Parry ML (1994) Potential impact of climate change on world food supply. Nature 367(6459):133–138
13. Zhao C, Liu B, Piao S, Wang X, Lobell DB, Huang Y, Asseng S (2017) Temperature increase reduces global yields of major crops in four independent estimates. Proc Natl Acad Sci 114(35):9326–9331
14. Kajikawa Y, Ohno J, Takeda Y, Matsushima K, Komiyama H (2007) Creating an academic landscape of sustainability science: An analysis of the citation network. Sustain Sci 2(2):221–231. https://doi.org/10.1007/s11625-007-0027-8
15. Carter CR, Rogers DS (2008) A framework of sustainable supply chain management: moving toward new theory. Int J Phys Distrib Logist Manag 38(5):360–387
16. Simon D (1987) Our common future: report of the world commission on environment and development (book review). Third World Plann Rev 9(3):285
17. Aguayo Gonzalez F, Lama Ruiz JR, Peralta Alvarez ME (2011) Ingeniería sostenible de la cuna a la cuna: una arquitectura de referencia abierta para el diseño c2c sustainable. DYNA Ingenieria E Industria 86(3):199–211
18. Markard J, Raven R, Truffer B (2012) Sustainability transitions: an emerging field of research and its prospects. Res Policy 41(6):955–967
19. Schaltegger S, Hansen EG, Lüdeke-Freund F (2016) Business models for sustainability: origins, present research, and future avenues. Organ Environ 29:3–10
20. Kajikawa Y (2008) Research core and framework of sustainability science. Sustain Sci 3:215–239. https://doi.org/10.1007/s11625-008-0053-1
21. Van Marrewijk M (2003) Concepts and definitions of CSR and corporate sustainability: between agency and communion. J Bus Ethics 44(2–3):95–105
22. Geissdoerfer M, Savaget P, Bocken NMP, Hultink EJ (2017) The circular economy—a new sustainability paradigm? J Clean Prod 143:757–768. https://doi.org/10.1016/j.jclepro.2016.12.048
23. Seuring S, Müller M (2008) From a literature review to a conceptual framework for sustainable supply chain management. J Clean Prod 16(15):1699–1710
24. Blanco-Mesa F, Merigó JM, Gil-Lafuente AM (2017) Fuzzy decision making: a bibliometric-based review. J Intell Fuzzy Syst 32(3):2033–2050
25. Buter RK, Raan AFJ (2012) Identification and analysis of the highly cited knowledge base of sustainability science. Sustain Sci 8(2):253–267
26. Cancino CA, Merigó JM, Torres JP, Diaz D (2018) A bibliometric analysis of venture capital research. J Econ Fin Admin Sci 23(45):182–195. https://doi.org/10.1108/JEFAS-01-2018-0016
27. Fetscherin M, Heinrich D (2015) Consumer brand relationships research: a bibliometric citation meta-analysis. J Bus Res 68(2):380–390
28. García-Orozco D, Espitia-Moreno IC, Alfaro-García V, Merigó JM (2020) Sustentabilidad en México un análisis bibliométrico de la investigación científica presentada en los últimos 28 años. Inquietud Empresarial 20(2):101–120

29. Merigó JM, Blanco-Mesa F, Gil-Lafuente AM, Yager RR (2017) Thirty years of the international journal of intelligent systems: a bibliometric review. Int J Intell Syst 32(5):526–554
30. Yılmaz B, Dinçol ME, Yalçın TY (2019) A bibliometric analysis of the 103 top-cited articles in endodontics. Acta Odontol Scand. https://doi.org/10.1080/00016357.2019.1621378
31. Zhang L, Zhong Y, Geng Y (2019) A bibliometric and visual study on urban mining. J Clean Prod 239. https://doi.org/10.1016/j.jclepro.2019.118067
32. Sarkodie SA, Strezov V (2019) A review on environmental kuznets curve hypothesis using bibliometric and meta-analysis. Sci Total Environ 649:128–145
33. Alfaro-García VG, Merigó JM, Alfaro-Calderón GG, Plata-Pérez L, Gil-Lafuente AM, Herrera-Viedma E (2020) A citation analysis of fuzzy research by universities and countries. Journal of Intelligent & Fuzzy Systems 38(5):5355–5367
34. Bilas Roy S, Basak M (2013) Journal of documentation: a bibliometric study. Library Philos Pract 945:1–10
35. Chabowski BR, Mena JA, Gonzalez-Padron TL (2011) The structure of sustainability research in marketing, 1958–2008: a basis for future research opportunities. J Acad Mark Sci 39(1):55–70
36. Gomezelj DO (2016) A systematic review of research on innovation in hospitality and tourism. Int J Contemp Hosp Manage 28(3):516–558
37. Martens ML, Carvalho MM (2017) Key factors of sustainability in project management context: a survey exploring the project managers' perspective. Int J Project Manage 35(6):1084–1102
38. Castro e Silva, M., & Teixeira, A. A. C. (2011) A bibliometric account of the evolution of EE in the last two decades. Ecol Econ 70(5):849–862
39. Dionisio M (2019) The evolution of social entrepreneurship research: a bibliometric analysis. Soc Enterp J 15(1):22–45
40. Li K, Rollins J, Yan E (2018) Web of Science use in published research and review papers 1997–2017: a selective, dynamic, cross-domain, content-based analysis. Scientometrics 115(1):1–20
41. Merigó JM, Gil-Lafuente AM, Yager RR (2015) An overview of fuzzy research with bibliometric indicators. Appl Soft Comput J 27:420–433
42. Liu W (2019) The data source of this study is web of science core collection? Not enough. Scientometrics 121:1815–1824
43. Harzing AW, Alakangas S (2016) Google scholar, scopus and the web of science: a longitudinal and cross-disciplinary comparison. Scientometrics 106(2):787–804
44. Kajikawa Y, Tacoa F, Yamaguchi K (2014) Sustainability science: the changing landscape of sustainability research. Sustain Sci 9(4):431–438
45. van Eck NJ, Waltman L (2010) Software survey: VOSviewer, a computer program for bibliometric mapping. Scientometrics 84(2):523–538
46. Waltman L, van Eck NJ (2013) A smart local moving algorithm for large-scale modularity-based community detection. Eur Phys J B 86(11):471
47. Waltman L, van Eck NJ, Noyons ECM (2010) A unified approach to mapping and clustering of bibliometric networks. J Informet 4(4):629–635
48. Merigó JM, Pedrycz W, Weber R, de la Sotta C (2018) Fifty years of information sciences: a bibliometric overview. Inf Sci 432:245–268
49. Carter CR, Easton PL (2011) Sustainable supply chain management: evolution and future directions. Int J Phys Distrib Logist Manage 41(1):46–62
50. Freeman RE (1999) Divergent stakeholder theory. Acad Manage Rev 24(2):233–236
51. Hart SL (1995) A natural-resource-based view of the firm. Acad Manage Rev 20(4):986. https://doi.org/10.2307/258963
52. Geels FW (2010) Ontologies, socio-technical transitions (to sustainability), and the multi-level perspective. Res Policy 39(4):495–510
53. Schaltegger S, Wagner M (2011) Sustainable entrepreneurship and sustainability innovation: categories and interactions. Bus Strateg Environ 20(4):222–237
54. Frias-Aceituno JV, Rodriguez-Ariza L, Garcia-Sanchez I (2013) The role of the board in the dissemination of integrated corporate social reporting. Corp Soc Responsib Environ Manage 20(4):219–233

55. Brandenburg M, Govindan K, Sarkis J, Seuring S (2014) Quantitative models for sustainable supply chain management: developments and directions. Eur J Oper Res 233(2):299–312
56. Varsei M, Soosay C, Fahimnia B, Sarkis J (2014) Framing sustainability performance of supply chains with multidimensional indicators. Supply Chain Manage Int J 19(3):242–257
57. Schoolman ED, Guest JS, Bush KF, Bell AR (2012) How interdisciplinary is sustainability research? Analyzing the structure of an emerging scientific field. Sustain Sci 7(1):67–80
58. Cruzen PJ (2002) Geology of mankind: the Anthropocene. Nature 415:23
59. Steffen W, Crutzen PJ, McNeill JR (2007) The anthropocene: are humans now overwhelming the great forces of nature. AMBIO: J Hum Environ 36(8):614–621
60. Alfaro-García VG, Merigó JM, Pedrycz W, Gómez Monge R (2020) Citation analysis of fuzzy set theory journals: bibliometric insights about authors and research areas. Int J Fuzzy Syst
61. Garcia AS, Mendes-Da-Silva W, Orsato R (2017) Sensitive industries produce better ESG performance: Evidence from emerging markets. J Clean Prod 150:135–147
62. Bocken NMP, de Pauw I, Bakker C, van der Grinten B (2016) Product design and business model strategies for a circular economy. J Ind Prod Eng 33(5):308–320. https://doi.org/10.1080/21681015.2016.1172124

Bibliometric Analysis of Sustainable Tourism Research for the Period 1991–2019

Abraham Nuñez-Maldonado and Martha Beatriz Flores-Romero

Abstract This text presents a bibliometric review of sustainable tourism, analysing 28 years of research, published between 1991 and 2019, within the Web of Science database, using the keyword "sustainable tourism". The analysis focuses on the results of bibliometric review software, Rstudio to perform different topics in bibliometrics and scientometrics, including the most cited paper and author. The results of the study indicate that sustainable tourism has achieved a complexity that is reflected in the exponential growth of scientific papers.

Keywords Bibliometrics · Scientometrics · Sustainable · Tourism

1 Introduction

The purpose of this study is to contribute to the academic discussion on sustainable tourism, presenting an approach across the bibliometric analysis, from the publications of the Web of Science (WoS). Tourism has been a characterized by huge innovativeness, due to its maturation as an area of study, where such articles are more and more common within the field's leading journals, thanks to the interest awakened in the academic world [1–3].

In recent years, the impacts of tourism have received crescent attention in discussions, negative social, and cultural impacts, environmental degradation, bibliometric journal reviews, and studies on related development because tourist activities impact directly and indirectly on ecosystems. The links between tourism and sustainability are complex. [4–9].

The concept of sustainable tourism has evolved through time, according to the United Nations World Trade Organization [10], in theirs development goals in a

A. Nuñez-Maldonado (✉) · M. B. Flores-Romero
Universidad Michoacana de San Nicolás de Hidalgo. Ave. Francisco J. Mujica S/N, Edificio A-IV, 3er. Piso, Col. Felicitas del Rio C.P., 58030 Morelia, Michoacán, México
e-mail: abraham.nunez@umich.mx

M. B. Flores-Romero
e-mail: betyf@umich.mx

E. León-Castro et al. (eds.), *Soft Computing and Fuzzy Methodologies in Innovation Management and Sustainability*, Lecture Notes in Networks and Systems 337,
https://doi.org/10.1007/978-3-030-96150-3_19

journey to 2030 has defining it as the tourism that fully takes into account its current and future economic, social, and environmental impacts, serving the needs of visitors, industry, the environment and host communities. A structured literature review like this could also provide an overview of the most developed areas of study in sustainable tourism [11].

2 Theoretical Framework

As a concept, [12] was the first to publish an article on sustainable tourism, in The Globe ´90 conference, which began a strategy for sustainable tourism development. The notion of sustainable tourism must be considered as one of the great success stories of tourism research and knowledge transfer [13], the relationship between sustainability and tourism is interesting because of the considerable role of tourism within the global economy [14]. Researchers and policymakers constantly question the efficiency of sustainable tourism and its applications, practices, and practical adoption [15]. The economic, social, and environmental crises of recent years have caused a world concern for the necessity to search out mechanisms that enable efficient development processes that lead countries towards sustainability [16].

3 Method

Bibliometric studies use a large range of methods. The foremost popular are those who take under consideration the number of publications and therefore the number of citations [17]. To approach the problem, this study empirically examines the recent sustainable tourism research published from 1990 to 2019. The research was limited to the papers published in the databases of the Web of Science, we used the words "Sustainable tourism" for the enquiry, there are: 2151 article, 26 articles; early Access, 43 articles; proceedings paper, 1 correction, 1 biographical-item, 74 book review, 55 editorial material, 1 editorial material; early Access, 3 letters, 1 letter; early Access, 1 meeting abstract, 92 reviews, 1 review; early Access, containing 85,139 cited references, from 500 sources.

Bibliometric review software (Rstudio), with the packages "Bibliometrix", provides several routines to import bibliographic data from Clarivate Analytics Web of Science databases, and to perform bibliometric analysis, and "Biblioshiny", performs science map analysis, which provides a collection of tools for quantitative research in bibliometrics and scientometrics [18], was used to perform.

4 Results

The study identified the key influential and also the most cited recent articles in sustainable tourism literature. Data statistics revealed the distribution of sustainable tourism articles, the country´s total production, country collaboration map, and the most productive authors. The number of research papers on sustainable tourism, published between 1991 and 2019 has only increased, Fig. 1 illustrates the exponential growth of annual publications in sustainable tourism, from 20 papers in 2005 to 429 in 2019, having an increase of 26.92% from the year before.

Table 1 illustrates the countries productive of research papers top ten, China being the country that generates the most research with 163 Single Country Publications (SCP) and 60 Multiple Country Publications (MCP), in second place are the United States of America (USA), with 169 SCP and 53 MCP, but Spain is the most productive country in single publications with 184, following by the USA. Also, we can see that some countries have a lot of papers in collaboration, such as the case of New Zealand with 47.1% of MCP.

Figure 2 shows us a heat map where the countries with the highest publications are in dark blue, while the countries that do not have publications are in grey, as we can see, there is also a relationship between the countries that they have at least 3 collaborations between them.

This bibliographic review also has the list of the 10 most cited articles, with Butler R. in first place with 153 citations with an article from 1999, followed by Hunter C with 147 citations and an article from 1997, the most recent article and with the highest number of citations is the 2012 Buckley R. with 141 citations (Table 2).

The 10 most cited authors in these 28 years of bibliometric review are listed in Table 3, where it can be seen that the most cited author over time is Gosslin S. with

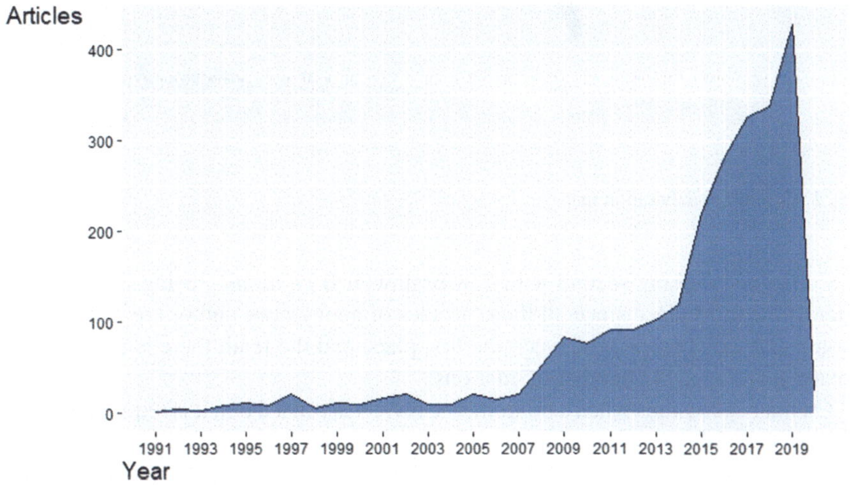

Fig. 1 Annual scientific production

Table 1 Production per country

	Country	Articles	SCP	MCP	MCP_Ratio
1	China	223	163	60	0.269
2	USA	222	169	53	0.239
3	Spain	220	184	36	0.164
4	Australia	215	162	53	0.247
5	United Kingdom	191	145	46	0.241
6	Italy	118	96	22	0.186
7	Canada	80	53	27	0.338
8	Turkey	67	59	8	0.119
9	Romania	66	54	12	0.182
10	New Zealand	51	27	24	0.471

Elaborated from the Web of Science database. *Abbreviations* SCP, Single country publications; MCP, Multiple country publications

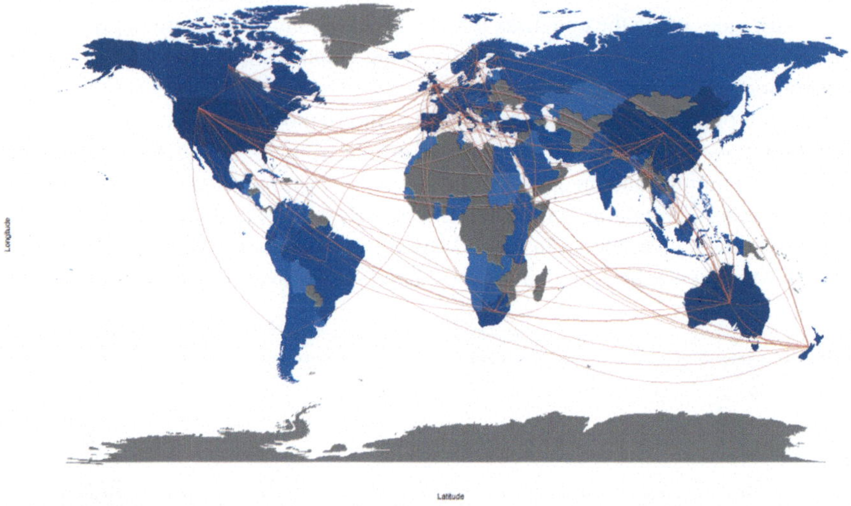

Fig. 2 Country collaboration map

828 citations, ranking second with 738 Bramwell B citations.., being of particular interest the third place since all those articles of anonymous author are collected, it should also be noted that between the first place and the tenth there is a difference greater than twice the number of citations.

On many occasions when a bibliometric review is carried out, it is required to have a clear vision of the production of the most prolific authors on the subject, as can be seen in Fig. 4, the authors with the most publications on sustainable tourism can be observed, where depending on the size of the circle, there is an annual production of

Table 2 Article most cited

	Article	Cited
1	BUTLER R., 1999, TOURISM GEOGR, V1, P7, DOI https://doi.org/10.1080/14616689908721291	153
2	HUNTER C, 1997, ANN TOURISM RES, V24, P850, https://doi.org/10.1016/S0160-7383(97)00036-4	147
3	LIU ZHENHUA, 2003, JOURNAL OF SUSTAINABLE TOURISM, V11, P459, https://doi.org/10.1080/09669580308667216	143
4	BUCKLEY R, 2012, ANN TOURISM RES, V39, P528, https://doi.org/10.1016/J.ANNALS.2012.02.003	141
5	BUTLER RW, 1980, CAN GEOGR-GEOGR CAN, V24, P5, https://doi.org/10.1111/J.1541-0064.1980.TB00970.X	139
6	SAARINEN J, 2006, ANN TOURISM RES, V33, P1121, https://doi.org/10.1016/J.ANNALS.2006.06.007	118
7	CHOI HC, 2006, TOURISM MANAGE, V27, P1274, https://doi.org/10.1016/J.TOURMAN.2005.05.018	113
8	JAMAL TB, 1995, ANN TOURISM RES, V22, P186, https://doi.org/10.1016/0160-7383(94)00067-3	103
9	MILLER G, 2001, TOURISM MANAGE, V22, P351, https://doi.org/10.1016/S0261-5177(00)00067-4	97
10	ANDERECK KL, 2005, ANN TOURISM RES, V32, P1056, https://doi.org/10.1016/J.ANNALS.2005.03.001	96

Elaborated from the Web of Science database

Table 3 Author most cited

	Author	Cited
1	Gossling S	828
2	Bramwell B	738
3	Anonymous	704
4	Hall CM	534
5	Hall C M	516
6	Buckley R	420
7	Unwto	417
8	Scott D	408
9	Sharpley R	367
10	Becken S	342

Elaborated from the Web of Science database

1 to 4 publications, while the colour that they have will represent the number of times these works were cited, in this way you can know what the trajectory of the authors with more publications, showing in a simple way what his career is through time, in which year were his most important works, either in a number of publications or citations.

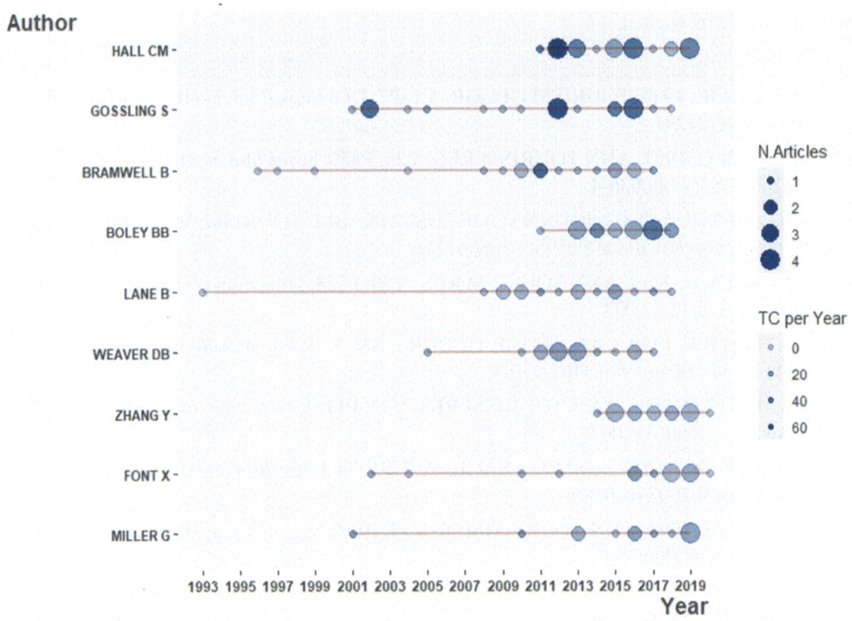

Fig. 4 Top-authors production over time

Table 4 Lotka's law coefficients

N.Articles	N.Authors	Freq	N.Articles	N.Authors	Freq
1	4120	0.8501857	11	4	0.0008254
2	467	0.0963681	12	1	0.0002064
3	134	0.0276517	13	2	0.0004127
4	47	0.0096987	14	1	0.0002064
5	15	0.0030953	15	2	0.0004127
6	18	0.0037144	16	1	0.0002064
7	10	0.0020636	17	1	0.0002064
8	8	0.0016508	18	1	0.0002064
9	8	0.0016508	22	1	0.0002064
10	4	0.0008254	23	1	0.0002064

Elaborated from the Web of Science database

The Lotka's law coefficients for scientific productivity [19], posits that there is an inadequate distribution of productivity among writers and that, no matter the area; most authors publish the least number of papers, while a few ones publish most of the pertinent literature on a given topic, and they form the most productive group [20], and this topic is not the exemption.

Table 5 Top ten sources

	Sources	Articles
1	Journal of sustainable tourism	388
2	Sustainability	203
3	Tourism management	139
4	Annals of tourism research	96
5	Current issues in tourism	49
6	Journal of cleaner production	49
7	Tourism geographies	37
8	Tourism management perspectives	34
9	Worldwide hospitality and tourism themes	34
10	International journal of tourism research	32

Elaborated from the Web of Science database

Table 4 illustrates Lotka's law, which shows the distribution of the number of publications per number of authors. While only one author published 23 research papers, giving us less than 2% has written 5 or more articles, the majority of authors (4120) have only written one paper in the field of sustainable tourism.

Of the 2449 results obtained from the WoS search for sustainable tourism, 500 publication sources were obtained and as shown in Table 5 of the top ten sources, the journal of sustainable tourism has the highest number of publications with 388 which represents 15.84% of the total, being the journal Sustainability the second place with 203 publications for 8.28%, among the ten sources with the most publications there is 43.32% which shows that these sources are the most important in this topic.

Is also important to know de h index of the authors from the topic, to know the quantity and quality of their work by the number of citation, being this index a balance between them, allowed us to know who researchers are the most distinguish and make comparisons between researchers [21], the g index has been well received in scientometrics, being an improvement of the h-index, since it is more similar to the feeling of visibility of an author [22], and the m index or metric index surpasses state-of-the-art precise research techniques in terms of Input/Output, computational costs, and response times to inquiries, having the ability to keep similar data close in the index [23] (Table 6).

5 Conclusions

This study was a bibliometric of sustainable tourism from the WoS databases, using two different packages from the program Rstudio (Bibliometrix and Biblioshiny), to explore the bibliometrics and scientometrics from this topic, we concluded that every day is more important the study of sustainable tourism, having an exponential growth year by year in the number of articles published and gives us a general overview of

Table 6 Top ten author index

Author	h_index	g_index	m_index	TC	NP	PY_start
Hall CM	16	23	1.6	1455	23	2011
Gossling S	20	22	1	2245	22	2001
Bramwell B	11	18	0.44	826	18	1996
Boley BB	13	17	1.3	514	17	2011
Lane B	11	16	0.3928571	598	16	1993
Weaver DB	13	15	0.8125	562	15	2005
Zhang Y	6	12	0.8571429	144	14	2014
Font X	8	14	0.4210526	457	14	2002
Miller G	8	13	0.4	669	13	2001
Ruhanen L	10	13	*NA*	429	13	2008

Elaborated from the Web of Science database. *Abbreviations* TC, times cited; NP, number of publications; PY_start, publication year start.

the importance in WoS. Being China the country that generates the most research, because it hosts more multiple country publications, with the United States, Spain, and Australia in second, third, and fourth respectively, the first country with single country publications was Spain with 184 articles and the country in the top ten with the most higher multiple country publications ratio was New Zealand with 47.1%. The source more important was the Journal of Sustainable Tourism followed by the journal Sustainability and Tourism Management, with research published in 500 sources.

The results of the study indicate that the academic implications of sustainable tourism have achieved a complexity that is reflected within the exponential growth of scientific papers. One of the limitations of this paper was that we used only the WoS database, new studies should include more databases, to expand the general vision of the research in sustainable tourism.

Acknowledgements We would like to thanks the Doctorate studies in Administration, Universidad Michoacana de San Nicolás de Hidalgo, and the National Quality Postgraduate Program of the National Council of Science and Technology (CONACyT) in México, is gratefully acknowledged.

References

1. Álvarez García J, Durán Sánchez A, Del Río Rama M, de la C. (2018) Scientific coverage in community-based tourism : sustainable tourism and strategy for social development. Sustainability 10:18. https://doi.org/10.3390/su10041158
2. Hjalager AM (2010) A review of innovation research in tourism. Tour Manage 31(1):1–12. https://doi.org/10.1016/j.tourman.2009.08.012

3. Ruhanen L, Weiler B, Moyle BD, McLennan C, lee J. (2015) Trends and patterns in sustainable tourism research: a 25-year bibliometric analysis. J Sustain Tour 23(4):517–535. https://doi.org/10.1080/09669582.2014.978790
4. Buckley R (2012) Sustainable tourism: research and reality. Ann Tour Res 39(2):528–546. https://doi.org/10.1016/j.annals.2012.02.003
5. Choi HSC, Sirakaya E (2006) Sustainability indicators for managing community tourism. Tour Manage 27(6):1274–1289. https://doi.org/10.1016/j.tourman.2005.05.018
6. Gössling S (2002) Global environmental consequences of tourism. Glob Environ Chang 12(4):283–302. https://doi.org/10.1016/S0959-3780(02)00044-4
7. Mihalič T (2000) Environmental management of a tourist destination: a factor of tourism competitiveness. Tour Manage 21(1):65–78. https://doi.org/10.1016/S0261-5177(99)00096-5
8. Saarinen J (2006) Traditions of sustainability in tourism studies. Ann Tour Res 33(4):1121–1140. https://doi.org/10.1016/j.annals.2006.06.007
9. Strandberg C, Nath A, Hemmatdar H, and Jahwash M (2018) Tourism research in the new millennium: A bibliometric review of literature in tourism and hospitality research. Tourism and Hospitality Res 18(3):269– 285. https://doi.org/10.1177/1467358416642010
10. UNWTO (2019) Tourism and the sustainable development goals-journey to 2030. https://doi.org/10.18111/9789284419401
11. Comerio N, Strozzi F (2019) Tourism and its economic impact: a literature review using bibliometric tools. Tour Econ 25(1):109–131. https://doi.org/10.1177/1354816618793762
12. Var T (1991) Sustainable tourism development. Ann Tour Res 18(2):327–329. https://doi.org/10.1016/0160-7383(91)90017-6
13. Hall CM (2011) Policy learning and policy failure in sustainable tourism governance: from first- and second-order to third-order change? J Sustain Tour 19(4–5):649–671. https://doi.org/10.1080/09669582.2011.555555
14. Niñerola A, Sánchez-Rebull MV, Hernández-Lara AB (2019) Tourism research on sustainability: a bibliometric analysis. Sustainability (Switzerland) 11(5):1–17. https://doi.org/10.3390/su11051377
15. Cocolas N, Walters G, Ruhanen L (2016) Behavioural adaptation to climate change among winter alpine tourists: an analysis of tourist motivations and leisure substitutability. J Sustain Tour 24(6):846–865. https://doi.org/10.1080/09669582.2015.1088860
16. Alfaro-Calderon GG, Godinez-Reyes NL, Gomez-Monge R, Alfaro-Garcia V, Gil-Lafuente AM (2019) Forgotten effects in the valuation of the social well-being index in Mexico'S sustainable development. Fuzzy Econ Rev 24(01):67–81. https://doi.org/10.25102/fer.2019.01.04
17. Merigó JM, Mas-Tur A, Roig-Tierno N, Ribeiro-Soriano D (2015) A bibliometric overview of the journal of business research between 1973 and 2014. J Bus Res 68(12):2645–2653. https://doi.org/10.1016/j.jbusres.2015.04.006
18. Aria M, Cuccurullo C (2017) bibliometrix: an R-tool for comprehensive science mapping analysis. J Informet 11(4):959–975. https://doi.org/10.1016/j.joi.2017.08.007
19. Lotka AJ (1926) The frequency distribution of scientific productivity. J Washington Acad Sci 16:317–323
20. Serrano L, Sianes A, Ariza-Montes A (2019) Using bibliometric methods to shed light on the concept of sustainable tourism. Sustainability (Switzerland) 11(24). https://doi.org/10.3390/SU11246964
21. Hirsch JE, Buela-Casal G (2014) The meaning of the h-index. Int J Clin Health Psychol 14(2):161–164. https://doi.org/10.1016/S1697-2600(14)70050-X
22. Egghe L (2006) Theory and practise of the g-index. Scientometrics 69(1):131–152. https://doi.org/10.1007/s11192-006-0144-7
23. Novak D, Batko M, Zezula P (2011) Metric index: an efficient and scalable solution for precise and approximate similarity search. Inf Syst 36(4):721–733. https://doi.org/10.1016/j.is.2010.10.002

Research Growth on Bioethanol: A Bibliometric Analysis

Adriana Paulina Aranzolo-Sánchez, Donaji Jiménez-Islas, and Miriam Edith Pérez-Romero

Abstract A bibliometric analysis of the research on bioethanol was carried out, using a comparative between the Web of Science and Scopus database from 1990 to 2019. The study includes the modeling of data using Logistic equation to describe quantitatively the growth of publication research with both databases. The results showed that Scopus database had a rate of publications of 0.31 years^{-1} with respect to WoS of 0.29 years^{-1}. The USA and China are countries with a high influence for research on bioethanol. This analysis showed that the logistic equation can be used to predict the evolution of publications on the field of bioethanol.

Keywords Bio-ethanol · Bioenergy · Biofuels · Scopus · WoS · Logistic equation

1 Introduction

At the moment, various environmental problems are addressed, such as pollution, in all its forms, deterioration of water quality, in its various phases, loss of biodiversity, caused by humans, and acceleration of damage to the planet and humanity itself; it has been suggested that the basis of environmental problems is a product of excessive consumption of resources, including energy resources [1]. Coupled with climate change, caused by the massive use of fossil energy sources, the aforementioned have had an impact on the human and natural condition [2]. Reducing anthropogenic CO_2 emissions is an environmental issue that has drawn attention due to the link between rising CO_2 levels in the atmosphere and increased global warming [3].

A. P. Aranzolo-Sánchez · D. Jiménez-Islas (✉)
División de Ingeniería en Energías Renovables, Instituto Tecnológico Superior de Huichapan, Huichapan, México
e-mail: djimenez@iteshu.edu.mx

M. E. Pérez-Romero
División de Ingeniería en Gestión Empresarial, Instituto Tecnológico Superior de Huichapan, Huichapan, México

E. León-Castro et al. (eds.), *Soft Computing and Fuzzy Methodologies in Innovation Management and Sustainability*, Lecture Notes in Networks and Systems 337,
https://doi.org/10.1007/978-3-030-96150-3_20

About 71% of global anthropogenic greenhouse gas emissions are caused by energy production and use [4], and dependence on fossil fuels has become an unworkable path due to its continued depletion in the world finite reserves [5, 6]. In order to mitigate the environmental problems that occur during the extraction, refining and combustion of energy resources, global efforts have been made to avoid the consequences of global warming by establishing international agreements that lead local policies to be adapted to the development of each signatory nation [7]. An example of this is the European Union, which has set a target of reaching 10% renewable energy in the transport sector by 2020 and 14% by 2030 [8], among renewable energies, we find wind, solar and biomass energies [9].

Biofuels are liquids, solids or gases derived from renewable biological sources [10], including biodiesel [11], biohydrogen [12], methanol [13], and bioethanol [14]. Bioethanol is expected to be the most widely used fuel in the world considering its, directly or mixed, use with fossil fuels, its rapid growth derives from the incorporation of pre-treating methods and technologies of raw materials that meet the demand, as long as they do not compete with materials used for human food [15].

Some countries have promoted the development of alternative energies including bioethanol [16], which brings with it significant advancement of scientific growth, Solano et al. [17] cites that there is a growing need to evaluate the processes of production and communication of the knowledge in the educational field, however, for the evaluation of scientific research is required to take into account the scientific work, institutions, countries and interrelationship of them through bibliometric indicators [18], among these, there are located documents ranked by type, language, years, journals, countries, institutions, index—H, authors, among others [19].

In recent years, bibliometric techniques have been frequently used in the literature, relate to the delivery of a complete general image of a research field [20]. The bibliometric analyses have been taken to various scientific research fields, where the area of renewable energies is no exception. Uzun [21] proposed a bibliometric analysis on priority patterns in renewable energy; also, Meriño et al. [22] conducted a bibliometric analytical of trends on biodiesel production from 2009 to 2016, finally, as well as, Azevedo et al. [23] reviewed the articles that have been published from 2005 to 2018 on the renewable energy value chain.

Many studies have been made to determine the increase of number of scientific publications in the biomass field, nevertheless no bibliographic analysis records were found on the subject of bioethanol, even when the bioethanol is considered as a gasoline substitute and the global market size is projected to 33.7 billons in 2020 to 64.8 billons by 2025. With the context of climate change and the knowledge of production of ethanol, it can be analysed to the currently state of the scientific research, also, to highlight hidden information such as publications trends, research areas of current interest and to identify scholar, countries and institutions with high productivity. This paper aims to analyse the evolution of scientific production of bioethanol through comparative literature from the Scopus and Web of Science (WOS) databases from 1990 to 2019.

The article consists of the following sections, materials and methods, section describing the activities, techniques, tools and formulas used to achieve the objective of research; followed by the results and discussion section, also, a section showing the results obtained and comparing these with other studies, finding similarities and differences, finally, it sets out the conclusions reached throughout this study.

2 Materials and Methods

The methodology followed in this bibliometric analysis consists of the following steps: identification of related bibliometric publications, delimitation of the study topic, selection of databases, establishment of the search algorithm, and selection of bibliometric indicators, obtaining and grouping of information in Microsoft Excel spreadsheets, and finally, analysis and discussion of information.

Step 1. Identification of related bibliometric publications. A search for articles and/or reviews was carried out on the basis of the keywords "bioethanol" and "bio-ethanol" and "Bibliometric analysis" in order to identify the existence of work on the subject or the like.

Step 2. Delimitation of the study theme. The previous step found that no bibliometric studies have been conducted on bioethanol-related publications, thus confirming the study topic for this biofuel that offers, inter alia, a solution to climate change.

Step 3. Selection of databases. WOS and Scopus databases were selected because both are considered for their quality as the most influential databases to classify academic research and index journals. Therefore, it is apparent that the information contained in these bases is the most representative and important in the topic of analysis.

Step 4. Establishing the search algorithm. Following Zhang et al. [19] bibliometric methodology, the search algorithm for both databases was defined, remaining as "bioethanol" OR "bio-ethanol", excluding the year 2020, document type "article" AND "review", search by "topic" with search date December 2019.

Step 5. Selection of bibliometric indicators. The two basic categories of bibliometric analysis according to Callon et al. [24] were addressed: Activity indicators and relationship indicators, the former provide data on the volume and impact of research activities, while the latter do so on links and interactions between researchers, fields of knowledge, etc. The selected publications were further analysed with respect to the type of document, publication language, number of publications per year, main journals, main countries, main institutions, Index -H, main authors, number of citations, mainly.

Step 6. Obtaining and grouping information in Microsoft Excel spreadsheets. The publications obtained from the indicated algorithm, were exported to text files and processed in Microsoft Excel 2016 to generate the corresponding tables and graphs.

Step 7. Analysis of scientific production. The selected publications were refined to the top 10 countries, authors, universities, journals and research areas, then their analysis and discussion was carried out.

Models were established to describe scientific production data, using the number of publications per year. The logistical (Eq. 1) and exponential equations were used to fit the data and the rate was found using the Excel solver function [25].

$$P(t) = \frac{P_{max}}{1 + \left(\frac{P_{max-}}{P_0} - 1\right)e^{-rt}} \tag{1}$$

$$(P_{SCOPUS} - P_{MODELO})^2 \tag{2}$$

3 Results and Discussion

This section presents the main bibliometric results found in WOS and Scopus for the search algorithm indicated above. For WOS, the first article dates from 1990, from that year to 2019 11,939 documents have been published between articles and reviews. As for Scopus, 9,545 documents have been registered in the same period.

3.1 Type of Document and Language

The top 10 countries of the scientific literature have been analysed with respect to the type of document and the most used language in the field. This compares by percentage to the overall production index which, in the search carried out on the 10th and 11th of December 2019, it was 9,545 documents for Scopus, consisting of 8,526 articles and 1,019 reviews. A concentration of 69% of the total publications in the world in this period was issued in the Top 10 countries. The top 10 countries and their scientific production are shown in Table 1.

For WOS, the number of publications was 11,939 in the period 1990–2019, 10,791 are articles and 1,148 reviews. About 72% of these publications are issued by the Top 10 countries producing scientific literature on bioethanol.

Both databases have the top five of the countries with the highest productivity in scientific publications in the field of bioethanol, with China and the United States topping the ranking. Yu and Meng [26] reviewed in the top three of the United States, China and Germany in a study of bibliometric indicators under the topic of searching for biomass, or biofuel, or biodiesel, or bioethanol, or biomass power, or biogas. Although the search was broader, the countries that dominate the field of energy through biomass continue to coincide with the present work, with the exception of Germany that has reduced its productivity. The reduction in the number of annual

Table 1 Top countries and document type

Ranking	Scopus				WOS			
	Country	Article	Review	Total	Country	Article	Review	Total
1	*China*	1098	113	1211	*China*	1612	118	1730
2	*United States*	1001	127	1128	*USA*	1389	159	1548
3	*India*	707	155	862	*Brazil*	922	94	1016
4	*Brazil*	636	73	709	*India*	818	176	994
5	*Japan*	556	44	600	*Japan*	676	48	724
6	*South Korea*	444	35	479	*Spain*	619	56	675
7	*Spain*	411	42	453	*South Korea*	567	41	608
8	*United Kingdom*	364	53	417	*Italy*	439	33	472
9	*Italy*	356	27	383	*England*	383	47	430
10	*Germany*	268	40	308	*Canada*	334	43	377
	Total Top 10	5841	709	6550	*Total Top 10*	7759	815	8574
	Global Total	8526	1019	9545	*Global Total*	10,791	1,148	11,939
	Global Average	69%	70%	69%	*Global Average*	72%	71%	72%

Source Own elaboration

publications may be associated with the fact that the production of liquid biofuels (including bioethanol) has grown very slowly and depends heavily on standards and regulations, which vary greatly from region to region [26].

In the Scopus database, the predominant type of documents is Articles, with a percentage of 89.18% and 10.82% of Reviews; the percentage found by WoS resembles the percentage of SCOPUS with 90.49% of articles and 9.51% of Reviews, Fig. 1a, b. Different result from publications in the biomass area for renewable energy production, which have reported 68.4% for Articles [28].

As for the language of publication registered in Scopus, there is a great diversity of languages in which they are published, the main languages are: English (95.02%) and Mandarin Chinese (2.03%), lower percentages are German (0.94%), Japanese (0.84%), Korean (0.61%), Spanish (0.26%), Portuguese (0.21%), Italian (0.06%), Turkish and French (both with 0.02%). These percentages do not directly demonstrate that publications are made in countries where the language of publication is the official language, but it does demonstrate that authors prefer to disseminate their research in English so that it has a greater scope or that it can be included in international journals [29].

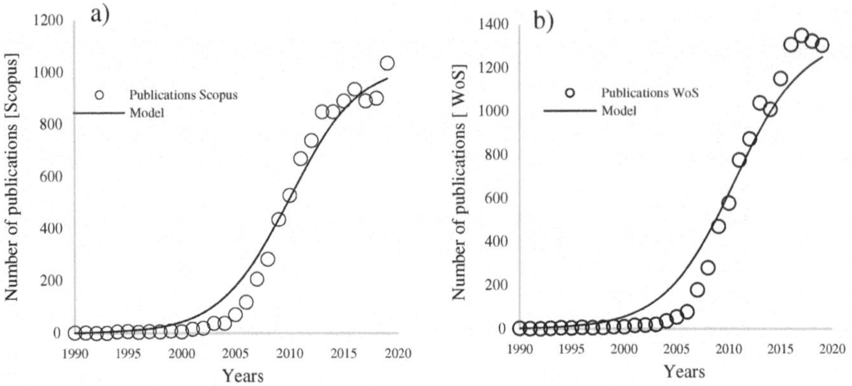

Fig. 1 Growth of total scientific production on bioethanol research. Scopus (1a) and WoS (1b)

The data obtained in WoS (99.02%) of the documents published from 1990 to 2019 is in English. This may be because WoS coverage is a database that pays special attention to journals that publish full texts in English. Yaoyang and Boeing [30] found in a mapping in the field of biofuels that 96.9% of scientific production was published in English. For its part, Perea-Moreno et al. [28] conducted an analysis of the trends of biomass research for renewable energy from 1978 to 2018, finding that 95.3% of the publications were written in English, followed by Chinese, German and Spanish. In addition, Valencia et al. [27] through an analysis of trends in liquid biofuel research from 2010 to 2018 reported that 98.8% of articles were written in English and the rest in Chinese, Polish, Serbian-Croatian, Portuguese and Spanish.

3.2 *Date of Publication*

The number of publications per year in each country in the top 10 was concentrated, and then the figures were compared with the total world production for the period 1990–2019 (Figs. 1 and 2).

From 1990 to 2005, the number of publications per year, both in the top 10 and in the world, did not have significant growth; however, after 2005 and until 2017 there was exponential growth in the number of publications per year in the top 10 and in the world's publications on the topic of bioethanol. As of 2013 in the United States, the country with the highest number of articles and reviews of the time, ranked a place below China, placing the latter as the first bioethanol information producer in the world. Although the change in the speed of publication in the United States began in 2013, the country experienced a marked reduction in its production in 2017. This behavior may be caused by budget changes in the administration of country [31], the budget in the United States fell from $22.6 million in 2016 to 2017 in the renewable energy research sector and continued to decline in later years.

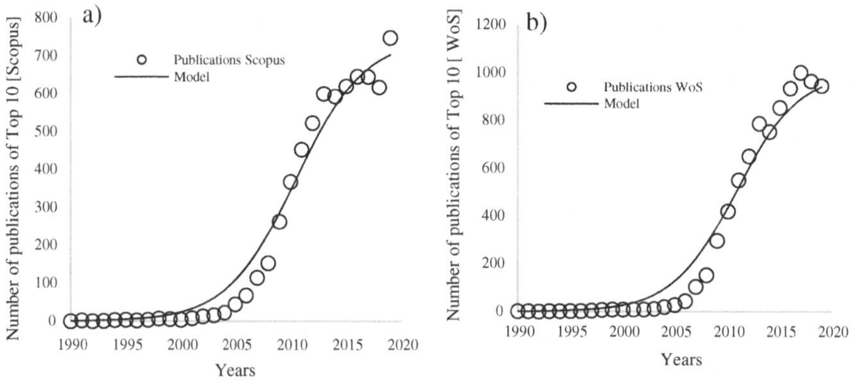

Fig. 2 Growth of total scientific production on bioethanol research, top 10 of countries, Scopus (2a) and WoS (2b)

In the quest to establish a comparative analysis between the WoS and Scopus databases, it was proposed through the logistical equation to estimate the speed at which articles and review are published in the field of bioethanol, to do so the experimental data were adjusted through the prediction of the model, Figs. 1 and 2.

The results of the logistic model tuning are shown in Table 2, where the Scopus database is observed to have higher speed compared to WoS; having the same behavior for the top 10 most influential countries in publications of the selected topic. One of the advantages of establishing comparisons through models and statistical analysis is that it not only considers the number in a specific year, it actually considers global evolution over the years, which allows to make a projection of what publications will look like in future years in the topic of bioethanol.

For Top 10 publications, it is important to note that the most influential countries in bioethanol publications maintain a higher speed compared to the global of the rest of the countries, so that the models proposed in this work could determine the projection of the number of scientific publications with a higher level of significance. In this context, there is model reports that are used to predict the behavior of scientific production. For example, the one reported by López-Muñoz et al. [32], where they apply in themes of bipolar disorder a linear adjustment: $y = 16.872x - 48.52$ with $R^2 = 0.7367$ and an exponential adjustment $y = 37.929e^{0.0942x}$ with $R^2 = 0.9019$. This

Table 2 Specific growth rate of scientific publications on bioethanol with WoS y Scopus

Publication/database	Specific growth rate ($year^{-1}$)	Coefficient of determination (R^2)
Total of publications Wos	0.298	0.9785
Total of publications Scopus	0.3123	0.9838
Top 10 Wos	0.3227	0.9817
Top 10 Scopus	0.3232	0.9807

Source Own elaboration

allows to compare the adjustment of the data through the models, this work (Table 2) has higher determination coefficient adjustments through the use of the proposed logistical model to determine the speed of publications in both databases.

3.3 Universities

Refining the 9545 results with the "Affiliation" indicator, the top ten institutions with bioethanol publications were selected (Table 3). The necessary information for each institution was the country of origin, the total number of documents published, the number of appointments it has accumulated, the H-index, the average number of citations per document and the category of appointments. In both databases, the universities of the countries that publish the most articles correspond to China (2), Brazil (2), and Denmark (1); the rest is not found in the same databases, this is the case of the United States (2), India (2), France (1), Japan (2), Sweden (1), and South Korea (1).

TC is influenced by the total number of documents (TP), the average number of citations per item (TC/TP) describes the average impact of each item. Therefore, the TC/TP index may better reflect the average influence of each item [26]. The TC/TP average of WoS and Scopus is 33.91 and 31.38 with the University Of Denmark, the Department Of Energy Doe, the Industrial Research CSIR (USA) and the Department Of Agriculture USDA; institutions that have a TC/TP > 40 in WoS for bioethanol issues in the Scopus database. The universities of Lunds and Danmarks Tekniske top the ranking in bioethanol publications. In addition, for both databases, Denmark is the country that has two universities with more than 100 quotations (Tables 3 and 4).

The number of products is related to the number of research institutions, the availability of research funds and the proportion of those focused on energy security; In addition, some countries, such as England, the United States and China, are very focused on energy security and cooperation with other institutions [33]. In the global context of biofuels, it has been described that the USA has a central role in international collaborations with the most productive countries, where the ranking is headed by the USA, Sweden, China and Brazil, where the leading institutions are the Academy of Sciences of China, the University of California, and Berkley [30].

3.4 Authors

The top 10 authors were listed by the number of publications establishing them as the main authors, Table 5. The indicators for each author were: university and country of origin, total publications, total subpoena, citation rate and H-index. The results include the presence of the Japanese university "Kobe University", with two authors from that university. It is also noteworthy, that universities from leading bioethanol countries, such as China, the United States and Brazil, are not included in this list,

Table 3 The top 10 most productive institutions in bioethanol research in WoS

Ranking	Country	University	TS	TC	H-index	Average citations per item (TC/TS)	>=100	>=75	>=50	>=25	<25	TOTAL
1	China	Chinese Academy Of Sciences	274	6408	44	23.4	9	5	25	43	192	274
2	Brazil	Universidade De Sao Paulo	237	5751	37	24.3	14	3	9	43	168	237
3	USA	United States Department Of Energy Doe	213	9784	54	45.9	27	9	24	40	113	213
4	India	Indian Institute Of Technology System IIt System	191	5270	34	27.6	6	5	9	23	148	191
5	India	Council Of Scientific Industrial Research CSIR India	185	8275	45	44.7	19	6	14	33	113	185
6	Brazil	Universidade Estadual De Campinas	176	4305	37	24.5	11	5	9	32	119	176
7	Denmark	Technical University Of Denmark	175	8549	50	48.9	21	5	24	29	96	175
8	USA	United States Department Of Agriculture USDA	152	6341	39	41.7	17	7	11	25	92	152
9	France	Centre National De La Recherche Scientifique CNRS	151	5183	38	34.3	13	4	11	32	91	151
10	China	Tianjin University	129	3073	33	23.8	5	5	11	23	85	129

Source Own elaboration

Table 4 The top 10 most productive institutions in bioethanol research in Scopus

Ranking	Country	University	TS	TC	H-index	Average citations per item (TC/TS)	>=100	>=75	>=50	>=25	<25	TOTAL
1	China	Chinese Academy of Sciences	179	4313	33	24.1	8	6	7	26	132	179
2	Brazil	Universidade de Sao Paulo	170	4538	35	26.7	11	3	5	33	118	170
3	Brazil	Universidade Estadual de Campinas	144	3458	31	24.0	9	4	6	19	106	144
4	Denmark	Danmarks Tekniske Universitet	138	6959	44	50.4	19	3	19	23	74	138
5	China	Ministry of Education China	135	1710	22	12.7	1	2	4	13	115	135
6	Sweden	Lunds Universitet	84	6244	39	74.3	19	6	9	20	30	84
7	Japan	National Institute of Advanced Industrial Science and Technology	78	1939	22	24.9	3	4	5	10	56	78
8	South Korea	Chonnam National University	72	1526	21	21.2	1	2	5	12	52	72
9	China	University of Chinese Academy of Sciences	70	1257	19	18.0	2	2	1	11	54	70
10	Japan	Kobe University	69	2590	28	37.5	5	6	11	8	39	69

Source Own elaboration

Table 5 The most influential authors in bioethanol research in WoS

Ranking	Author(s)	University	Country	TS	TC	TC/TS	H-Index
1	Kondo, A	Kobe University	Japan	73	2,529	34.64	27
2	Karimi, K	Isfahan University of Technology	Iran	60	1,737	28.95	19
3	Ballesteros, M	Research Centre for Energy, Environment and Technology	Spain	54	4,137	76.61	27
4	Pandey, A	Indian Institute of Toxicological Research	India	52	3,917	75.33	27
5	Taherzadeh, M.J	University of Borås	Sweden	48	1,768	36.83	20
6	Sun, RC	Dalian Polytechnic University	China	47	1,483	31.55	20
7	Liu, Y	University of Chinese Academy of Sciences	China	45	777	17.27	18
8	Li, Y	Qilu University of Technology	China	44	1,173	26.66	18
9	Zacchi, G	Lund University	Sweden	44	4,602	104.59	31
10	Zhang, Y	Jilin University	China	42	413	9.83	11

Source Own elaboration. *Note* The university of affiliation of the researcher is obtained by consulting his most recent article published on the subject under investigation. The TS/TC/H-index data correspond to the author´s studies on the subject.

while universities from countries not in the top 10 are listed (Sweden, Malaysia and Portugal).

The author Kondo, ranked first with 63 publications as the first author, it is also the author with a higher H-index (27), although Zacchi, G. of Sweden has a higher publication rate than either in the list of both databases with TC/TS of 104.59 for WoS and 125.71 in Scopus, Tables 5 and 6. Four of the most cited authors in bioethanol issues in the WoS database belong to universities in China, two to Universities of Sweden and the rest of Japan, Iran and Spain.

Note: The university of affiliation of the researcher is obtained by consulting his most recent article published on the subject under investigation. The TS/TC/H-index data correspond to the author´s studies on the subject.

The Scopus database identifies, among the most influential authors besides Zacchi, Ballesteros of the University of the Research Centre for Energy, Environment and Technology, with a TC/TS of 107.85, followed by Pandey of the Indian Institute of Toxicological Research with TC/TS of 84.04 in the field of bioethanol.

The diversity of universities indicates the global knowledge of the issue and the need of countries to promote ethanol as biofuel; Zhang and Meng [34] suggest that research in the field of biofuels has been disseminated globally in many research centres and university institutions.

Table 6 The most influential authors in bioethanol research in Scopus

Ranking	Author(s)	University	Country	TS	TC	TC/TS	H-Index
1	Kondo, A	Kobe University	Japan	63	2492	39.56	27
2	Pandey, A	Indian Institute of Toxicological Research	India	48	3986	83.04	26
3	Taherzadeh, M.J	University of Borås	Sweden	40	1487	37.18	17
4	Bae, H.J	Chonnam University	South Korea	33	840	25.45	17
5	Ballesteros, M	Research Centre for Energy, Environment and Technology	Spain	33	3559	107.85	18
6	Lee, K.T	Universiti Sains Malaysia	Malaysia	33	1754	53.15	20
7	Hasunuma, T	Kobe University	Japan	32	1776	55.50	20
8	Sukumaran, R.K	National Institute for Interdisciplinary Science and Technology	India	31	2424	78.19	20
9	Zacchi, G	Lund University	Sweden	31	3897	125.71	26
10	Domingues, L	University of Minho	Portugal	30	1204	40.13	17

Source Own elaboration

3.5 *Journals*

The published magazines were purged in the top 5; the indicators were total publications, total subpoena, citation rate and H-index. Bioresource Technology tops the ranking of magazines with H index of 85 for both databases, followed by Biotechnology for Biofuels and Biomass and Bioenergy, see Tables 7 and 8. In addition,

Table 7 The top 5 most productive journals in bioethanol to biofuels (WoS)

Ranking	Source Titles	TS	TC	TC/TS	H-Index
1	*Bioresourse Technology*	1060	42,281	39.88	85
2	*Biotechnology for Biofuels*	320	9324	29.13	48
3	*Biomass and Bioenergy*	285	10,804	37.90	51
4	*Industrial Crops and Products*	243	4557	18.75	37
5	*Renewable Sustainable Energy Reviews*	238	12,160	51.09	60

Source Own elaboration

Table 8 The top 10 most productive journals in bioethanol to biofuels (Scopus)

Ranking	Source Titles	TS	TC	TC/TS	H-Index
1	*Bioresource Technology*	712	33,467	47.00	85
2	*Biotechnology for Biofuels*	257	7533	29.31	46
3	*Biomass and Bioenergy*	197	9632	48.89	48
4	*Renewable and Sustainable Energy Reviews*	173	9476	54.77	56
5	*Industrial Crops and Products*	155	3343	21.56	34

Source Own elaboration

in the Scopus database, the subpoena rate (TC/TS) is greater than 50 for Applied Energy, Renewable, and Sustainable Energy Reviews; for the WoS database, the subpoena that rated greater than 39 was for Renewable and Sustainable Energy Reviews and Bioresource Technology. It is noteworthy that the setbacks that publish on bioethanol are mainly from the fields of biotechnology, energy and industrial processes. The identification of journals and the area of research is essential for students and professionals when conducting a literature review and helps readers to get better information [35].

The most frequently cited articles in most cases proved to be the pioneers of research and marked a milestone in the history of research and from these follow-up articles were created, thus reinforcing the importance in pioneering the research area [36]. Valencia et al. [27] through an analysis of liquid biofuels report to Renewable and Sustainable Energy Reviews in the ranking 1 of the period 2010 to 2018, in this work is ranked 4 and 5 for Scopus and WoS.

3.6 Citation by Countries

For country citations, the elected indicators were the number by rank of citations (>−100, > −75, >−50, >−25, and <25), total studies and citations, Tables 9 and 10. The United States tops the ranking of publications >100 citations, followed by China in both databases. China stands out for the number of published articles; however, the TC/TS ratio is higher for the USA because its articles are referent in the field and have a higher number of citations. For its part, Brazil is the only Latin American country in the top 5 with a TC/TS of 19.41 and 19.50 for WoS and Scopus, respectively.

In both databases, the USA stands out in the bibliometric indicators, the reason why is found in the works cited by Tahamtan et al. [37], where he describes that authors who are affiliated to certain countries get more or less citations for having the privilege and scientific background, additionally to the financial support to conduct the research and therefore publish articles with higher quality. In this case, the US institutions are the ones that receive the most citations from other countries [37].

Table 9 Citation of the top 10 most productive countries in bioethanol research in WoS

Ranking WoS	Country	>=100	>=75	>=50	>=25	<25	Total studies	Total citation	TC/TS
1	USA	127	52	104	251	1014	1548	55,897	36.11
2	China	42	27	84	226	1351	1730	31,999	18.50
3	India	39	23	45	104	783	994	23,266	23.41
4	Spain	29	22	56	111	457	675	19,862	29.43
5	Brazil	31	27	48	129	781	1016	19,721	19.41
6	Japan	18	13	33	95	565	724	13,584	18.76
7	England	21	13	23	81	290	428	11,744	27.44
8	South Korea	7	8	24	100	467	606	10,808	17.83
9	Italy	15	19	13	87	336	470	10,795	22.97
10	Canada	12	9	15	57	282	375	8,373	22.33

Source Own elaboration

Table 10 Citation of the top 10 most productive countries in bioethanol research in Scopus

Ranking scopus	Country	>=100	>=75	>=50	>=25	<25	Total studies	Total citation	TC/TS
1	United States	103	39	74	183	729	1128	43,853	38.88
2	China	34	17	45	134	981	1211	20,449	16.89
3	India	37	17	37	90	681	862	19,943	23.14
4	Spain	24	16	31	76	306	453	14,136	31.21
5	Brazil	27	13	28	86	555	709	13,889	19.59
6	United Kingdom	26	14	21	67	289	417	13,511	32.40
7	Japan	17	16	30	77	460	600	11,792	19.65
8	Italy	18	13	13	62	277	383	9,072	23.69
9	South Korea	6	10	25	76	362	479	8,549	17.85
10	Germany	15	6	10	35	242	308	7,037	22.85

Source Own elaboration

The article "Pretreatment technologies for an efficient bioethanol production process based on enzymatic hydrolysis: A review" by Alvira et al. [38] is positioned as the most cited article in the field, with a total of quotations of 2208 to the date that it was searched, the article covers the subject of the pretreating of lignocellulosic to improve accessibility to enzymes improving the digestibility of cellulose [38]. Valencia et al. [27] found, through a bibliometric analysis of trends in liquid biofuels from 2010–2018, that there is a coincidence of the article by Alvira et al.

[38] as the most influential in the field of liquid biofuels. The first published articles are from 1990, however, the articles with the highest number of citations are from 2004 to 2011 in both databases, which coincides with the exponential stage found in Figs. 1 and 2. This can be attributed to the increase in the development of renewable energy, which is of particular relevance to global warming, energy security and laws promoting clean energy [39, 40].

4 Conclusion

The present work offers the first bibliometric analysis of scientific publications in the field of bioethanol. In addition, a way of measuring the rate with which it is published in the field of bioethanol is established in this work. A logistical model was applied to estimate the specific growth rate of scientific publications, where Scopus database had a high rate of publications in the field of bioethanol, and the future behavior of field publications can be projected using the model. The USA and China are the largest contributors to the biofuels literature. Also, the journal Bioresource Technology is the most published in the field of bioethanol where the most dominant areas of research are energy and engineering. In future research we will carry out the analysis of keywords and cooperation between institutions through the VOSviewer software to determine the leading research groups in the field of bioethanol.

Acknowledgements We thank the Michoacan University of San Nicolas of Hidalgo for the data base access.

References

1. Arias BN (2016) El consumo responsable: educar para la sostenibilidad ambiental. Aibi Revista de Investigación, Administración e Ingeniería 4(1):29–34. (*spanish*). https://doi.org/10.15649/2346030x.385
2. Kahia M, Ben Jebli M, Belloumi M (2019) Analysis of the impact of renewable energy consumption and economic growth on carbon dioxide emissions in 12 MENA countries. Clean Techn Environ Policy 21(4):871–885. https://doi.org/10.1007/s10098-019-01676-2
3. Mendiara T, García-Labiano F, Abad A, Gayán P, de Diego LF, Izquierdo MT, Adánez J (2018) Negative CO_2 emissions through the use of biofuels in chemical looping technology: a review. Appl Energy 232:657–684. https://doi.org/10.1016/j.apenergy.2018.09.201
4. Mahbub N, Gemechu E, Zhang H, Kumar A (2019) The life cycle greenhouse gas emission benefits from alternative uses of biofuel coproducts. Sustain Energy Techn Assess 34:173–186. https://doi.org/10.1016/j.seta.2019.05.001
5. Ahmad Dar R, Ahmad Dar E, Kaur A, Gupta Phutela U (2018) Sweet sorghum-a promising alternative feedstock for biofuel production. Renew Sust Energ Rev 82:4070–4090. https://doi.org/10.1016/j.rser.2017.10.066
6. Miller RG, Sorrell SR (2014) The future of oil supply. Philos Trans A Math Phys Eng Sci 2014(372):20130179. https://doi.org/10.1098/rsta.2013.0179

7. Martins F, Felgueiras C, Smitkova M, Caetano N (2019) Analysis of fossil fuel energy consumption and environmental impacts in European countries. Energies 12(6):964. https://doi.org/10.3390/en12060964
8. Jåstad EO, Bolkesjø TF, Rørstad PK (2020) Modelling effects of policies for increased production of forest-based liquid biofuel in the Nordic countries. Forest Policy Econ 113:102091. https://doi.org/10.1016/j.forpol.2020.102091
9. Jiménez ID, Abreu CA, López y López VE, Téllez JA, Gracida RJN (2012) Fermentable sugar production through acid hydrolysis of *Beta vulgaris* L. Rev Int Cont Amb 28(2):151–158
10. Falcone PM, Lopolito A, Sica E (2017) Policy mixes towards sustainability transition in the Italian biofuel sector: dealing with alternative crisis scenarios. Ener Res Socl Sci 33:105–114. https://doi.org/10.1016/j.erss.2017.09.007
11. Reddy ER, Sharma M, Chaudhary JP, Bosamiya H, Meena R (2014) One-pot synthesis of biodiesel from high fatty acid *Jatropha curcas* oil using bio-based basic ionic liquid as a catalist. Curr Sci 106(10):1394–13400
12. Stamenković OS, Siliveru K, Veljković VB, Banković-Ilić IB, Tasić MB, Ciampitti IA, Dalovic IG, Mitrović PM, Sikova VS, Prasad PVV (2020) Production of biofuels from sorghum. Renew Sust Energ Rev 124:109769. https://doi.org/10.1016/j.rser.2020.109769
13. Eom T, Chaiprapat S, Charnnok B (2019) Enhanced enzymatic hydrolysis and methane production from rubber wood waste using steam explosion. J Environ Manag 235:231–239. https://doi.org/10.1016/j.jenvman.2019.01.041
14. Ebrahimian F, Karimi K (2020) Efficient biohydrogen and advanced biofuel coproduction from municipal solid waste through a clean process. Bioresour Technol 300:122656. https://doi.org/10.1016/j.biortech.2019.122656
15. Yu KL, Chen W-H, Sheen H-K, Chang J-S, Lin C-S, Ong HC, Show PL, Ling TC (2020) Bioethanol production from acid pretreated microalgal hydrolysate using microwave-assisted heating wet torrefaction. Fuel 279:118435. https://doi.org/10.1016/j.fuel.2020.118435
16. Chuck-Hernández C, Pérez-Carrillo E, Heredia-Olea E, Serna-Saldívar SO (2011) Sorghum as a multifunctional crop for bioethanol production in Mexico: technologies, advances and improvement opportunities. Rev Mex Ing Quim 13(3):529–549
17. Solano LE, Castellanos QS, López RRM, Hernández FJ (2009) La bibliometría: una herramienta eficaz para evaluar la actividad científica postgraduada. MediSur **7**(4):59–62
18. Belter CW (2015) Bibliometric indicators: opportunities and limits. J Med Libr Assoc: JMLA 103(4):219–221. https://doi.org/10.3163/1536-5050.103.4.014
19. Zhang M, Gao Z, Zheng T, Ma Y, Wang Q, Gao M, Sun X (2016) A bibliometric analysis of biodiesel research during 1991–2015. J Mater Cycles Waste Manag 20(1):10–18. https://doi.org/10.1007/s10163-016-0575-z
20. Valenzuela-Fernández L, Merigó JM, Nicolas C (2018) The most influential countries in market orientation: a bibliometric analysis between 1990 and 2016. Int J Eng Bus Manag 10:1–9. https://doi.org/10.1177/1847979017751484
21. Uzun A (2002) National patterns of research output and priorities in renewable energy. Energy Policy 30(2):131–136. https://doi.org/10.1016/s0301-4215(01)00084-2
22. Meriño Stand LI, Valencia GE, Benavides AE (2018) Trend analysis of research results in biodiesel production from 2009 to 2016. Prospectiva 16(2):89–93. https://doi.org/10.15665/rp.v16i2.1656
23. Azevedo SG, Santos M, Antón JR (2019) Supply chain of renewable energy: a bibliometric review approach. Biomass Bioenerg 126:70–83. https://doi.org/10.1016/j.biombioe.2019.04.022
24. Callón M, Curtial JP, Penan H (1995) Cienciometría. Ediciones Trea, Gijón España
25. Jiménez-Islas D, Páez-Lerma JB, Soto-Cruz ON, Gracida-Rodríguez JN (2014) Modelling of ethanol production from reed beet juice by *Saccharomyces cerevisiae* under thermal and acid stress conditions. Food Tech Biotech 52:93–100
26. Yu D, Meng S (2018) An overview of biomass energy research with bibliometric indicators. Energy Environ 29:576–590. https://doi.org/10.1177/0958305x18756304

27. Valencia GE, Duarte JE, Obregón LG (2018) Trend analysis of liquid biofuels research during the years 2010 to 2018. Int J Appl Eng Res 13:1104–1111
28. Perea-Moreno M-A, Samerón-Manzano E, Perea-Moreno A-J (2019) Biomass as renewable energy: worldwide research trends. Sustainability 11:863. https://doi.org/10.3390/su11030863
29. Niño-Puello M (2013) El inglés y su importancia en la investigación científica: algunas reflexiones. RECIA 5(1):243–254. https://doi.org/10.24188/recia.v5.n1.2013.487
30. Yaoyang X, Boeing WJ (2013) Mapping biofuel field: a bibliometric evaluation of research output. Renew Sust Energ Rev. 28:82–91. https://doi.org/10.1016/j.rser.2013.07.027
31. Gallagher KS, Anadon LD (2018) DOE Budget Authority for Energy Research, Development, and Demonstration Database. Fletcher School of Law and Diplomacy, Tufts University; Department of Land Economy, University of Cambridge; and Belfer Center for Science and International Affairs, Harvard Kennedy School
32. López-Muñoz F, Vieta E, Rubio G, García-García P, Alamo C (2006) Bipolar disorder as an emerging pathology in the scientific literature: a bibliometric approach. J Affect Disord 92(2–3):161–170. https://doi.org/10.1016/j.jad.2006.02.006
33. Zhou W, Kou A, Chen J, Ding BA (2018) Retrospective analysis with bibliometric of energy security in 2000–2017. Energy Rep 4:724–732. https://doi.org/10.1016/j.egyr.2018.10.012
34. Zhang M, Meng H (2019) The scientometric evaluation on the research of biofuels based on citespace. Adv Eng Res 184:6–11
35. Cheng Z, Kong W, Cheng Z, Qi H, Yang S, Zhang A, Niu S (2020) A bibliometric-based analysis of the high-value application of Chlorella. 3 Biotech 10(3). https://doi.org/10.1007/s13205-020-2102-0
36. Schwartz FW, Fang YC, Parthasarathy S (2005) Patterns of evolution of research strands in the hydrologic sciences. Hydrogeol J 13(1):25–36. https://doi.org/10.1007/s10040-004-0423-4
37. Tahamtan I, Safipour Afshar A, Ahamdzadeh K (2016) Factors affecting number of citations: a comprehensive review of the literature. Scientometrics 107(3):1195–1225. https://doi.org/10.1007/s11192-016-1889-2
38. Alvira P, Tomás-Pejó E, Ballesteros M, Negro MJ (2010) Pretreatment technologies for an efficient bioethanol production process based on enzymatic hydrolysis: a review. Bioresour Technol 101(13):4851–4861. https://doi.org/10.1016/j.biortech.2009.11.093
39. Coelho MS, Barbosa FG, da Rosa Andrade Zimmermann de Souza M (2019) A bibliometric analysis of top-cited papers in the biogas field. Environ Earth Sci 78:318. https://doi.org/10.1007/s12665-019-8303-3
40. Neij L, Heiskanen E, Strupeit L (2017) The deployment of new energy technologies and the need for local learning. Energy Policy 101:274–283

The Interplay of Management Information Systems in Industry 4.0: A Bibliometric Review

Jorge Lerma Beltrán and Eleazar Gónzalez Álvarez

Abstract In this study, a review of the interaction literature of Management Information Systems (MIS) in Industry 4.0 is carried out. It is done through a bibliometric analysis of citation, keywords and authorship of the relevant literature contained in the Science Direct database. Tools such as Microsoft Excel, Mendeley and VOSviewer are used, applied to a content review of 55 documents resulting from selective search. The results corroborate a high degree of interaction MIS and contribution in the implementation in industry 4.0 in companies; in addition, the close relationship with information and communication technology projects, innovation and performance improvement within organizations.

JEL Codes M1 · M15 · O33

Keywords Management Information Systems (MIS) · Industry 4.0 · Innovation · Performance

1 Introduction

The fast and Big technological advances in industrialization methods and use of Information and Communication Technologies (ICT), have caused tremendous progress in development of new generation of manufacturing technology [58]. Since 2011, when Industry 4.0 concept was introduced, at the Hannover fair, Germany, the Fourth Industrial Revolution is being experienced, placing the Germans in the role of pioneers of industrial sector, by announcing their German Strategic Initiative in 2013.

The fourth industrial revolution, known as Industry 4.0, and underlying digital transformation, is a cutting- edge research topic across various disciplines. Industry 4.0 literature is growing exponentially, overexpanding current understanding of digital industrial revolution through thousands of academic publications [23]. Many

J. L. Beltrán (✉) · E. G. Álvarez
Universidad Autónoma de Sinaloa, Facultad de Contaduría Y Administración, Ciudad Universitaria Edificio de Posgrado, Cubículo 7, Colonia Villa Universidad, 80010 Culiacán, Sinaloa, México

E. León-Castro et al. (eds.), *Soft Computing and Fuzzy Methodologies in Innovation Management and Sustainability*, Lecture Notes in Networks and Systems 337,
https://doi.org/10.1007/978-3-030-96150-3_21

countries are adopting programs and strategies that seek to improve development and adoption of Industry 4.0 technologies, starting in Germany; followed by countries such as USA that establishes the Advanced Manufacturing Association, in China, Made in China 2025 and in France: *La Nouvelle France Industrielle* and many others countries [10].

Reviewing the literature, it´s observed that the Internet of things (IoT) and industrial systems are closely related to each other, which means that evolution of MIS has caused development of new technologies, starting from systems such as: Computer Numerical Control (CNC), Computer-Aided Design (CAD), Computer-Aided Processing Planning (CAPP), evolving and developing new systems implemented in Industry 4.0, such as: Cyber-Physical Systems (CPS), IoT, Cloud Computing, Blockchain, Industrial Information Integration, Artificial Intelligence (AI), to the creation of smart cities and development of technological innovation projects in different disciplines and fields [4, 24, 27, 32, 44, 45,47, 51, 57, 58, 39, 49].

The literature review allows the analysis of scientific articles, units, journals and institutions which disseminate and publish them, parameters, number of citations between articles, authors who generate them and various related subjects, among others that are no less important [19]. Likewise, it contributes to analysis of scientific works that reveal their purposes, advantages, and/or limitations, allows enriching production of new knowledge, finding networks or clusters, identifying strengths, weaknesses and opportunities within scientific dissemination [15].

Research by authors such as López et al., [36], [21, 32, 40, 51–53] among others, suggest importance of literature review studies and their many contributions to scientific heritage. From start to finish, this literary review follows this structure: Sect. 2 introduces theoretical framework, background, evolution, impact of MIS and its relationship with Industry 4.0. Section 3 talks about the method and tools to carry out the literature review. Section 4 shows information and results of the literary review study. Finally, in Sect. 5 conclusions of study are given and references are shown.

2 Theoretical Framework

2.1 MIS Background

MIS origins go back to emergence of digital age in the 1970s, thanks to computers invention; It was then that people conceived of its use in organizations [13]. Later, in the 80's, research carried out made it possible to unite electronics, computing and telecommunications, making possible an interconnection between networks, moving from Electronic Data Processing era to Information Systems Engineering era. [13, 59]. ICT can be described, in general terms, as a set of elements made up of tools, practices and techniques used for treatment, processing, storage and transmission of data; in order to structure them into useful information that leads to problem solving and knowledge generation [46].

MIS are described by different authors as: a set of organized processes that, when carried out, provide important information in decision-making and control of an organization, where information can be tangible or intangible, that is, physical or virtual [30, 33, 51]. In the same way, interest in ICT investments in organizations has increased considerably, which translates into a significant change in the way business is done, both inside and outside organizations [41], p. 13).

As well known, new technologies and innovation go hand in hand with changes and progress in the world environment. The interest and importance that these future technologies should be given is fundamental, since historically they have given rise to revolutions and/or industrial eras from the steam era, through computer age and reaching the fourth era called industry. 4.0.

2.2 MIS Evolution

When talking about MIS, it is almost inevitable not to link it with technological advances and computing, since these as is well known, go hand in hand with the history of ICT [35]. For practical study purposes, these technological advances, like industrial revolutions, can be divided into eras, remaining as the author [35] lists them as follows:

(1) Computing in main frame and minicomputers. (2) Personal computers. (3) Client/Server Networks. (4) Business computing. Main contribution of this fourth era is improvement of services and networks within companies. With high-speed networks, performance of various areas such as finance, sales, accounting, marketing and even human capital management was substantially improved. (5) Cloud computing.

With industry 4.0 arrival, MIS continues to have solid validity within all processes, including manufacturing, it is crucial to follow them and be aware of their evolution in the coming years, as they have been doing since the 1960s, it is very likely that they will make great contribution to following eras and/or industrial revolutions, perhaps, in not so distant future.

2.3 Impact of MIS on Companies

Within companies, MIS are tools used to support processes, operations, intelligence, and information technology. MIS tools move and manage important information within organizations. These are the core of information science management and are often considered the first systems of the computer age [9]. In contrast, contribution of other authors is shown, with respect to MIS, such as [33] who mention that the term MIS also designates a specific category of information systems that serve mid-level management, provide managers at this level with reports on the current performance of organization, such information is used to monitor and control the company, in addition to predicting it's future performance.

Some research mentions that information is considered a strategic resource of great importance, it's management turns out to be an essential process for modern companies [56], p. 91, in the same way, results obtained by [3] demonstrate that information systems have a positive impact on productivity, and it is exposed to failures when there is a deficiency in training or the systems have very poor performance (p. 1). Among great contributions that authors such as [6, 8, 30, 33, 51] have made, is MIS classification, listing them as follows: MIS: Management Information Systems, ES: Expert Systems, NN: Neural Networks, EIS: Executive Information Systems, GDSS: Group Decision Support Systems, ESS: Executive Support Systems, OAS: Office Automatization Systems, TPS: Transactions Process Systems, IS: Intelligent Systems, KWS: Knowledge Work Systems.

Given the modernity and necessity and integrating concept of MIS, systems such as Enterprise Resources Planning (ERP) and Manufacturing Execution System, arise. In conjunction, industry 4.0 tries to connect in one, the ERP and MES systems, which in turn connects production and products with intelligent information and communication systems. [6, 38]. Resuming, the wide variety of MIS, which can be used for many of the operations carried out inside and outside companies, is more clearly visualized. They obtain a number of benefits in performance and efficiency, from reports, information management and archiving, to transactions, customers, suppliers, networks and decision-making support systems; MIS have been and will be present in organizations doing the important job of managing enormous amounts of information, filtering it, processing it and making it reach right people at the right time, in order to make better and more assertive decision making.

2.4 MIS and Industry 4.0: MIS 4.0

Transformations in the environment, which come to modify and improve the way in which goods and services are produced, is due in large part to the digitization of production [16]. New revolution takes what started with adoption of computers and automation, projecting to intelligent and independent systems driven by data and machine learning [16]. According to what has been described, MIS 4.0 can be seen as prerequisite form systematic application of digital world linked to Industry 4.0. MIS 4.0 are based on modern ERP and MES systems. All information of both products, suppliers and customers is managed in MIS and is clearly displayed. In addition, MIS contains relevant data on infrastructure of companies such as: facilities, machines and components, all managed in real time to optimize planning, strategy and execution performance [16, 43]. Analyzing information above, it can be distinguished that MIS still has a lot to offer, after all these years of use and adoption by companies. Industry 4.0, came to rescue, amplify, integrate and make intelligent the best that left the third revolution, information age, much needed and as a prerequisite for a better of Industry 4.0 implementation.

3 Methodology

Information and data to carry out this bibliometric study, were obtained from Science Direct database, whose website is www.sciencedirect.com, it was selected because it is one of the largest databases, which contains different quality formats of scientific dissemination, such as: magazines, articles, books, reviews, conference information, among many other additional formats. Steps proposed by Gräßler et al., [25] were followed to realize the bibliometric study, which are Identify, Locate and Select, Critical Appraise, Analyze and Synthesize, Disseminate.

Step 1. Identifying the review question. The question should be as clear, precise and meaningful. It is defined as follows: which studies address MIS, industry 4.0, performance in organizations, their relations and interplays?

Step 2. Localization and selection of relevant studies. Literature research was made with Science direct website. As search terms, articles published from 2016–2021 were analyzed. First, bibliometric studies about industry 4.0, MIS and performance were examined to identify previous works that focused on the industry 4.0 and the interplay of MIS and performance as shown in Table 1. The result of the search was a total of 659 papers on Science Direct.

Step 3. Critical appreciation of studies. As a crucial step, any systematic review is that the studies are critically assessed in terms of quality and review protocol. After reading titles, abstract and keywords of selected papers, unrelated papers were excluded. 31 documents of 2021 were selected, for being on the border of most current. Results of selection are shown in Table 2.

Step 4. Analyzing and synthesizing findings from studies. All articles related to review question are collected, analyzed and synthesized. A total of 55 papers were shortlisted and analyzed. Articles were sorted accordingly to MIS interplay on

Table 1 Searching strategies and selection

	Science direct
Search strategies	1. Advanced search
	2. "Performance" OR "productivity" AND "management" AND "information system" AND "industry 4.0" were applied in the title field
	3. "Performance" and "productivity" and "management information system" and "industry 4.0"
	4. Language: English
	5. Time span: 2017–2021
Filtered selection	1. Selection of most recent articles from 2021
	2. Analizing revelant articles

Table 2 Searching keywords and results

	Science direct	
Search terms	Total hits	Selected
"Performance" OR "productivity" AND "management" AND "information system" AND "industry 4.0"	659	31
"Performance" AND "productivity" AND "management information system" AND "industry 4.0"	55	24
Total	714	55

industry 4.0 and performance in organizations. The total 55 articles were uploaded to Mendeley software, which is a program to manage references, whose basic functions is archiving of literature in different text formats, it is used for citation, to insert bibliography and to export bibliographic data. From this database, RIS files were extracted, which were used to generate files necessary to make relationship maps that were generated on hand of VOSviewer program.

Microsoft Excel is also used, it is a spreadsheet tool that allows to import database files, manage them, manipulate them and generate tables and graphs for a better analysis and visualization of the information. This study was done through database use, in certain period of time May 15–22, 2021. Therefore, results obtained may vary little or significantly, depending on criteria and keywords used and largely updates on regarding topics and articles within the database itself.

Step 5. Sharing review findings. Main purpose of bibliometric reviews is to share researched information and make it accessible to all researchers. Next section contains results of analyzing and researching processes, additionally some recommendations and opportunities.

4 Results of Studies

As mentioned above, a total of 55 articles were obtained, published in a range from 2017 to 2021, which are composed by 1 book chapter, 30 conference articles, 21 Journal articles and 3 Magazine articles. Research related to industry 4.0 starts from approximately on 2011, publications obtained can be estimated as low production in 2017 with only one article, and 2021 as high with 31 publications (Fig. 1).

Sciences related to research on MIS, Industry 4.0, performance and other concepts, research have been found manufacturing, engineering and technology, health sciences, robotics, aerospace science, petrochemicals, mathematics, economics, manufacturing, information and computer science and administrative sciences among others. The results are seen in Fig. 2.

Following the analysis through VOSviewer software tool, compiled data and keywords, citations and authors were used; results were obtained for keywords with highest match, as shown in Fig. 3.

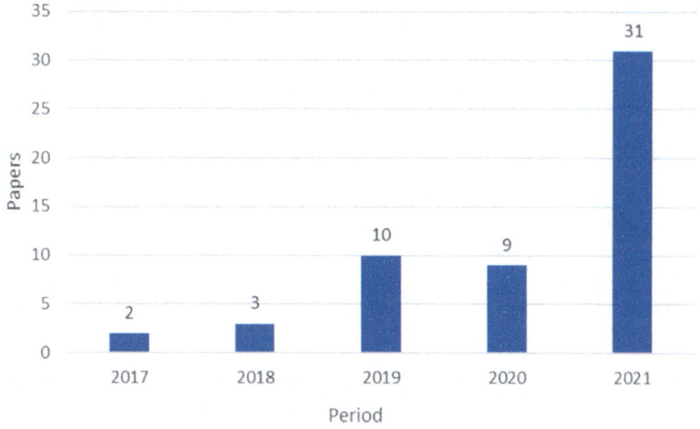

Fig. 1 Papers for period. *Source* Science Direct database 2021

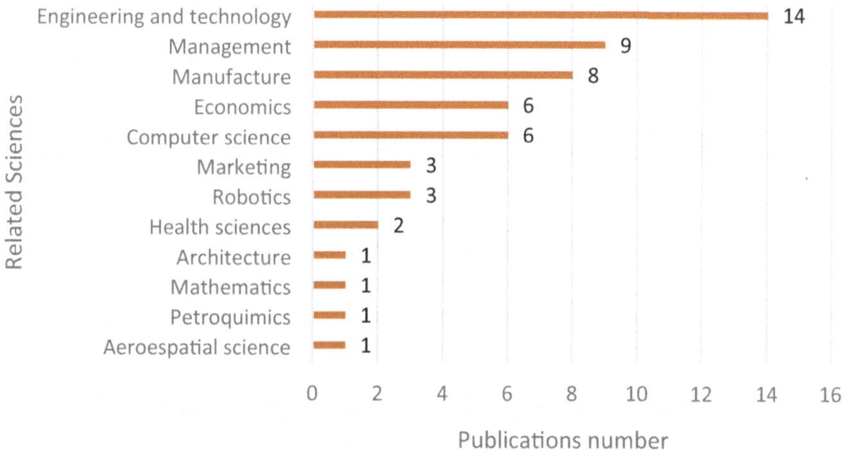

Fig. 2 Numbers of publications per science. *Source* Science Direct database 2021

It can be seen, many interesting clusters that are formed, Industry 4.0, Information science, Management, IoT, Big data and Supply chain are strongly related to the ecosystem. The number of citations on papers are shown in Table 3.

According with articles analysis and review, Internet of Things (IoT) paradigm enables interconnection, intercommunication and interaction among supply chain (SC) actors that allow for the dynamic management of global network operations, Industry 4.0's theme is Smart Manufacturing for the Future. It´s goal is simple and similar to previous revolutions: To increase productivity and achieve mass production using innovative technology, being as a prerequisite the correct implementation of MIS in organizations. Systems use, like Project Portfolio Management (PPM),

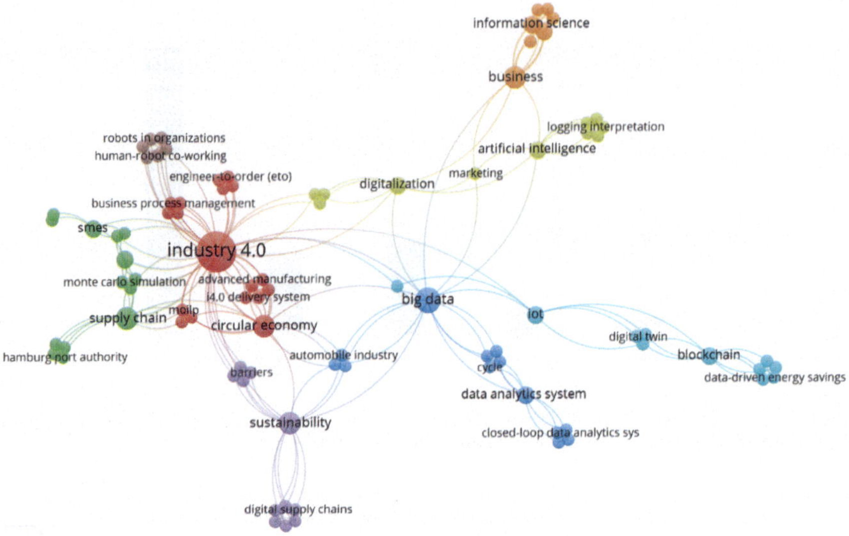

Fig. 3 Keyword map. *Source* Science Direct database 2021

computer-aided design (CAD), Enterprise Performance Management (EPM), Enterprise Resource Planning (ERP) and recent building information modeling software (BIM), have changed the way the construction industry works [2, 12, 17].

Industry 4.0 technologies can be used to effectively aid digital transformations of an organization to achieve sustainable development goals, aims to link disruptive technologies to manufacturing systems, combining smart operations and supply chain management (OSCM). With a few more development in enabling technologies such as 5G developments, IoT, standardization, AI and blockchain 3.0 utilization, it is but a matter of time that the industry will transition towards the digital twin-based approach. There is a relationship between quality management practices, MIS and Industry 4.0 technologies that improve quality. Industry 4.0 and digital age have dramatically influenced both ITC job characteristics and ITC labor demand [1, 5, 7, 52, 54].

Development of information technology has significantly changed the functioning of contemporary organizations in different areas as marketing, advertising, logistics, sales or banking also in the scope of work organization or project management. Use of Machine Learning and Data Mining techniques are well established in manufacturing because intelligently analyzed data is a valuable resource since it gains new insights and can provide a significant competitive advantage [14, 28, 42].

Therefore, MIS implementation in companies is essential to take the next step into the industrial revolution 4.0, it is necessary to have a vigorous implementation and use, as a requirement to integrate into ecosystem, to adapt companies to effectiveness and efficiency of production and/or manufacturing that is demanded in present and future of all businesses. Additionally, other authors assure that ICTs, MIS and new

Table 3 Citation on documents

Rank	Title	Authors	Cites	Year
1	Intelligent Autonomous Vehicles in digital supply chains: A framework for integrating innovations towards sustainable value networks	Bechtsis, Dimitrios; Tsolakis, Naoum; Vlachos, Dimitrios; Srai, Jagjit Singh	53	2018
2	Industry 5.0 and Human–Robot Co-working	Demir, Kadir Alpaslan; Döven, Gözde; Sezen, Bülent	47	2019
3	The Paradigm of the Integration of Different Types of Management Information Systems in Investment and Construction Company Implementing the Project Approach	Eroshkin, S. Yu; Koryagin, N. D.; Kovkov, D. V.; Panov, D. V.; Sukhorukov, A. I	36	2017
4	Industry 4.0 adoption and 10R advance manufacturing capabilities for sustainable development	Bag, Surajit Gupta, Shivam Kumar, Sameer	35	2021
5	A fuzzy rule-based industry 4.0 maturity model for operations and supply chain management	Caiado, Rodrigo Goyannes Gusmão Scavarda, Luiz Felipe Gavião, Luiz Octávio Ivson, Paulo Nascimento, Daniel Luiz de Mattos Garza-Reyes, Jose Arturo	20	2021
6	Recent advances on industrial data-driven energy savings: Digital twins and infrastructures	Teng, Sin Yong; Touš, Michal; Leong, Wei Dong; How, Bing Shen; Lam, Hon Loong; Máša, Vítězslav	16	2021
7	Assessment of Leadership Effectiveness Dimensions in Small and Medium Enterprises (SMEs)	Madanchian, Mitra Taherdoost, Hamed	16	2019
8	An exploratory study of digital workforce competency in Thailand	Siddoo, Veeraporn; Sawattawee, Jinda; Janchai, Worawit; Thinnukool, Orawit	15	2019
9	Opportunities and threats presented by social media in project management	Hysa, Beata Spalek, Seweryn	13	2019
10	Machine Learning and Data Mining in Manufacturing	Dogan, Alican Birant, Derya	12	2020

Source Science Direct database 2021

digital technologies have been a positive influence in recent years in performance, professions, support of decision-making, employment, business and at social level. Other capabilities also mentioned are business strategy alignment, business support, organizational agility, and an overall promise of improved performance and increased agility in organizations [18, 18–29, 37, 50, 51, 52, 53, 55].

5 Conclusions

This bibliometric study presents the selection, analysis and synthesis of literature research related to the interaction of MIS with Industry 4.0 in a period of time, being the year 2021 with the most production of scientific literature on Science Direct website. The obtained visual maps show close relationship between information sciences, MIS and industry 4.0, resulting in a constant interaction between them.

These types of studies are important to publicize scientific panorama and visualize the opportunities that arise to carry out research on related topics. Among sciences most attached to the topics of MIS and Industry 4.0 are engineering and manufacturing, computer science, administration, economics, among others no less important.

The methodology used in this research makes the task of collecting and displaying the information simple so that any researcher can access it. Reports in this study are filtered and synthesized information to visualize ecosystem and relationship of the MIS and Industry 4.0, which sciences are more and better involved, which authors are the most cited and in which year more literature has been generated.

The future of industry 4.0, is imminent arrival of industry 5.0, in which the interaction, co-working, between machine and man is predicted, which is related to interesting advances in artificial intelligence, robotics and the systems that make it up, including the use of ICT and MIS and whose relationship and interaction will continue to be in force, which gives an opportunity to grow the scientific heritage on these interesting topics that will be approached in the future no far away.

References

1. Bag S, Gupta S, Kumar S (2021) Industry 4.0 adoption and 10R advance manufacturing capabilities for sustainable development. International Journal of Production Economics, 231(December 2019):107844. https://doi.org/10.1016/j.ijpe.2020.107844
2. Bechtsis D, Tsolakis N, Vlachos D, Srai JS (2018) Intelligent Autonomous Vehicles in digital supply chains: a framework for integrating innovations towards sustainable value networks. Journal of Cleaner Production, 181:60–71. https://doi.org/10.1016/j.jclepro.2018.01.173
3. Bere A, Brijlal P (2014) The impact of information systems usage on productivity: a retrospective analysis and an empirical study in Cape Town tourism of South Africa. In: African Journal of Hospitality, Tourism and Leisure, vol 3, issue 1. www.ajhtl.com
4. Bogoviz AV (2019) Industry 4.0 as a new vector of growth and development of knowledge economy. Stud Syst Dec Control 169:85–91. https://doi.org/10.1007/978-3-319-94310-7_8
5. Caiado RGG, Scavarda LF, Gavião LO, Ivson P, Nascimento DLdeM, Garza-Reyes JA (2021) A fuzzy rule-based industry 4.0 maturity model for operations and supply chain management. Int J Prod Econ 231. https://doi.org/10.1016/j.ijpe.2020.107883
6. Camacho H, Rubén CR (2018) SISTEMAS DE INFORMACIÓN EMPRESARIAL. EVOLUCIÓN HISTÓRICA Y ACTUALIDAD
7. Carvalho AV, Enrique DV, Chouchene A, Charrua-Santos F (2021) Quality 4.0: An Overview. Procedia Computer Science, 181(2019):341–346. https://doi.org/10.1016/j.procs.2021.01.176
8. Cohen Karen D, Asín Lares E (2014) Tecnología de la información: Estrategia y transformación en los negocios (6a ed). McGraw-Hill

9. Cummings M (2012) The role of management information systems|smartsheet. https://www.smartsheet.com/management-information-systems
10. Dalenogare LS, Benitez GB, Ayala NF, Frank AG (2018) The expected contribution of Industry 4.0 technologies for industrial performance. Int J Prod Econ 204:383–394. https://doi.org/10.1016/j.ijpe.2018.08.019
11. Dallasega P, Rauch E, Linder C (2018) Industry 4.0 as an enabler of proximity for construction supply chains: a systematic literature review. Comput Ind 99:205–225. https://doi.org/10.1016/j.compind.2018.03.039
12. Demir KA, Döven G, Sezen B (2019) Industry 5.0 and Human-Robot Co-working. Procedia Computer Science, 158:688–695. https://doi.org/10.1016/j.procs.2019.09.104
13. Dietz JLG, Mulder HB (2020) Enterprise ontology: a human-centric approach to understanding the essence of organisation. In: Enterprise ontology
14. Dogan A, Birant D (2020) Machine learning and data mining in manufacturing. Expert Syst Appl 114060. https://doi.org/10.1016/j.eswa.2020.114060
15. Dávila RM, Guzmán SR, Macareno AH, Piñeres HD, de la Rosa BD, Caballero-Uribe CV (2009) Bibliometría: conceptos y utilidades para el estudio médico y la formación professional. Salud Uninorte 25(2):319–330
16. Edirisinghe S (2020) What is the role of MIS in an era of IR 4.0?
17. Eroshkin SY, Koryagin ND, Kovkov DV, Panov DV, Sukhorukov AI (2017) The paradigm of the integration of different types of management information systems in investment and construction company implementing the project approach. Proc Comput Sci 103:605–608. https://doi.org/10.1016/j.procs.2017.01.076
18. Fadhilah AN, Subriadi AP (2019) The role of IT on firm performance. Proc Comput Sci 161:258–265. https://doi.org/10.1016/j.procs.2019.11.122
19. Fernanda L, Jaramillo C (2020) Trazabilidad en la cadena de suministro alimentaria: Un estudio bibliométrico. 76:277–297
20. Fernández-Rovira C, Álvarez Valdés J, Molleví G, Nicolas-Sans R (2021) The digital transformation of business. Towards the datafication of the relationship with customers. Technological Forecasting and Social Change, 162(September 2020):120339. https://doi.org/10.1016/j.techfore.2020.120339
21. Franco-Paredes K, Díaz-Reséndiz FdeJ, Pineda-Lozano JE, Hidalgo-Rasmussen CA (2016) Bibliometric analysis of scientific production of Mexican journal of eating disorders, 2010–2014. Revista Mexicana de Trastornos Alimentarios 7(1):9–16. https://doi.org/10.1016/j.rmta.2016.03.001
22. Gallo T, Cagnetti C, Silvestri C, Ruggieri A (2021) Industry 4.0 tools in lean production: A systematic literature review. Procedia Computer Science, 180(2019): 394–403. https://doi.org/10.1016/j.procs.2021.01.255
23. Ghobakhloo M, Fathi M, Iranmanesh M, Maroufkhani P, Morales ME (2021) Industry 4.0 ten years on: a bibliometric and systematic review of concepts, sustainability value drivers, and success determinants. J Cleaner Prod 302:127052. https://doi.org/10.1016/j.jclepro.2021.127052
24. Gilchrist A (2016) IIoT reference architecture. In industry 4.0. https://doi.org/10.1007/978-1-4842-2047-4_4
25. Gräßler I, Roesmann D, Pottebaum J (2021) Model based integration of human characteristics in production systems: a literature survey. Procedia CIRP, 99: 57–62. https://doi.org/10.1016/j.procir.2021.03.010
26. Guo Y, Wang N, Xu ZY, Wu K (2020) The internet of things-based decision support system for information processing in intelligent manufacturing using data mining technology. Mech Syst Signal Process 142:106630. https://doi.org/10.1016/j.ymssp.2020.106630
27. Helo P, Hao Y, Toshev R, Boldosova V (2021) Cloud manufacturing ecosystem analysis and design. Robot Comput-Integ Manuf 67:102050. https://doi.org/10.1016/j.rcim.2020.102050
28. Hysa B, Spalek S (2019) Opportunities and threats presented by social media in project management. Heliyon, 5(4):e01488. https://doi.org/10.1016/j.heliyon.2019.e01488

29. Joppen R, Enzberg S, Kühn A, Dumitrescu R (2019) Data map—Method for the specification of data flows within production. Proc CIRP 79:461–465. https://doi.org/10.1016/j.procir.2019.02.127
30. Kendall KE, Kendall JE (2011) ANÁLISIS Y DISEÑO DE SISTEMAS (8va ed). PEARSON EDUCACIÓN. http://cotana.informatica.edu.bo/downloads/ld-Analisis%20y%20Diseno%20de%20Sistemas_Kendall-8va.pdf
31. Kim J, Lee JY (2021) Server-Edge dualized closed-loop data analytics system for cyber-physical system application. Robotics and Computer-Integrated Manufacturing, 67(August 2020):102040. https://doi.org/10.1016/j.rcim.2020.102040
32. Kim J, Lee JY (2021) Server-edge dualized closed-loop data analytics system for cyber-physical system application. Robot Comput-Integ Manuf 67:102040. https://doi.org/10.1016/j.rcim.2020.102040
33. Laudon KC, Laudon JP (2016) Management information systems, 14th ed. Pearson Education Limited
34. Li Q, Tang Q, Chan I, Wei H, Pu Y, Jiang H, Li J, Zhou J (2018) Smart manufacturing standardization: architectures, reference models and standards framework. Comput Ind 101:91–106. https://doi.org/10.1016/j.compind.2018.06.005
35. Luenendonk M (2017) Management Information Systems (MIS): definition and how it works|Cleverism. https://www.cleverism.com/management-information-systems-mis/
36. López K, Alvarado S, Fong C, González E (2019) Capital social en la perspectiva de la teoría de recursos y capacidades: un análisis bibliométrico/Social capital in the perspective of the theory of resources and capabilities : a bibliometric analysis. Inquietud Empresarial, XIX(1):63–77
37. Majeed A, Zhang Y, Ren S, Lv J, Peng T, Waqar S, Yin E (2021) A big data-driven framework for sustainable and smart additive manufacturing. Robot Comput-Integr Manuf 67:102026. https://doi.org/10.1016/j.rcim.2020.102026
38. Marík V, Kadera P, Rzevski G, Zoitl A, Anderst-Kotsis G, Tjoa Am, Khalil I (2019) Industrial Applications of Holonic and Multi-Agent Systems. In Industrial Applications of Holonic and Multi-Agent Systems: Vol. 11710 LNAI. https://doi.org/10.1007/978-3-030-27878-6
39. Morgan J, Halton M, Qiao Y, Breslin JG (2021) Industry 4.0 smart reconfigurable manufacturing machines. Journal of Manufacturing Systems, 59(November 2020):481–506. https://doi.org/10.1016/j.jmsy.2021.03.001
40. Muñoz AL, Rodríguez BMP, Hernández López AM (2017) Análisis bibliométrico sobre la implementación de las tecnologías de la informacion y counicación en las administraciones públicas: Aportaciones y oportunidades de investigación. https://doi.org/10.15446/innovar.v26n63.60674
41. Pereira JL, Varajão J, Sá JOe, Silva A (2019) Performance evaluation in IST projects: a case study. Lect Notes Inf Syst Organ 31:13–27. https://doi.org/10.1007/978-3-030-14850-8_2
42. Piechnicki F, Loures E, Santos E (2017) A conceptual framework of knowledge conciliation to decision making support in RCM deployment. Proc Manuf 11(June):1135–1144. https://doi.org/10.1016/j.promfg.2017.07.235
43. PlenarIT-Wholistic and Lean Information Systems (2020) Knowledge in the Fourth industrial revolution and the digital WorldenarIT AG. https://www.plenarit.com/knowledge/?lang=en
44. Qin J, Liu Y, Grosvenor R (2016) A categorical framework of manufacturing for industry 4.0 and beyond. Proc CIRP 52:173–178. https://doi.org/10.1016/j.procir.2016.08.005
45. Ramos I, Quaresma R, Silva P, Oliviera T (2019) Information systems for industry 4.0. https://doi.org/10.1007/978-3-030-14850-8
46. Reyes Echeagaray DA, Bribiesca Correa G, Carrillo López VH, Corona Cabrera A, Cruz Quiroz RE, Ramírez Munive YA, Ramírez Chavero MO, Torres Garibay R (2016) Tecnologías de Información y Comunicación en las organizaciones. UNAM. http://publishing.fca.unam.mx/tic/TIC-Organizaciones.pdf
47. Ruiz MÁG, Ruiz BÁE, Calvillo BS (2019) México Y La Industria 4.0
48. Salas G, Ponce FP, Méndez-Bustos P, Vega-Arce M, de los Ángeles Pérez M, López-López W, Cárcamo-Vásquez H (2017) 25 años de Psykhe: Un análisis bibliométrico. Psykhe, 26(1):1. https://doi.org/10.7764/psykhe.26.1.1205

49. Salvadorinho J, Teixeira L (2021) Organizational knowledge in the I4.0 using BPMN: a case study. Procedia Computer Science, 181(2019):981–988. https://doi.org/10.1016/j.procs.2021.01.266
50. Sartal A, Llach J, Vázquez XH, de Castro R (2017) How much does lean manufacturing need environmental and information technologies? J Manuf Syst 45:260–272. https://doi.org/10.1016/j.jmsy.2017.10.005
51. Sharma D, Sharma D, Prakash Sharma J (2015) Information systems. In: International Journal of Scientific Research Engineering & Technology (IJSRET). http://www.ijsret.org/pdf/EATHD-15015.pdf
52. Siddoo V, Sawattawee J, Janchai W, Thinnukool O (2019) An exploratory study of digital workforce competency in Thailand. Heliyon 5(5):e01723. https://doi.org/10.1016/j.heliyon.2019.e01723
53. Silviana BG (2019) Product acceptance model-case study for a nanotechnology research and development project. Proc Manuf 32:1058–1063. https://doi.org/10.1016/j.promfg.2019.02.321
54. Teng SY, Touš M, Leong WD, How BS, Lam HL, Máša V (2021) Recent advances on industrial data-driven energy savings: digital twins and infrastructures. Renew Sustain Energy Rev 135. https://doi.org/10.1016/j.rser.2020.110208
55. Theron P (2018) Through-life cyber resilience in future smart manufacturing environments. A research programme. Proc Manuf 16:193–207. https://doi.org/10.1016/j.promfg.2018.10.157
56. Tundidor MDOL, Nogueira RD, Medina LA (2018) ARTÍCULO ORIGINAL Organización de los sistemas informativos para potenciar el control de gestión empresarial information systems Organ Strength Bus Manag Control 12(1). http://scielo.sld.cu/pdf/cofin/v12n1/cofin07118.pdf
57. Wang L, Törngren M, Onori M (2015) Current status and advancement of cyber-physical systems in manufacturing. J Manuf Syst 37:517–527. https://doi.org/10.1016/j.jmsy.2015.04.008
58. Xu LDa, Xu EL, Li L (2018) Industry 4.0: state of the art and future trends. Int J Prod Res 56(8):2941–2962. https://doi.org/10.1080/00207543.2018.1444806
59. Zenteno I (2012) El origen de las TICS. https://prezi.com/kg1xnovnpwuy/el-origen-de-las-tics/

The manufacturer's authorised representative in the EU is Springer Nature Customer Service Centre GmbH, Europaplatz 3, 69115 Heidelberg, Germany. If you have any concerns regarding our products, please contact ProductSafety@springernature.com

Printed and bound by CPI Group (UK) Ltd, Croydon, CR0 4YY
30/09/2026
02225493-0015